ETHYL ALCOHOL PRODUCTION

AND USE AS A MOTOR FUEL

ETHYL ALCOHOL PRODUCTION AND USE AS A MOTOR FUEL

Edited by J.K. Paul

NOYES DATA CORPORATION
Park Ridge, New Jersey, U.S.A.
1979

Published in the United States of America by
Noyes Data Corporation
Noyes Building, Park Ridge, New Jersey 07656

Library of Congress Cataloging in Publication Data

Main entry under title:

Ethyl alcohol production and use as a motor fuel.

(Chemical technology review ; no. 144) (Energy
technology review ; no. 51)
"A companion volume to ... [editor's] Methanol
technology and application in motor fuels."
Bibliography: p.
1. Alcohol - - Patents. 2. Alcohol as fuel - - Patents.
I. Paul, J. K. II. Series. III. Series: Energy
technology review ; no. 51.
TP593.E83   621.4'023   79-22900
ISBN 0-8155-0780-1

FOREWORD

Ethanol is available from so-called "renewable raw materials," i.e., agricultural crops such as corn, grain, potatoes, sugar beets and sugar cane, or other biomass and suitable garbage.

In this book an attempt has been made to present an economic assessment of possible modes of preparation of ethanol from various forms of biomass, natural resources and their waste materials or by-products. A chapter on current technology is also included. The present and potential availability of biomass from sugar crops, grains and grasses and silviculture is considered.

Current crop production, proposed crops grown specifically for energy production and crop wastes and residues are discussed. Finally, to determine the actual practicality of fueling motor vehicles with ethanol, either 100% or in blends, several sets of engine test data are reviewed. The results seem favorable, 10 to 20% ethanol blends performing very similarly to straight gasoline with slight gains in octane rating and mileage.

This book is a companion volume to our *Methanol Technology and Application in Motor Fuels.*

Because the information in this book is taken from multiple sources, it is possible that certain portions of this book may disagree or conflict with other parts of the book. This is especially true of monetary values and opinions of future potential. We chose to include these, however, in order to make the book more valuable to the reader.

Cost figures provided are those given in the report cited, the date of which is always given. When the dates of the cost figures themselves are given, we have included them.

Advanced composition and production methods developed by Noyes Data are employed to bring these durably bound books to you in a minimum of time. Special techniques are used to close the gap between "manuscript" and "completed book." Industrial technology is progressing so rapidly that time-honored, conventional typesetting, binding and shipping methods are no longer suitable. We have bypassed the delays in the conventional book publishing cycle and provide the user with an effective and convenient means of reviewing up-to-date information in depth.

The expanded table of contents is organized in such a way as to serve as a subject index. It provides easy access to the information contained in this book which is based on various studies produced by and for diverse governmental agencies under grants and contracts. These primary sources are listed at the end of the volume under the heading Sources Utilized. The titles of additional publications pertaining to topics in this book are found in the text.

CONTENTS AND SUBJECT INDEX

INTRODUCTION

Ethanol has been identified as being a possible candidate for the partial replacement of petroleum-based transportation fuels. In particular, ethanol fuels may play a role in the critical transition period between today's natural petroleum derivatives and the synthetic fuels of the future. Due to their relative availability and moderately advanced production and resource technology, alcohol fuels seem positioned to play a role in alleviating the effects of a sudden petroleum shortfall such as the 1973–1974 oil embargo or the more recent Iranian political crisis.

Since 1974, the requirement to conserve petroleum resources has become more pronounced, and the investigation of conservation measures, through the introduction of alternative transportation fuels, more intense.

Throughout the world, motor vehicle operators are becoming aware of ethanol-gasoline blends. Brazil has for some time been fueling many of its vehicles with a 20% ethanol-80% gasoline blend (and experimenting with 100% ethanol), and anticipates converting the entire country within a few years; and France is carrying out feasibility studies.

The Nebraska gasohol (10% ethanol-90% gasoline) supporters have seen the availability of gasohol at the pump start with one Nebraska service station in February, 1978 and grow to more than 800, in over half of the United States, a year later. Incentives, such as the exemption of the 4¢/gal excise tax on motor fuels containing at least 10% ethanol produced from renewable resources, have helped make this possible. Loan guarantees for future pilot plants and an expected Department of Energy R&D funding program for fiscal 1980 are expected to brighten the picture even more.

Through the mid-1980s, there are expected to be sufficient surplus grain and waste raw materials for the conversion facilities now operating; however, to prepare for the future, efforts will need to be concentrated on more and improved production facilities and processes as well as the development of the most efficient forms of biomass for conversion. While the use of ethanol won't completely resolve the U.S. energy problems, it certainly presents a possible interim solution.

In this book an attempt has been made to present an economic assessment of possible modes of preparation of ethanol from various forms of biomass, such as renewable natural resources and their waste materials or by-products. A chapter on current technology is also included. The present and potential availability of biomass from sugar crops, grains and grasses and silvicultural forms is considered. Current crop production, proposed crops grown specifically for energy production and crop wastes and residues are discussed.

Finally, to determine the actual practicality of fueling motor vehicles with ethanol, either 100% or in blends, several sets of engine test data are reviewed. The results seem favorable, 10 to 20% ethanol blends performing very similarly to straight gasoline with slight gains in octane rating and mileage.

In summary, this book reviews the latest available data on the production and technological potential for ethanol, and the effects of its use in motor vehicle fuels.

NEAR-TERM POTENTIAL
FOR BIOMASS-BASED ETHANOL

The information in this chapter is based on *Biomass-Based
Alcohol Fuels: The Near-Term Potential for Use with Gasoline*,
prepared by W. Park, G. Price and D. Salo of The Mitre Corporation, Metrek Division, for the U.S. Department of Energy
(DOE Report HCP/T4101-03), August 1978.

INTRODUCTION

Considerable attention has been focused on the near-term possibility of blending
biomass-derived alcohol with gasoline for transportation fuels. Conflicting reports and recommendations have appeared concerning the wisdom of such ventures.

Because of the intensifying interest in biomass-derived alcohol fuels, the Fuels
from Biomass Systems Branch of the Department of Energy requested the Metrek
Division of the Mitre Corporation to assess the near-term potential of biomass-
based alcohol-gasoline fuel systems. This report presents an overview of biomass-
based alcohol fuels and analyzes the requirements and prospects for the development of a nationwide alcohol-gasoline fuel system by the year 1990.

Potential of Biomass-Based Alcohol Fuels

There are compelling reasons for carefully considering alcohols produced from
biomass as potential fuels. Some of the most important are presented below.

Alcohols produced from biomass are renewable fuels. This is not the case for
the petroleum products which are consumed today in this country. Biomass-
derived alcohol fuels could provide an alternate source of energy that would reduce the nation's reliance on nonrenewable petroleum resources for liquid fuels.

The biomass resources available for alcohol production today are wood and various sugar, grain and other crops such as wheat, corn, grain/sweet sorghum, sugar

cane, sugar beets and potatoes. As these resources are produced within the United States, alcohol fuels produced from biomass can reduce the dependence on imported energy.

The technology for alcohol production is well-known. The U.S. chemical industry has had a long and successful history in the production of alcohols. The production of beverage alcohol from grain is a time-honored process. The establishment of a nationwide biomass-based alcohol production system should, therefore, not be technically difficult.

Alcohol fuels are promising alternatives to fossil-based liquid fuels for the transportation sector. Alcohols blend well with gasoline and, when present in relatively small amounts, they provide a fuel which has been used successfully in automobiles today.

Reason for the Absence of an Alcohol Fuel System in the United States Today

The primary barrier to alcohol fuel production and use to date has been the cost of production. Inexpensive fuels derived from petroleum have essentially prevented the development of alcohol fuels. The use of alcohol fuels has generally occurred only when unusual political or economic conditions have existed. For example, alcohol fuels emerged in Europe during World War II when gasoline shortages were widespread. In Brazil, in recent years, excess sugar cane has been purchased by the government and converted to alcohol for addition to gasoline both as a sugar price support mechanism and more recently as a hedge against energy imports.

Today, in the United States, cost is still the barrier to alcohol fuel production. As we look to the future, however, the economic prospects for alcohol fuel are more promising because of increasing prices of transportation fuels. It, therefore, is reasonable to ask whether the current cost differences have been reduced to the point that the development of a United States biomass alcohol-gasoline fuel system is in the national interest.

There are some impediments in addition to production cost in establishing a United States alcohol-gasoline fuel system. These problems are not barriers, however, and simply result in additions to the final cost of the fuel product. These problems are discussed later in more detail.

Biomass-Based Alcohols

Ethanol and methanol are the only alcohols which are currently considered suitable for extensive use as transportation fuels. Most likely production routes for biomass derived alcohol fuels are identified in Figure 1.1. The upper route is for methanol, the lower for ethanol.

Ethanol is the well-known intoxicating beverage alcohol. All beverage alcohol is produced from the fermentation of grains and other sugar or starch feedstocks. However, ethanol used for industrial purposes today is primarily made from ethylene, a gas derived from petroleum.

As shown in Figure 1.1, the most likely biomass to ethanol production process for fuel products is the well-known fermentation and distillation process used for beverage alcohol. This process can accept any biomass feedstock whose carbohydrate content can be easily fermented. This limits biomass resources to food sources such as grains, sugar crops, potatoes, etc. Other biomass resources, such as wood or agricultural residues, could be used for the generation of ethanol. However, to convert these cellulosic materials to alcohol requires an additional front-end solubilization step called hydrolysis. Interesting developments in the hydrolysis of cellulose to fermentable sugars have been made recently and the DOE Fuels from Biomass Program is actively supporting research in this area to identify ways of improving process economics. However, as of this time (1978), the costs of these processes still appear too high to expect any significant production by 1990. This route to ethanol is, therefore, shown in dotted boxes in Figure 1.1.

Figure 1.1: Biomass to Alcohol Fuel Routes

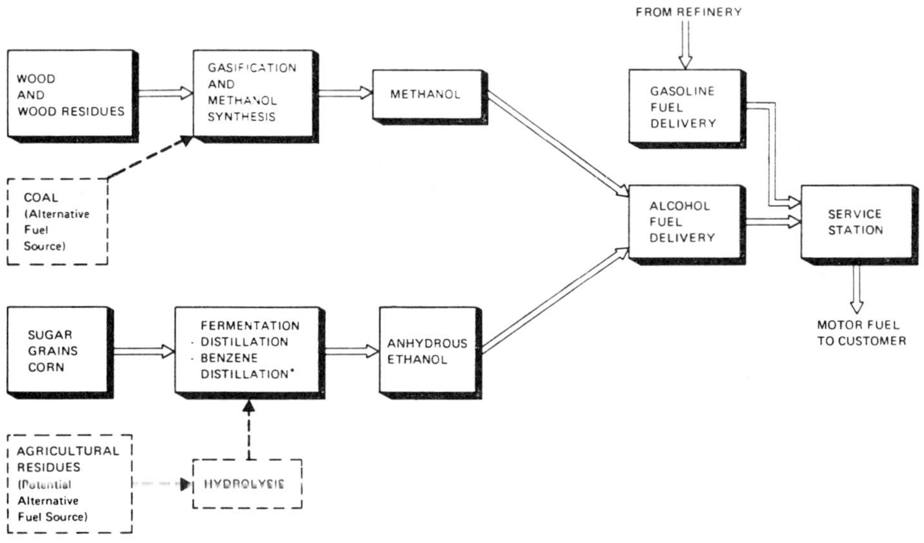

Source: DOE HCP/T4101-03

Potential Use of Alcohol Fuel in Automobiles

Ethanol could be used directly as a substitute fuel in automobile engines, however, carburetion systems that are in use today would require some significant changes to accommodate the use of alcohol. An engine designed to run on gasoline will not operate on pure alcohol fuel without these conversions. The conversion to an alcohol fuel system would require a large-volume fuel-injection system plus some types of induction heating system to improve the vaporization of the alcohol fuel.

Ethanol can be mixed with gasoline in any desired proportions. Present automobile engines designed for gasoline use have been run on alcohol-gasoline blends with alcohol contents as high as 30%. As the percentage of alcohol in the fuel blend increases, reduction in driving performance occurs. Stalling and hesitation on warm-up are the most common problems. The extent of performance loss depends on the type of vehicle, its age and state of tune-up. These performance problems are usually not severe as long as the percentage of alcohol is less than about 15%.

An advantage to the use of alcohols in gasoline relates to fuel octane rating. When added to gasoline, ethanol boosts the octane value of the original gasoline in much the same way as tetraethyllead and no-lead additives in gasoline.

There are some problems that may occur with the use of alcohol-gasoline blends for automobile fuels. None of these appears to be a technical barrier; however, the solution of these problems adds a cost to the consumer. Estimates of these costs are given in the next section. Problems identified with the use of alcohol fuel blends in present automobiles are:

> Performance—The use of alcohol-gasoline blends in cars tuned for 100% gasoline utilization can result in some performance problems, usually detected as slight hesitation on acceleration during the warm-up period. For other cars, there are no reported performance difficulties.
>
> Fuel Phase Separation—When small percentages of water are added to alcohol-gasoline fuel mixtures, the alcohol is drawn out of the mixture to blend with the water. The alcohol-water combination is heavier than the gasoline and settles to the bottom of the fuel tank. An automobile engine will not run on this alcohol-water solution. Water is present in virtually all of the present gasoline storage systems as a small layer at the bottom of the tank. To accommodate alcohol-gasoline fuel mixtures, these tanks would have to be dried and efforts made to prevent water contamination. Because of this, a lower cost alternative has been suggested that alcohol used as a supplement to gasoline be distributed separately to retail filling stations. At the station, it would be kept in separate tanks and mixed with gasoline only at the last stage of pumping into automobile tanks. (This distribution method was used in the costing performed in the analysis of this paper.)
>
> Corrosion Problems—Alcohols are well-known solvents. Much of the market for them is dependent on their use as solvents. It has been shown that alcohol-gasoline blends have corrosive effects on fuel system components of present day automobiles. Plastic components in fuel systems might have to be replaced with other materials and possibly different materials would have to be used for fuel tanks, lines, and carburetors.
>
> Vapor Pressure—Vapor pressure for gasoline is limited by state law throughout the country. In those cases where vapor pressure limitations are exceeded, the gasoline makeup would have to be changed.
>
> Emissions—Exhaust emissions do not seem to be a major problem for small percentage blends of alcohol in gasoline. Aldehydes and unburned alcohols do appear in exhaust gases as new potential pollutants when alcohol fuels are used.

COSTS OF BIOMASS-BASED ALCOHOL FUELS

The costs of biomass-derived alcohol fuels are presented both in terms of production cost and cost to the consumer at the service station.

Production Costs

The energy content (gross or higher heating values) of a gallon of gasoline is nearly twice that of methanol and 1.5 times that of ethanol. The theoretical effects of these energy differences on the distances a car would travel on each of the three fuels are presented in Figure 1.2.

Figure 1.2: Relative Energy Content of Gasoline and Alcohol Fuels

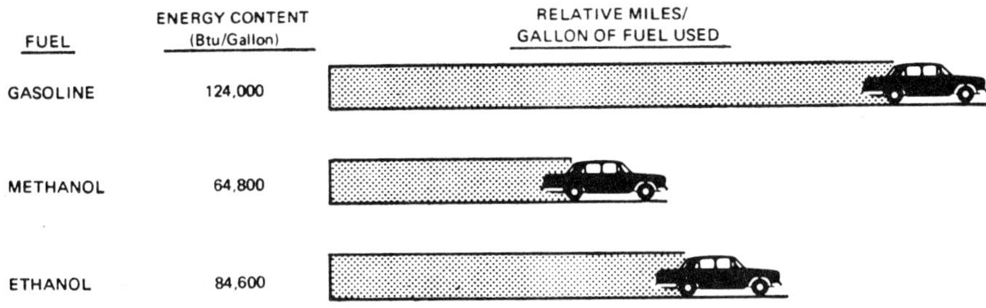

Source: DOE HCP/T4101-03

Illustrative production costs for gasoline and biomass-derived alcohol fuels in 1982 and 1990 have been abstracted from a number of recently completed studies and are presented in Figure 1.3. All costs are given in constant 1976 dollars so that inflation effects are removed.

These costs are also presented on a constant energy ($/10^6$ Btu) rather than a constant volume ($/gallon) basis because of the energy differences in the fuels. Alcohol production costs are shown to decrease over time because new plants are assumed to become increasingly large and, therefore, result in improved economies of scale.

The primary conclusion that can be drawn from this figure is that alcohol fuels will cost more than gasoline on an equivalent energy basis over the time period in question. Ethanol costs are about 3 to 4 times those of gasoline.

Consumer Costs

Estimates of the price of alcohol-gasoline blends relative to the price of no-lead gasoline are presented in Figure 1.4. These prices are "at the pump" estimates and are given in 1976 dollars. The no-lead gasoline price of $0.74 per gallon is based on the cost of imported crude oil at $14.00 per barrel. Marginal gasoline costs, derived from imported oil costs, are used since replacement of gasoline with alcohol will likely apply to the highest cost gasoline source, namely, gasoline from imported oil.

Figure 1.3:   Estimated Fuel Production Costs for 1982 and 1990

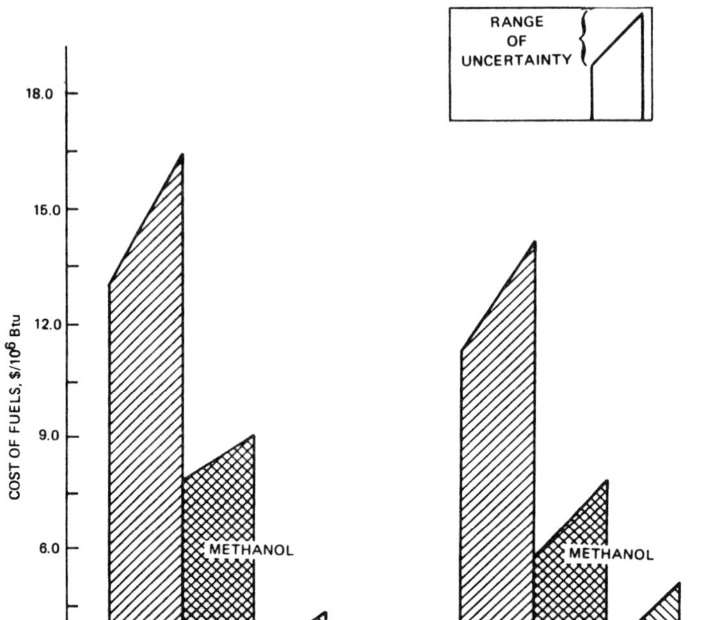

Figure 1.4:   Consumer Costs (Based on Gasoline from Imported Crude Only)

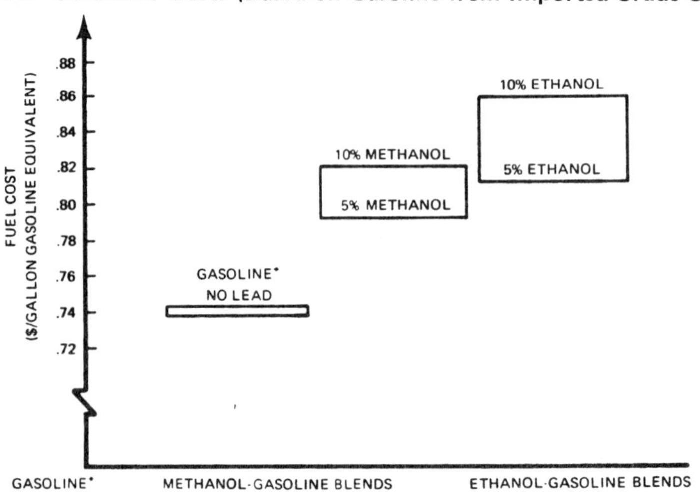

*Gasoline prices derived from imported crude at $14.00/barrel

Source: DOE HCP/T4101-03

As seen in Figure 1.4, the total cost addition to the consumer for a 5% methanol-gasoline blend is about $0.05 per gallon. It is about $0.08 per gallon for a 10% methanol-gasoline blend. The cost addition is $0.07 per gallon for a 5% ethanol-gasoline blend. For a 10% blend, the increase is $0.12 per gallon.

In brief, the costing is based on methodology whereby additional costs are assigned to alcohol-gasoline blends for fuel processing and delivery, construction of the alcohol fuel distribution system, and conversion of automobile fuel systems. Credit is given to the blends for the octane benefits of the alcohol-gasoline fuel.

A graphical breakdown of the consumer costs of gasoline and the two 5% alcohol-gasoline blends is presented in Figure 1.5. In this figure, it is important to note the relatively small impact the price of the alcohol has on the total fuel price for 5% blends.

Figure 1.5: Breakdown of the Consumer Costs (Based on Gasoline from Imported Crude Only)

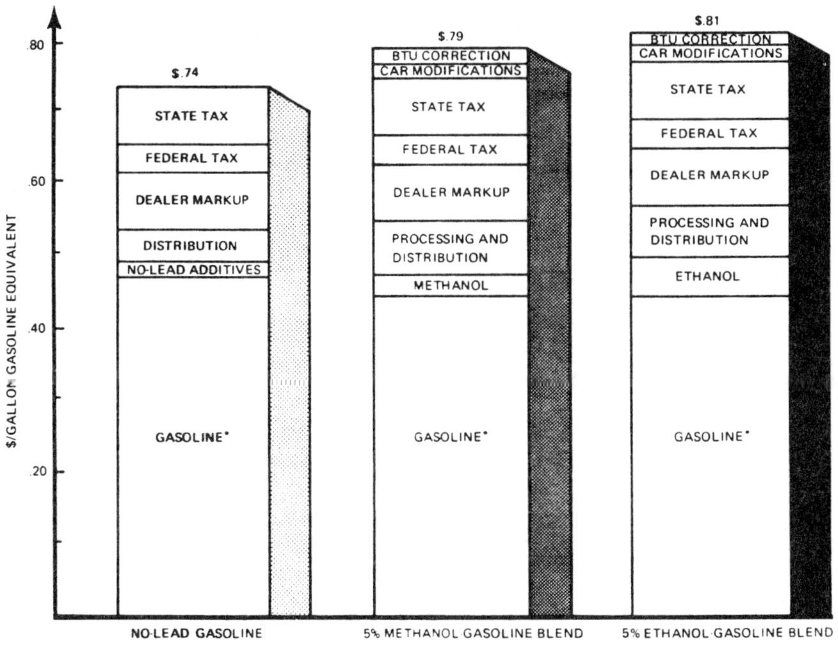

*Gasoline price relative to imported crude only ($14.00/barrel)

Source: DOE HCP/T4101-03

PRODUCTION OF BIOMASS-BASED ALCOHOL FUELS

The analysis presented in this section is based upon the MOPPS (Market Oriented Program Planning Study—DOE 1977) year 1990 gasoline demand estimate of 115 billion gallons (14.3 quads). A larger volume of fuel would be required to meet this energy demand when using alcohol-gasoline blends because the energy content of the alcohols is less than that of gasoline. The actual differences in volumetric demand among 5% alcohol-gasoline blends and gasoline are shown in Figure 1.6. The total fuel demand is held constant at 14.3 quads in each column.

Figure 1.6: Gasoline-Alcohol Fuel Volumes Required to Meet 1990 Energy Demand

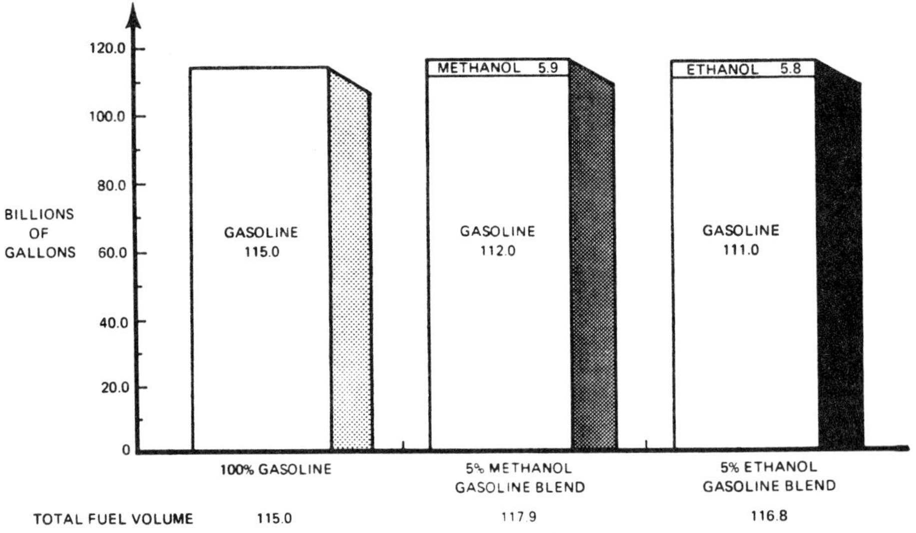

Note: All cases assume constant 14.3 quads for 1990 fuel demand

Source: DOE HCP/T4101-03

Resources Required for Nationwide 5% Ethanol-Gasoline Blend by 1990

It was shown in Figure 1.6 that 5.8 billion gallons of ethanol would be required to produce 116.8 billion gallons of a 5% ethanol-gasoline blend in 1990. This is approximately 29 times the current national production of industrial ethanol. An ethanol plant consuming 12,200 tons of raw sugar juice per day would produce 207,600 gallons of ethanol per day or 67.3 million gallons per year. A total of 86 such ethanol plants would be required to produce enough ethanol for a 5% fuel blend.

The capital cost of this size ethanol plant has been estimated to be 126.8 million dollars (1). Eighty-six plants would cost approximately 10.9 billion dollars.

Although sugar cane is used here as a feedstock, corn, wheat, sorghum, molasses, etc. can also be used for ethanol production at similar facilities with similar capital costs.

Figure 1.7 graphically presents these production requirements. It is reasonable to assume that if production facilities were increased at an effective annual growth rate of about 35%, starting with 2 new plants in 1982, a maximum attainable market penetration of 5% ethanol in gasoline could be achieved by 1990.

Figure 1.7: Production Requirements for Nationwide 5% Ethanol-Gasoline System in 1990

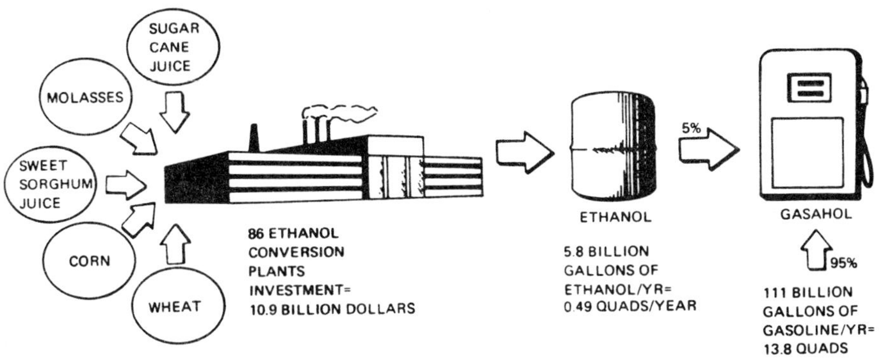

Source: DOE HCP/T4101-03

Feasibility of Attaining Ethanol from Biomass Production by 1990

It has been reported that 100 million acres of unused cropland are currently available in the United States (2). It is unlikely that all of this could be put to effective food production for alcohol production.

Battelle-Columbus Laboratories have estimated that the maximum potential sugar cane production in the continental United States could be nearly 26 million dry tons per year by 1990 (1). This corresponds to about 1.6 billion gallons of anhydrous ethanol per year or just 28% of the amount required to obtain a 5% ethanol-gasoline blend in the country. In other words, only 25 of the required 86 ethanol plants could use sugar cane juice as feedstock. As the continental United States could support just one sugar cane crop per year, the storage of feedstocks for the ethanol plants would also be a major factor in the feasibility of this option.

Grains have the advantage that long-term crop storage is not as serious a problem. Battelle has suggested that in a critical need situation, 20 million acres could be added to our national corn production capacity (1). As a bushel of corn can be converted to approximately 2.8 gallons of ethanol at 83 bushels per acre (average U.S., 1973-75), the expected annual yield from 20 million acres would be 4.6 billion gallons of ethanol. This is about 80% of that required for a nationwide 5% ethanol-gasoline blend.

Production of sweet sorghum, sugar beets, wheat and other grains could also be increased to provide ethanol. The additional land requirements for these crops would overlap those estimated for the additional corn production mentioned above.

It does appear, however, that sufficient additional feedstock could be produced from sugar crops, corn and other grains to make the ethanol required for a 5% alcohol-gasoline blend, nationally. Potential feedstock contributions for 1990 ethanol production are shown in Figure 1.8.

Figure 1.8: Resources Available for Ethanol Production in 1990

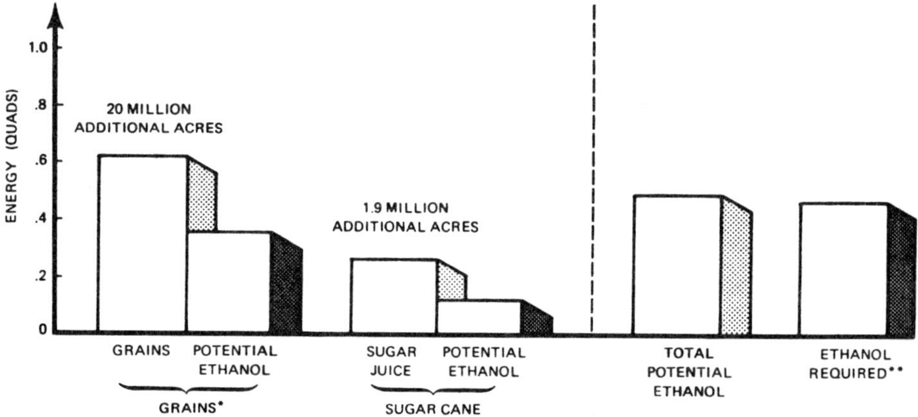

Source: DOE HCP/T4101-03

GOVERNMENT ROLE IN NATIONWIDE BIOMASS-BASED ALCOHOL-GASOLINE FUEL SYSTEM

It has been shown that the combined effects of a relatively low cost for gasoline, relatively high cost for alcohols, and the additional costs of an alcohol-gasoline system would result in higher consumer prices for alcohol-gasoline blends.

To guarantee that an alcohol-gasoline system would be economically viable for the country, it is evident that the prices of gasoline and alcohol-gasoline blends must be competitive. When two fuel types are equivalent in all performance aspects, their prices should be approximately equal. The cost differences of alcohol-gasoline blends and gasoline are small on a cents per gallon basis. On a national scale, however, these small differences will be magnified significantly.

The annual expense of converting from no-lead gasoline to a 5% ethanol-gasoline blend was calculated by multiplying the cost difference for the two fuels by the amount of blended fuel consumed. The annual governmental cost to balance the market for the two fuels was determined to be approximately 6.9

billion dollars by 1990. The imported oil saved by the conversion to the blended fuel would be 0.49 quad.

It was assumed that only one-half of the nation's annual gasoline consumption would be blended fuel when the costs of balancing markets for 10% alcohol blends and no-lead gasoline were determined. This assumption was made because of the uncertainty surrounding the production of fuel-alcohol in quantities greater than about 6 billion gallons per year by 1990.

The 1990 governmental cost to balance the costs of a 10% ethanol-gasoline blend and no-lead gasoline was determined to be approximately 5.9 billion dollars. The government cost is less than in the 5% cost since there are less conversion costs in applying the alcohol system to just one-half the national gasoline system. Import savings would again be 0.49 quad.

To obtain a better feel for the magnitude of the nationwide costs of these programs, total costs were divided by the estimated number of barrels of imported oil saved in each case. These values are presented in Figure 1.9. Net additional costs per barrel range from $67 to $79 or 3.6 to 4.3 times the cost of an imported barrel of oil in 1990 ($18.44 in 1976 dollars).

Figure 1.9: Net Additional Costs of Each Barrel of Imported Oil* Replaced by Alcohol Fuel System Options (in 1990)

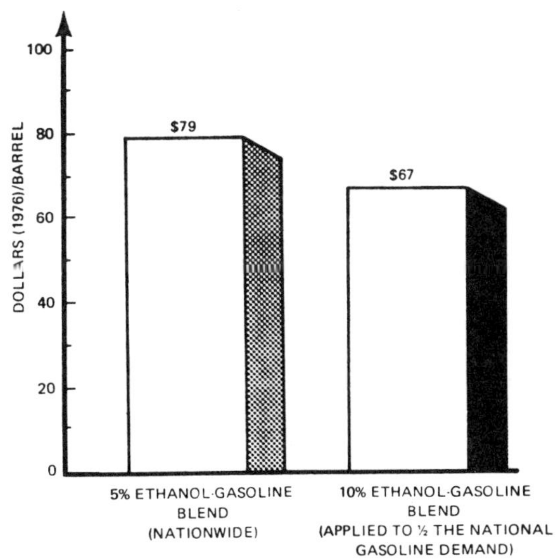

*Assumed cost of barrel of imported crude in 1990 is $18.44 (1976 dollars)

Source: DOE HCP/T4101-03

Although assumptions made in this analysis were based upon current information, there is still significant uncertainty regarding the economic benefit of octane improvement for alcohol-gasoline blends and the requirements and costs for automobile conversion. Total annual costs and costs per barrel for the alcohol-gasoline options considered are quite sensitive to the assumptions made.

For the alcohol-gasoline options assessed, the table below presents the variations in cost per barrel for each of the following separate alternate assumptions:

> (1)   There is no octane benefit from alcohol-gasoline blends;
> (2)   No automobile conversion is required; and
> (3)   There is no octane benefit from alcohol-gasoline and no automobile conversion is required.

Sensitivity of Alcohol-Gasoline Option Costs to Critical Assumptions
(Dollars/Barrel of Replaced Crude Oil)

	Base Case	No Octane Improvement	No Automobile Conversion	No Octane Improvement and No Automobile Conversion
5% Ethanol-gasoline blend	79	106	49	77
10% Ethanol-gasoline blend (½ the national gasoline demand)	67	81	53	66

Under promising assumptions (No. 2) for alcohol-gasoline blends, the net additional costs per barrel are at least 2.5 times the cost of imported oil in 1990. Under less promising assumptions (No. 1), costs per barrel increase to more than 4 times the cost of imported oil.

Further investigation is in order to confirm or improve the accuracy of the cost assumptions. Still, it is evident from this analysis that the relative costs of alcohol and gasoline are not the only factors that constrain the market penetration of alcohol-gasoline blends. The cost of the transportation and processing systems required to develop and maintain an alcohol-gasoline fuel system in the country is significant.

With the small percentages of alcohol in the alcohol-gasoline blends, there would have to be a significant change in the price of the alcohols to effect even a relatively small change in the price of the blend. Government incentives such as investment tax credits for biomass alcohol production, therefore, would have little effect in reducing the price of an alcohol-gasoline blend.

SUMMARY

This preliminary assessment is summarized in the following conclusions:

- There are sufficient biomass resources available in the United States to support alcohol production for a nationwide 5% ethanol-gasoline blend automotive fuel system in 1990.

- The technology required for production of biomass-based alcohols is available today.

- It would be possible, but difficult, to construct sufficient biomass conversion facilities to support a 5% alcohol automotive fuel system by 1990.

- It is technically feasible to convert automotive fuel transportation and utilization systems to accommodate an alcohol-gasoline fuel system.

- Cost to the taxpayer of implementing any national alcohol-gasoline fuel system would be substantial, ranging to 6.9 billion annually by 1990.

- The import savings resulting from an alcohol-gasoline system would range from 68 to 88 million barrels per year by 1990.

REFERENCES

(1) Lipinsky, E.S., et al, *Fuels from Sugar Crops,* Battelle-Columbus Laboratories, Columbus, OH (1976).

(2) McElroy, A.D., "Utilization of Land with Limited Capabilities," Midwest Research Institute, paper presented at Biomass—A Cash Crop for the Future Conference, Kansas City, MO (March 1977).

ETHANOL FROM BIOMASS COMPARATIVE ECONOMIC ASSESSMENT

MITRE STUDY—28 PROCESSES

The information in this section is based on *Comparative Economic Assessment of Ethanol from Biomass*, prepared by the Mitre Corporation, Metrek Division for the U.S. Department of Energy (DOE Report HCP/ET-2854), September 1978.

Introduction

The production of alcohol fuels from biomass is frequently mentioned as one way to reduce the nation's reliance on petroleum. Numerous studies have been conducted to assess the potential of biomass-based alcohol fuels and several others are either currently underway or are planned. Results of these analyses, however, vary enough that the economic viability of an alcohol-from-biomass program remains in doubt.

Ethanol is the alcohol which has received the most attention recently because: it can be produced from several surplus agricultural commodities; feedstock could be grown on set-aside agricultural land; commercial biomass-to-ethanol conversion systems are available; and the fuel properties of gasoline-ethanol blends are often considered to be better than those of gasoline-methanol blends.

Metrek has reviewed fourteen of the ethanol-from-biomass studies which were conducted during 1976–1978 to explain some of the important differences in economic conclusions among them. No attempt was made to critically evaluate any of these assessments. Instead, economic data were examined in the context of a common analytical framework so that comparisons among studies could be made and some of the larger differences could be explained. A life cycle cost model was used for this purpose. Financial data were converted to 1977 dollars and all other data were converted to common units. The results of the analysis are presented below following a brief background discussion of biomass-based ethanol fuels.

History of Ethanol Production

The knowledge to produce alcohol from grains and fruits has existed since the earliest days of recorded history. Processes to brew beer are depicted by the Mesopotamians and the Egyptians as early as 2500 B.C. Pictures of grape presses in Egyptian towns exist from 2000 B.C.

Despite this ancient and almost universal capability to convert sugar and starches to ethyl alcohol, modern methods were developed only as recently as the mid-19th century. At that time the German botanist F.T. Kutzing and French chemist Louis Pasteur demonstrated that fermentation was due to the action of yeast cells which produce the enzymes necessary to convert sugars into ethyl alcohol. Later research showed that the complete yeast cell is not necessary for fermentation to occur but that an appropriately prepared extract will serve equally well. It was also discovered that certain acids and enzymes could be used to convert the complex nonfermentable cellulose molecule into the fermentable glucose molecule via a process called hydrolysis. Glucose could then be converted into alcohol via fermentation.

The earliest systematic studies of ethanol production through the hydrolysis of cellulose were carried out in Germany during World War I. Two commercially important processes were developed: a weak acid method (Scholler) and a strong acid method (Rheinau). The weak acid process was further researched and improved in the United States and became the basis for the commercial development of the synthetic route of ethanol production from petroleum-derived ethylene.

The historical use of ethanol has been primarily as a beverage, but it had other uses as well. Before the discovery of petroleum in the last century it was widely used for cooking, heating and lighting purposes. An early enthusiast for using alcohol as a motor fuel was Henry Ford. Between 1935 and 1937 he sponsored three conferences in Detroit on the industrial uses of such farm products as grains, soybeans and peanuts. When questioned about the availability of fuel for automobiles, he replied, "Our roads are lined with the energy to run our cars." The Model A was usually equipped with an adjustable carburetor designed to permit the burning of alcohol, gasoline or any mixture of the two.

Another well-known figure of the time who advocated the use of alcohol fuel was Alexander Graham Bell. In 1922 he wrote, "The world's consumption of fuel has become so enormous as to show that our present supplies cannot possibly last for many generations more. . .Alcohol is a beautifully clean and efficient fuel which can be produced from vegetable matter of almost any kind. The waste products from our farms are available, even the garbage of our cities."

The production of industrial ethanol from agricultural crops was widespread in Europe until the mid-1940s. The main crops used in the process were corn, potatoes and sugar beets. After 1945 ethanol from petroleum-based sources replaced fermentation of crops in the industrial chemical market because it could be produced at lower cost.

In Europe between the Wars, there was extensive experimentation with, and use of, a variety of substitute fuels. Alcohol blends were used with success in over four million vehicles at this time. Blends were generally in mixtures of up to 25% with gasoline.

In Great Britain after World War I, there was a clause inserted in the Finance Act of 1920 legalizing the use of ethyl alcohol for fuel purposes. This fuel was free of all duty and restrictions in Great Britain, and the same regulations applied throughout most of the countries of the British Empire.

Some of the countries and territories throughout the world which used ethanol blends in gasoline during the decades of the 1920s and 1930s include: Argentina, Australia, Cuba, Japan, Natal, New Zealand, the Philippines, South Africa and Sweden.

In 1936 a study by the U.S. Department of Agriculture predicted that fossil fuel supplies would eventually be exhausted. It concluded, however, that replacement by alcohol fuels would not be economically feasible without government subsidies.

There were indeed heavy costs involved in the European government-sponsored programs for the development of an ethanol-production industry. Imported oil at that time sold for about $0.09/gallon while the cost of alcohol averaged about $0.44/gallon. The expense of the program was justified, however, on grounds of the need: (1) to achieve independence from imported petroleum in the event of war, and (2) to establish a secure source of ethanol for munition industry requirements. Added benefits to national interests were seen as: (3) the stimulation of home agriculture, labor and industry during the depression years, and (4) reduced national trade deficits.

During the 1930s the major interest for the United States in alcohol fuels lay in the foreign export markets. Chrysler Motor Corporation produced cars which were slightly modified to accommodate shipments to New Zealand, a country which was using 100% alcohol fuel at the time. International Harvester also made trucks for export to the Philippines which were powered with engines designed to burn ethanol.

During the Second World War, the German army fueled most of its vehicles on alcohol made from potatoes. The United States built an ethanol plant in Omaha, Nebraska to produce motor fuel for the Army, and the potential for petrochemical products from grain-derived ethanol was recognized in the large-scale synthetic rubber production for wartime needs. Throughout the war there were service stations in Kansas, Illinois and Nebraska which sold an alcohol-gasoline blend called "Agrol." By 1944 the biomass-produced alcohol industry, aided by government incentives, had registered a sixfold increase in production in five years.

The end of the war in 1945 gradually restored the cheap availability of petroleum to the world. Governments withdrew their subsidies to national ethanol production programs, and by the end of the 1940s most facilities had been closed. Several ethanol-fuel plants are said to continue in operation in Eastern Europe.

Interest in the use of surplus or distressed grain for alcohol production has persisted in the midwestern United States because such a program would benefit farmers. In 1971, two years before the Arab oil embargo provided additional incentives, the Nebraska legislature passed several bills which established programs to aid in the development of a grain alcohol industry. An automotive fuel, known as "Gasohol," consisting of a blend of 10% agriculturally-derived ethyl alcohol and 90% unleaded gasoline, was developed and tested.

Many countries around the world are looking into the feasibility of converting local crops into energy. Australia is actively investigating the development and manufacture potential of ethanol from its eucalyptus forests. But it is Brazil which has implemented the largest program to use biomass feedstocks for the production of alcohol fuel. (The Brazilian experience and the Nebraska Gasohol program are reviewed in later chapters.)

Summaries of Studies Reviewed

In order to develop a comparison between various estimates of the costs of producing ethanol, a variety of reports, papers and studies were read and appropriate material extracted from each. These reports, all prepared between late 1976 and mid-1978, consist, either in whole or in part, of descriptions and analyses of the economic aspects of ethanol manufacture from biomass. The reports reviewed are the following:

(1)   U.S. Department of Agriculture, Economic Statistics and Cooperative Service, "Gasohol from Grain—the Economic Issues," January 19, 1978, ESCS-11.

(2)   Battelle-Columbus Laboratories, F.C. Schaffer and Associates, Inc., "Economic Study of Alcohol from Cane Juice," April 15, 1978.

(3)   Battelle-Columbus Laboratories, Louisiana State University, Texas A&M University, University of Florida, F.C. Schaffer and Associates, U.S. Department of Agriculture, "Second Quarterly Report on Fuels from Sugar Crops," October 31, 1977.

(4)(5)   Battelle-Columbus Laboratories, "Systems Study of Fuels from Sugarcane, Sweet Sorghum, Sugar Beets and Corn, Volume II," March 31, 1977; "Systems Study of Fuels from Sugarcane, Sweet Sorghum and Sugar Beets, Volume III: Conversion to Fuels and Chemical Feedstocks," December 30, 1976.

(6)   University of California/Berkeley, Lawrence Berkeley Laboratory, "Process Design and Economic Studies of Alternative Fermentation Methods for the Production of Ethanol," G.R. Cysewski and C.R. Wilke, August 1977.

(7)   Development Planning and Research Associates, University of Nebraska, "Gasohol Economic Feasibility Study, Progress Reports No. 1-4," January-May 1978. (Material contained in the DP&RA final report, dated July 1978, was not available for inclusion in this study at the time the analysis was carried out.)

(8)   The Mitre Corporation, Metrek Division, "Silvicultural Biomass Farms, Vol. V: Conversion Processes and Costs," C. Bliss and D.O. Blake, May 1977.

(9)-(11)   University of Nebraska, A. Scheller, Department of Chemical Engineering, "Grain Alcohol—Process, Price and Economic Information," September 1976; "The Production of Industrial Alcohol from Grain," October 1976; "The Production of Ethanol by the Fermentation of Grain," November 1977.

(12)   Stanford Research Institute (SRI), Menlo Park, California, "Preliminary Economic Evaluation of a Process for the Production of Fuel Grade Ethanol by Enzymatic Hydrolysis of an Agricultural Waste," January 1978.

(13)   Stanford Research Institute, Menlo Park, California, "An Evaluation of the Use of Agricultural Residues as an Energy Feedstock—a Ten Site Survey (Vol. II)," July 1977.

(14)   Stanford Research Institute, Menlo Park, California, "Mission
        Analysis for the Federal Fuels from Biomass Program," June 12,
        1978.

Prior to summarizing these reports, several general comments should be made
about them:

- The level of sophistication of economic analysis varies greatly from report to report. This is because in some of the work the economic
  aspects of ethanol manufacture were of secondary importance, while
  process design considerations received more emphasis.
- The approach taken to obtain product prices varies. In some cases, life
  cycle costs were obtained using a revenue requirements approach. In
  others, discounted cash flow analysis was used; and in still others, return on investment was computed for an assumed product price.
  While in most of the studies the time value of money was taken into
  consideration, this was not always the case.
- Due to both the differing methodologies used and the varying assumptions concerning feedstock costs, by-product credits, and financing of
  plants, the ethanol selling prices which are given in each study are not
  comparable and do not necessarily reflect the relative merits of each
  process.
- A large variety of feedstocks was used in the studies reviewed. Some of
  these, such as wood and corn, require pretreatment to convert cellulosic or starchy material to sugars; others, such as molasses and
  sugar juice do not. These differences are reflected in the feedstock
  costs, capital and operating costs and process efficiencies used.
- The facility sizes reviewed vary from 4 to 68 million gallons of ethanol
  per year. The median plant size, however, was about 20 million gallons per year.

Brief descriptions are provided below of the reports used in this study. The
descriptions are not intended to be critiques or evaluations of the work. They
summarize the economic aspects of the studies, the major assumptions made,
omissions in data supplied, and any other aspects of the work which, in the context of this evaluation, should be considered.

While some data are included in the summary descriptions, a complete data set
obtained from the studies is used in the subsequent analyses.

(1) U.S. Department of Agriculture, "Gasohol from Grain": In this report, the
major issues involved in producing ethanol from grain and in mixing ethanol with
gasoline to yield gasohol are discussed. The report is presented in a question-
and-answer format.

Capital costs developed by Battelle are used for a 20-million gallon per year
ethanol facility. A range of operating costs is taken from studies by Battelle,
the U.S. Department of Agriculture, and Scheller. Ethanol prices are computed.
for corn costing between $1.00 and $3.50/bushel in $0.5 increments and for high
and low debt financing of the plants. The value of distillers' dried grain (ddg) by-
product is not given in terms of a fixed value per ton but is allowed to vary with
the price of corn. Because the ddg has a feed value that is 135% of corn, the
monetary value assigned to it is also given as 135% of the price of corn.

The report does not take credit for by-products (other than ddg), such as fusel oil, esters, aldehydes and carbon dioxide. It is shown that there is uncertainty as to whether these by-products can be sold; and if so, that these credits would only amount to one-half cent per gallon of alcohol. For these reasons, the comparative economic analysis by Mitre does not assume by-product credits for these products.

The costs per gallon for ethanol obtained in the USDA report vary from a low of $0.74/gallon for corn costing $1.00/bushel (using the lowest range of both conversion and capital costs), to a high of $1.54/gallon for corn costing $3.50/bushel (using the high range of conversion and capital costs). For a generally used corn price of $2.50/bushel, ethanol prices of $1.07 to $1.31/gallon were obtained.

(2) F.C. Schaffer and Associates, "Alcohol from Cane Juice": This report presents a detailed cost estimate, as well as an operating income and earnings forecast, for existing ethanol-from-sugarcane technology. The facility is expected to process 9,000 tons of raw cane feedstock per day (designed for 10,000, assuming 10% lost time) from which the juice is extracted by a typical sugarcane milling tandem.

There will be four 150,000 lb/hr steam boilers designed to use the bagasse which remains from the milling operation. The system will provide all of the energy needs of the facility, estimated to be an average of 568,000 lb/hr.

After the cane milling and squeezing, the raw juice is processed by conventional clarification and rotary vacuum filtration. During this phase, the juice is partially treated with lime, not for neutralization (as would be the case in sucrose crystallization), but to enhance the juice clarification which assists in yeast recycling and increasing ethanol recovery. The expected extraction of fermentable sugars is about 99% of that in the sugar juice and 92% of that in the raw sugarcane.

The clarified juice is then concentrated through evaporation to about 20% total sugars to reduce the steam requirements in later process steps. The concentrated juice undergoes batch fermentation for 18 hours in ten 250,000 gallon steel tanks. Finally, about 80% of the yeast is recycled after centrifuge separation. The remaining 20% of the yeast is available for sale, but no by-product credit is assumed. Distillation of the fermented juice in a stopping/rectifying column produces a 95.5% ethanol solution. Afterwards, benzene is added to produce 99.5% (absolute) ethanol with a final yield of 140,000 gallons/day. In addition, almost 19,000 lb/hr of 60°Brix concentrated stillage is produced, but neither credits nor disposal costs were included in the financial analysis.

In this study, the economic analysis was made on an assumed selling price of $1.20/gallon for all cases, based on the current price of industrial ethanol. As such, there was no explicit assumption regarding the desired return on equity which, along with other costs over the project life, is often used to derive a product cost through calculating total revenue requirements.

A total of six cases were developed by varying the number of operating days per year (90, 180, 330) based on the sugarcane grinding season, and varying the purchase price of raw cane feedstock ($10.00/ton, $13.50/ton). While most of the cost components are fairly uniform, the sugarcane grinding season varies greatly from less than three months in Louisiana, to around five months in Florida, Texas and Puerto Rico, to a full year-round crop in Hawaii. Therefore, of the six cases, only four were considered sufficiently economical to warrant comparison with other ethanol technologies. The two 90-day operations were omitted. Finally, each case assumed a 60/40 debt/equity ratio, with 9% interest over a 10-year amortization schedule, and no maximum economic life was specified for the cane milling/ethanol fermentation.

(3) Battelle-Columbus Laboratories, F.C. Schaffer Associates, "Ethanol from Molasses": In this study, the detailed economic data of producing ethanol from molasses under varying conditions were provided to Battelle by F.C. Schaffer and Associates. Unlike most studies on ethanol production where financial parameters are the main variables, the two cases presented here represent fundamentally different operations posing a major economic tradeoff.

Molasses, the syrupy residue from sucrose crystallization, has an advantage over raw sugar juice in that it can be stored during the off-season. Since a prime drawback of using sugar juice has been an inability to spread the facility's capital cost over sufficient output (given the limited processing season), the use of purchased molasses in the off-season allows a more efficient use of plant and equipment. Furthermore, as a by-product of sucrose production, molasses is substantially less expensive as an ethanol feedstock than sugar juice on an equivalent Btu basis, with only a minor loss in process conversion efficiency.

The two cases presented in this report are variations of 10,000 tpd sugar mills such as those which exist in Florida or Texas. Specifically, these cases call for: (1) production of ethanol only during a sugarcane grinding season of 150 days; and (2) year-round production of ethanol, using purchased molasses for operation during the off-season.

As in the Schaffer study which examined cane juice, the economic analysis was conducted with an assumed ethanol selling price of $1.20 per gallon.

In the first operation, the facility is assumed to use only the molasses that it generated by sucrose crystallization during a sugarcane processing season of 150 days. Standard cane processing practices are used as described in the Schaffer report on sugarcane. The sugar juice is processed to sucrose, after which the residue, 60,000 tpd of molasses, is purified with sulfuric acid and lime to prevent contamination. The purified molasses is then subjected to fermentation for about 30 hours in seven 50,000-gallon fermentation tanks. After yeast separation and recycling, the mash is distilled in a two-column beer still and rectifying column. Product yield is 22,223 gpd of 95% ethanol. In addition, about 24,800 lb/hr of fermentation stillage at 60°Brix and a small quantity of unrecyclable yeast are produced. Credits for these by-products and disposal costs are not considered in the analysis.

In the second operation, the same processes occur except that stored molasses from other sugar mills is purchased to enable the ethanol plant to operate 330 days a year. However, in order to operate the distillery year-round, it is assumed that fuel oil must be purchased for steam generation at $0.30/gallon, or about 15% of total operating costs. Although it may be argued that stored bagasse could displace this expensive fuel, industry practice (i.e., the design of mills to use all of the bagasse they produce) limits this possibility at present. Since daily output is assumed to be the same, the difference is represented by an annual output of more than double that of the first operation for only a 12% increase in capital cost. One big advantage of year-round operation is that it reduces the capital expenditure from about $2/annual gallon of ethanol to only $1/gallon.

It is impossible to specify the precise capital utilization because, as in the Schaffer study on ethanol from cane juice, no maximum economic life of the facility is assumed. Some advantage exists for the year-round facility under most probable circumstances, extending well beyond a 50-year economic life. Using the common assumptions of the 18-year tax life as the effective economic life, $1.20/gallon selling price of ethanol and 100% debt financing at 9% for 10 years, the year-round facility was estimated to have an annual return on investment of 20% as compared to 11% for the 150-day facility.

In addition to the two Battelle cases, Schaffer and Associates also provided data for two other cases applicable to Louisiana. These included a 70-day operation based on the indigenous cane-growing season, and a year-round operation based on the same type of molasses purchase as the previous case. Although the year-round case was relatively more economical, both of the Louisiana cases were estimated to result in negative net returns on investment.

(4) Battelle-Columbus Laboratories, "Conversion to Fuels and Evaluation of Cane": Production of ethanol via the fermentation of raw sugarcane juices is one of four microbiological conversion processes examined in this first report. Three feedstock sources of fermentable sugars were considered: juices extracted from the stalk or root of the sugar crop; molasses derived as the syrupy residue from the crystallization of sucrose from sugarcane; and the total saccharides derived from hydrolysis of the cellulose and hemicellulose of sugar crop fiber. Of the three, the only economic analysis presented is for fermentation of cane juice, although the processing costs of sweet sorghum juice in cane mills is considered comparable. Molasses is suggested as a supplemental feedstock during the off-season. It has not been explicitly considered in this evaluation, however, because of its greater value as animal feed and its general price instability. Finally, derivation of total saccharides by hydrolysis was judged less attractive as an ethanol feedstock in the near future.

The ethanol facility purchases 8,900 tpd of raw cane juice which is extracted through the squeezing of sugarcane in the same roller mill system as is used in the first step of sucrose crystallization. The juice is then clarified. The clarification process used here does not employ lime neutralization as is the standard industry practice, because the fermentation is designed to run at an approximate pH of 4. The clarified juice is concentrated through evaporation and then undergoes batch fermentation.

Fermentation is followed by centrifugation for yeast separation and distillation to separate the ethanol from the water. The facility is expected to operate 330 days/year with an output of 207,000 gpd of 95% ethanol. In addition, a stillage recycle system is included to dewater the fermentation residue to a 9% water mixture of stillage and yeast. The estimated stillage by-product, including that derived from sugarcane tops and leaves, was estimated at 615 tpd.

Economic parameters were varied in this study mainly to assess the impact of high and low debt financing. The two cases assumed, respectively: 60/40 and 30/70 debt/equity ratios; 14 and 15% returns on equity; an 8.75 and 8.5% interest rate on debt to be paid over the 20-year project life. Common financial parameters not already mentioned include $13.4 million in annual by-product credits, a 24% investment tax credit, and working capital requirements that include a half-year (167 out of 330 operating days) inventory of feedstock cane juice.

(5) Battelle-Columbus Laboratories, "Conversion to Fuels and Evaluation of Corn": This study by Battelle evaluates the production of ethanol from the fermentation of corn. As an old and well-known technology in the distilled beverage industry, corn-to-ethanol is considered to have an advantage over sugar-cane juice in that it is easily stored and readily available in commodity markets. This enables a distillery to purchase its grain as needed, rather than having to store a half-year's supply (i.e., sugar juice in the form of molasses) to tide it over until next season.

Corn may be converted to ethanol in a number of different ways. The minimum processing method includes coarse grinding of the whole grain, conversion of the starch to glucose by enzyme treatment, fermentation of the glucose by yeast, and distillation. This simple process for which the technology is readily available, is not presently being used in the U.S. because of economic constraints.

Other corn-to-ethanol techniques involve some integration with the processing of corn for food products through two basic approaches, dry milling and wet milling. In dry milling, the germ is separated from the endosperm and the hull is processed to make corn oil. The endosperm, and hull are ground into cereal products, and the remaining cake is sold for animal feed. This technique has the advantage of obtaining corn oil, a valuable food commodity which helps to defray the cost of ethanol production. Wet corn milling, on the other hand, is a more complicated process with numerous by-products. It involves the loosening of kernel components by soaking and grinding while still wet to break the germ loose, and centrifuging to separate the starch from the gluten.

Several milling plants in recent years have integrated the production of ethanol for beverage use with this technology, by grinding less vigorously, and centrifuging so as to remove only that starch which is easily separated. A noncommercial modification of the wet milling system could achieve the separation of starch from gluten by enzyme treatment.

Despite the advantages of integrating ethanol with a food product conversion operation, the ethanol output amounts to only 22% of that from a conventional ethanol plant of the same corn grinding capacity. Furthermore, the net cost of producing ethanol is dependent on the price of corn and the prices assigned to related products such as starches, corn or fructose syrups and animal feeds. The only detailed economic analysis provided in this report was for the minimum processing plant, described above. That facility assumed a corn input of 24 million bushels at $2.50/bushel, and an annual output of 95% ethanol of 68.5 million gallons, based on 330 days of operation. Economic parameters based on high and low debt financing were respectively as follows: 60/40 and 30/70 debt/equity ratios; 14 and 15% return on equity; an 8.75 and 8.5% interest rate on debt to be paid over the 20-year project life. Common financial parameters not already mentioned include $24.5 million in annual by-products credits (233 tons of stillage at $105/ton), a 24% investment tax credit, and working capital requirements that include about 10% of the year's inventory of corn.

(6) University of California, Berkeley, "Alternative Fermentation Methods": This report describes four methods for producing ethanol from molasses. Molasses is delivered to a fermentation facility as a 50% sugar solution and is fermented using yeast. The ethanol produced is then concentrated to 75% by a single distillation step. The four fermentation systems used to produce ethanol are:

(1)   Batch: Fermentors are operated in batches with a 16-hour fermentation time and another 6 hours to fill, drain and sterilize each fermentor.

(2)   Continuous: Fermentors are operated with a 16-hour fermentation time and another 6 hours to fill, drain and sterilize each fermentor.

(3)   Continuous with cell cycle: This is identical to the above process except that some of the cell concentrate from the centrifuge is returned to the fermentors. This increases the cell mass concentration in the fermentor, allowing a higher ethanol production. As a result, fewer fermentors are needed than in the continuous fermentation method without the cell recycle system.

(4)   Vacuum fermentation with cell recycle: Molasses and mineral supplements are fed to a vacuum fermentor with oxygen sprayed through the fermentor to satisfy trace oxygen requirements of the yeast. Ethanol and water are boiled away from the fermentation broth. Fermented beer is pumped to atmospheric pressure and fed to continuous centrifuges to remove the yeast concentrate, some of which is returned to the fermentor. Clarified beer is distilled and the ethanol concentrated to 95%. The remaining yeast is dried and sold.

Process economics data are given for each of these four fermentation systems. Capital costs and production costs exclusive of feedstocks are given for a plant producing 78,000 gpd of 95% ethanol. Capital costs are highest for the batch process ($225/gallon ethanol/day). They are about half the cost for continuous fermentation and are lowest for the vacuum and continuous-with-cell-recycle systems. To obtain capital costs on an annualized basis, capital costs are multiplied by a fixed charge rate of 19% and then divided by the annual production rate in gallons to obtain a quantity called "investment-related costs." Ethanol production processes obtained are strongly dependent on the price of sugar, which is given at $0.70 to $0.75/gallon of ethanol produced, and which accounts for 77 to 82% of the total cost of producing ethanol.

Assuming that the ethanol produced would be sold at $1.10/gallon, returns on investment were computed for each process and for both yeast by-product credits of $0.10/lb and for no yeast by-product credit. Return on investment was defined as yearly profit divided by total capital investment. With a yeast credit, return on investment varied from 18.5% for batch fermentation to 81.5% for vacuum fermentation. Without yeast credits, return on investment ranged from 3.3% for batch fermentation to 64.3% for vacuum fermentation.

(7) Development Planning and Research Associates, "Gasohol Economic Feasibility Study": These progress reports describe work carried out for the U.S.

Department of Energy and the University of Nebraska. The reports provide some data concerning the costs of producing ethanol from corn, potatoes and sugar beets. The data, however, are preliminary and incomplete and it is therefore not possible to use the information provided in the progress reports to obtain ethanol prices where potatoes and sugar beets were used as feedstocks. Based on data provided in the Development Planning and Research Associates (DP&RA) report, only the cost of producing ethanol from corn could be computed. While a complete set of data will appear in the Gasohol Economic Feasibility Study final report, this information was not available for use in this study of the economics of ethanol production.

In the DP&RA reports, three model ethanol plants were considered:

(1) A 9.7 million gallon/year plant using sugar beets in which 4,000 tons of beets are processed a day for 120 days/year and in which the processed sugar juice is converted to ethanol for 300 days/ year at a rate of 32,300 gpd. While the sugar beets are, at $26/ ton, the most expensive feedstock of the three, the by-product obtained is similar in value to distillers dried grains ($0.42/gallon of ethanol vs $0.41/gallon of ethanol).

(2) A 10 million gallon/year plant using corn in which 3.85 million bushels of corn are used a year and in which a ddg by-product is obtained. Raw material costs for corn are lower than for sugar beets ($0.89/gallon of ethanol vs $1.29/gallon of ethanol) and the by-product has a somewhat higher value.

(3) A 7.2 million gallon/year plant using potatoes as a feedstock. This plant has the lowest raw material cost but produces a by-product that is only worth about $6/ton or $0.67/gallon of ethanol produced.

The break-even price of ethanol for all three facilities is computed as the net ethanol production cost in dollars per year, divided by the quantity of ethanol production in gallons per year. The cost of producing ethanol is computed by using a financial assessment model in which investment costs, the cost of investment recapture, annual operating and feedstock costs and by-product credits are taken into account.

The DP&RA model appears to be similar in structure to the Mitre life cycle cost model used in this study. As noted, because certain data are not available in the DP&RA progress reports, not all of their model plants could be used in the economic assessment.

(8) The Mitre Corporation, Metrek Division, "Silvicultural Biomass Farms": Production of ethanol from wood is one of many products (fuel and nonfuel) from silvicultural biomass farms covered in this report. Conversion of wood to ethanol involves two distinct process steps—the hydrolysis of wood to sugars and the fermentation of sugars to ethanol. Of the many technologies which integrate these two steps, the one considered for economic analysis in this report is a composite of the Scholler and the Madison processes. This method is in use in the Soviet Union and Japan.

In this process, the biomass feedstock is discharged from the storage hopper of the standard front-end unit into a digester. Steam is introduced, followed by

sulfuric acid and a recycled dilute prehydrolyzate stream, and finally, hot water.
The contents are maintained at 135° to 150°C for 30 minutes. The prehydrolyzate is drained, and dilute sulfuric acid is introduced at the top of the digester
with high-pressure steam to raise the temperature to 150°C in order to begin the
main hydrolysis. The main hydrolysis is carried on for three hours and during
this time the reaction temperature is raised from 150° to 190°C.

Methanol and furfural contained in the vapors resulting from "flashing" the solution to lower pressure are separated in a distillation tower. The hot hydrolyzate
is neutralized with lime and the calcium sulfate precipitate is separated. A wash
cycle is used on this precipitate to recover a small amount of sugar. The sugar
is passed to fermenting tanks for alcohol or yeast production, although concentration to crude molasses is another alternative. The dilute alcohol stream passes
to distillation towers which produce 95% ethanol.

Seven cases of ethanol production are developed in this study by employing four
variations of input prices for wood biomass and three variations of plant costs
by feed/output capacity. Of the seven cases, four were selected for comparison
with other ethanol technologies.

In terms of bracketing the range of acceptable wood feedstock costs, the $1/
MMBtu and $2/MMBtu cases were selected. Regarding plant size, the low and
intermediate plants were selected with feed capacities of 850 and 1,700 oven
dry tons (ODT)/day and outputs of 84,400 and 168,800 gpd of 95% ethanol,
respectively. Based on experience in the pulp and paper industry, the maximum
size facility of 3,400 ODT capacity, although technically feasible, is not considered a realistic option in the near-term. Waste products in all facilities include
residual sugars, especially pentose, and calcium sulfate. The 850 ODT/day facilities will produce approximately 54 tpd residual sugars and 104 tpd calcium sulfate. No by-product credits are given in any of the plant configurations which
give all process steam from wood biomass residues.

Economic parameters in this study include: financing at 53% debt, 12% preferred equity, and 35% common equity; interest on debt of 8% amortized over
30-year life; return on preferred and common equity of 15 and 25%, respectively;
and 10% working capital requirements without specification of inventory size.

*(9) through (11) University of Nebraska, W.A. Scheller, "Grain Alcohol—Process,
Price and Production," "The Production of Ethanol by the Fermentation of
Grain," and "The Production of Industrial Alcohol from Grain":* In the three
papers reviewed, Scheller describes the production of ethanol from corn, distressed grain, and grain sorghum. In all cases, the grain is ground and cooked to
solubilize and gelatinize the starch which is then converted to dextrose via batch
enzymatic hydrolysis. Yeast fermentation converts the dextrose to ethyl alcohol.
The alcohol and residual grains are sent to a distillation section where the alcohol-
water solution is separated from residual grains. The grains are dried to yield
ddg and solubles, while the alcohol-water solution is further distilled to obtain
190 or 200 proof ethanol.

The major soluble by-product produced is ddg, which is used as cattle feed.
Based on the Btu content of the ethanol produced and the digestible energy of
the ddg, high conversion efficiencies, in the 84 to 90% range, are achieved. In
computing profit, Scheller does not treat capital costs as an expense. Rather,

the straight-line depreciation of capital costs and the simple interest on the loan used to finance part of the plant are subtracted from profit before taxes and depreciation. In this case, "profit" is the surplus of income from ethanol and by-products over annual feedstock and operating expenses. Net cash flow is computed as the sum of depreciation and net cash balance (after interest, depreciation, taxes and loan repayment reserve are subtracted). The measure of project feasibility used is the return on private capital, defined in the net cash balance divided by the quantity of private capital invested in the ethanol plant.

The facilities postulated by Scheller each produce 20 million gal/yr of ethanol and substantial quantities of ddg (741 x 10^9, 1,084 x 10^9 and 1,171 x 10^9 Btu/yr for grain sorghum, corn and milo, respectively.)

(12) Stanford Research Institute, "Fuel Grade Ethanol by Enzymatic Hydrolysis": The process described in this study is one which produces 25 million gal/yr of 95% ethanol from a sugar solution obtained by enzymatic hydrolysis of wheat straw. Processing is divided into six sections, all of which operate concurrently except for the fermentation of sugar to ethanol.

In the substrate pretreatment section, the wheat straw is milled and treated with a 1% solution of sulfuric acid at 212°F for about one hour. The pretreatment produces xylose and arabinose which are separated from the reactor effluent by filtration and become part of an aqueous acid extract which also includes other solubles. A wet filter cake is also obtained from the filtration process and this contains the cellulosic substrate for sugar production. The aqueous acid extract is conveyed to mixing tanks in the enzyme recovery section. Within the enzyme recovery section and the hydrolysis section there is a continuous process which recycles sugar and enzyme solutions to produce a sugar solution which is recovered by rotary filtration. [The enzyme production section, meanwhile, yields an enzyme from a 1% cellulose (newsprint) slurry.] The enzyme is fed to the hydrolysis section. The enzyme solution is recovered from the hydrolysis effluent through centrifugation and filtration. The resulting wet cake contains unhydrolyzed solids and is removed in the waste stream.

The enzyme-free sugar solution from the enzyme recovery section is concentrated from about 4 to about 11% sugar in the sugar solution concentration section by multiple-effect evaporation. This concentrate is then fed to the batch fermentors in the ethanol production section. Fermentation is carried out at a pH of 4 to 5 and a temperature of 86°F for about 40 hours. Yeast is cultivated aerobically in a series of step-up culture tanks and prefermentors. Broth is continuously filtered to remove yeast cells. The filtrate is purified and concentrated to 95% ethanol in three distillation columns.

The investment costs of plant sections described here have been estimated at $70.5 million. Utilities, tankage, waste treatment, general facility start-up costs and working capital would bring the total capital investment, excluding land, to $121 million. The plant gate manufacturing cost, which includes the wheat straw feedstock, utilities, maintenance and labor, operating labor and supplies, control laboratory, plant overhead, taxes, insurance and depreciation, is $3.34/gal produced ($44.14/MMBtu).

(13) Stanford Research Institute, "Agricultural Residues as an Energy Feedstock": This study evaluates the use of agricultural residues as an energy feedstock in ten

study areas selected for on-site analysis. Each of the ten site reports is treated as a complete and independent analysis and contains the following sections: introduction and summary, study area description, residue characteristics, residues most feasible for energy production, energy supply and demand, and feasibility of residue conversion to energy. The report does not include a description of energy conversion processes or facilities.

The site which is reviewed for this paper is the Hendry, Florida area where the main crop is sugarcane, produced on approximately 300,000 acres.

The economics of the conversion of sugarcane to blackstrap molasses or cane juice and then to ethanol is addressed in this report. Economic feasibility is dependent both on future prices of synthetic ethanol derived from petroleum products and on alternative uses for, and sales costs of molasses. The cost of sales is dependent on the cost of molasses fermentation, capital equipment, and plant operations conducted at an existing Florida raw sugar mill location. Long-term projections of these factors are discussed in the report.

The process economics of manufacture of neutral spirits (95% alcohol) from molasses has been examined on the basis of a 1974 study for Alexander and Baldwin (Brown, G., "Technoeconomic Study of Selected Potential Sugarcane By-Products," for Hawaiian Development Company, Ltd., Honolulu, Hawaii; Stanford Research Institute, Menlo Park, CA). Assumptions from this earlier study have been modified to reflect Florida operating conditions. The assumptions are as follows:

- 250 days of molasses storage required when the sugar refinery is not in operation;
- Domestic water system requires a $300,000 capital improvement to supply alcohol plant with process water requirements;
- Plant to run at 0.9 stream factor or 328.5 days per year;
- No capital improvements needed to supply cooling water;
- Steam to be supplied by sugar refinery boilers for the 4-million gal/yr alcohol plant;
- Existing effluent water system adequate for small alcohol plants;
- Product storage of 30 days;
- No by-product recovery or processing equipment is included;
- No contingency factor for inflation has been added;
- No interest charges on construction or land costs are included;
- Working capital includes provisions for product inventory, raw materials inventory, accounts receivable and payable, operating cash, and spare parts;
- Start-up cost includes cost of training operators, initial start-up inefficiencies and minor plant modifications.

Total capital investments are based on standard factors for installed costs. For a 4-million gal/yr ethanol plant, the fixed capital investment is $3,986,000 and total capital requirement is $5,249,000.

Production costs for the 4-million gal/yr plant are calculated from these capital data, known materials and labor requirements, local materials and utilities costs, and certain industry average ratios for other costs.

Basic assumptions, in addition to the quantity of molasses required to produce one gallon of alcohol (2.5 gal/gal 95% ethanol) are as follows:

- Operating labor rate of $6.00 per man-hour;
- Local utility rates are supplied;
- General and administrative expenses estimated at 3% of fixed capital;
- Ten-year straight-line depreciation of fixed capital investment; and
- Interest on working capital and sales and research expenses are not included.

The total estimated production cost is $1.18/gal of ethanol or $15.64/MMBtu. This assumes that there is no return on investment. For a discounted cash flow return on investment of 20%, the required selling price is $1.49/gal ($19.69/ MMBtu).

(14) Stanford Research Institute, "The Federal Fuels from Biomass Program": The commercialization potential of fifteen biomass feedstock-to-production missions was investigated in this study for the Department of Energy, Fuels from Biomass Systems Branch. A "mission" is defined as a possible technical route from a biomass feedstock, through conversion, to a specific product. Of the fifteen missions considered, three involve production of ethanol using the following feedstocks: algae, sugarcane and wheat straw. The processes involved are as follows:

(1) Algae: The algae are converted to ethanol via acid hydrolysis and fermentation. Approximately 25 million gal/yr of 95% ethanol are produced from 1,126 tpd of dry algae (16 x 10^6 Btu/ton). The overall process efficiency is 32%, total capital investment for the sugar and ethanol plants is $53.7 million, annual operating costs are $12.2 million and feedstock costs are $27.9 million/year. The cost of producing ethanol is $26.90/MMBtu. The estimated commercialization date for this plant is post-1990.

(2) Sugarcane: Sugarcane is converted to a 10.7% sugar solution in a plant operating 165 days/yr. This solution is converted to ethanol in a plant operating 330 days/yr. A total of 25 million gal/yr of 95% ethanol is produced from 2,756 tons of dry sugarcane per day. Overall process efficiency is 26%, total capital investment is $53.9 million, annual operating costs are $11.8 million, and annual feedstock costs are $29.56 million. Assuming that it is feasible to store the sugar solution, this plant could be in commercial operation at this time. The cost given for ethanol production is $32.20/ MMBtu.

(3) Wheat Straw: Wheat straw is converted to a 4% sugar solution and then to ethanol by enzymatic hydrolysis. A total of 25 million gallons of 95% ethanol is produced per year from 3,270 dried tons of wheat straw per day. The cost of the wheat straw is given as $15/ton. The plant operates 330 days/yr. Overall process efficiency, including by-products is 23.6%, capital costs are $115.8 million, annual operating costs are $62.6 million and annual feedstock costs are $16.1 million. In this process, significant quantities of a usable by-product, pentose sugar, are produced (34,000 lb/hr). The product is valued at $0.03/lb and thus accounts for $8 million in revenue per year. The cost given for ethanol pro-

duction is $52.60/MMBtu. The available date for this plant is
assumed to be post-1990.

The life cycle costs of ethanol production for all three of the ethanol plants in
this report were computed in the same manner using the same financial parameters. Economic plant lifetime was assumed to be 30 years, depreciation tax
life was 20 years, construction time was 2 years, the return on equity was 15%
and return on debt was 9%. The plant financing was 65% debt and 35% equity.
Combined federal and state income tax rate was 52%. Capital costs included
working capital and interest during construction. When used in Metrek's life
cycle costing model, capital costs without interest during construction were used
because the latter interest during construction is computed by the model. The
price computed for ethanol in $/MMBtu was based on the total revenue require-
ments of the plants over their lifetimes and their total lifetime Btu production.

Economic Data Base

In all, 28 separate ethanol facility configurations were investigated, making use
of eight feedstocks and employing a variety of methods for converting the feedstocks to ethanol. The ethanol technology performance, process economics, and
financial data base used in this study is given in Tables 2.1 and 2.2. Table 2.1
presents the technical performance and process cost information obtained from
the 28 ethanol configurations reviewed in this study. For each one, the following data were obtained:

- Capital cost: This includes cost of land, working capital and start-up
  costs. It does not include interest during construction (idc). Where
  idc was a part of the capital costs given in the reports reviewed, it
  was extracted. This was done to avoid double counting since idc is
  computed as part of the Metrek life cycle costing model. Capital and
  other costs are given in various years' dollars in the reports reviewed.
  In order to compute ethanol costs that are expressed on a common
  basis, the dollar figures in the reports have been converted to 1977
  dollars using the GNP Implicit Price Deflator. Consequently, the
  cost data used in this analysis are expressed in the same (1977)
  year's dollars. The capital costs given in Table 2.1, however, are
  given in the same year's dollars as they appear in the reports reviewed. The capital costs in Table 2.1 are presented in three ways:
  dollars, the ratio of dollars to ethanol production, and the ratio of
  dollars to rate of total production in MMBtu per day. The latter
  figure provides a measure of the comparative costs to produce one
  million Btu per day of ethanol and salable by-products in each facility.
- Operating costs: These are the annual costs of operating the ethanol
  facility and include maintenance materials and supplies, operating
  labor, administrative and support labor, plant supervision, overhead
  expenses, utilities, annual property taxes and insurance. They do not
  include the feedstock cost. Operating costs are given both in terms
  of dollars per year and dollars per million Btu produced. As with
  capital costs, they are expressed in 1977 dollars in the analysis but
  are presented in the table as they were given in the reports reviewed.

Table 2.1: Summary of Cost/Performance Parameters Used

Feedstock name and other data	Schaffer				Schaffer/Battelle		Battelle	
	sugarcane 330 days/yr low price	sugarcane 330 days/yr high price	sugarcane 180 days/yr low price	sugarcane 180 days/yr high price	molasses half-year	molasses full year	corn	cane juice
Feedstock cost								
$MM/yr	29.70	40.10	16.20	21.87	1.62	3.67	60.73	65.13
$/MMBtu	2.47	3.33	2.47	3.33	5.03	5.18	6.65	9.90
Capital cost								
$MM	59.5	59.5	59.5	59.5	6.53	7.45	108.43	118.00
$/MMBtu produced/day	5,941.38	5,941.38	10,843.08	10,843.08	9,441.75	4,922.54	4,898.92	7,955.25
$/gal ethanol/yr	1.29	1.29	2.35	2.35	1.96	1.02	1.58	1.73
Operating cost								
$MM/yr	5.60	5.60	4.64	4.64	0.25	1.41	30.28	21.46
$/MMBtu produced*	1.53	1.53	2.32	2.32	0.99	2.55	3.75	3.96
Process efficiency								
Btu out/in*	0.306	0.306	0.306	0.306	0.783	0.783	0.885	0.823
Ethanol production								
MM gal/yr	46.12	46.12	25.27	25.27	3.33	7.30	68.50	68.20
Btu/yr x 10^9	3,655.27	3,655.27	2,002.91	2,002.91	252.24	552.41	5,183.40	5,160.30
By-product production								
Name	—	—	—	—	—	—	stillage	stillage
Btu/yr x 10^9	—	—	—	—	—	—	2,895.3	253.7
Market value, $MM/yr	—	—	—	—	—	—	24.46	12.18
Operating availability	0.90	0.90	0.49	0.49	0.41	0.90	0.90	0.90
Year dollars	1977	1977	1977	1977	1977	1977	1976	1976
Year available	1977	1977	1977	1977	1977	1977	1976	1976

*Includes by-product Btus.

(continued)

Table 2.1: (continued)

Feedstock name and other data	Mitre.					SRI.		mission analysis	
	850 tpd wood low price	850 tpd wood high price	1,700 tpd wood low price	1,700 tpd wood high price	wheat straw	10-site study molasses	algae	wheat straw	sugarcane
Feedstock cost									
$MM/yr	4.22	8.44	8.44	16.88	9.71	2.60	27.90	16.11	29.56
$/MMBtu	1.00	2.00	1.00	2.00	0.601	7.26	4.70	1.00	4.34
Capital cost									
$MM	127.46	127.46	217.73	217.73	120.92	5.25	53.70	115.80	53.90
$/MMBtu produced/day	24,946.58	24,946.58	21,307.50	21,307.50	23,237.58	6,330.92	10,470.33	11,136.96	10,398.25
$/gal ethanol/yr	5.17	5.17	4.42	4.42	4.84	1.31	2.15	4.63	2.16
Operating cost									
$MM/yr	22.75	22.75	38.00	38.00	64.17	2.13	12.20	62.60	11.80
$/MMBtu produced*	12.20	12.20	10.19	10.19	33.92	7.05	6.46	16.50	6.53
Process efficiency									
Btu out/in*	0.442	0.442	0.442	0.442	0.146	0.845	0.319	0.236	0.265
Ethanol production									
MM gal/yr	24.64	24.54	49.29	49.29	25.00	4.00	24.96	25.00	25.00
Btu/yr x 10^9	1,864.90	1,864.90	3,729.70	3,729.70	1,892.00	302.68	1,872.00	1,892.00	1,892.00
By-product production									
Name	—	—	—	—	—	—	—	pentose	not sig-
Btu/yr x 10^9	—	—	—	—	—	—	—	1,903.2	nifi-
Market value, $MM/yr	—	—	—	—	—	—	—	8.04	cant
Operating availability	0.80	0.80	0.80	0.80	0.90	0.90	0.90	0.90	0.90
Year dollars	1976	1976	1976	1976	1977	1977	1977	1977	1977
Year available	1977	1977	1977	1977	1990	1977	1990	1990	1978

*Includes by-product Btus.

(continued)

Table 2.1: (continued)

	Scheller			Lawrence Berkeley Laboratories.				DP&RA	USDA (ESCS)
Feedstock name and other data	corn & dis-tressed grain	grain sorghum	milo	molasses batch	continuous	cell recycle	vacuum	corn	corn
Feedstock cost									
$MM/yr	17.55	15.27	15.37	20.20	21.51	21.51	20.40	7.77	9.23
$/MMBtu	6.12	5.72	5.21	7.04	7.04	7.04	7.04	5.85	6.65
Capital cost									
$MM	21.00	35.00	27.00	17.60	7.50	5.58	5.21	34.30	29.00*
$/MMBtu produced/day	2,951.00	5,382.00	3,671.21	2,851.13	1,139.64	847.90	840.73	7,726.00	4,211.38
$/gal ethanol/yr	1.05	1.75	1.35	0.64	0.27	0.20	0.19	3.43	1.45
Operating cost									
$MM/yr	6.00	9.32	6.00	5.50	5.03	4.94	3.29	4.40	7.50**
$/MMBtu produced***	2.31	4.13	2.24	2.44	2.09	2.05	1.45	3.30	2.99
Process efficiency									
Btu out/in***	0.897	0.840	0.910	0.778	0.787	0.787	0.781	0.920	0.869
Ethanol production									
MM gal/yr	20.00	20.00	20.00	27.62	27.62	27.62	27.62	10.00	90.00
Btu/yr x 10^9	1,513.40	1,513.40	1,513.40	2,089.70	2,089.70	2,089.70	2,089.70	756.70	1,513.40
By-product production									
Name	ddg	ddg	ddg	yeast	yeast	yeast	yeast	ddg	ddg
Btu/yr x 10^9	1,084.0	741.0	1,171.0	163.7	312.3	312.3	172.1	575.3	1,000.0
Market value, $MM/yr	8.60	6.03	9.20	2.182	4.165	4.165	2.295	4.14	7.88†
Operating availability	0.90	0.90	0.90	0.97	0.97	0.97	0.97	0.82	0.90
Year dollars	1976	1976	1977	1975	1975	1975	1975	1977	1976
Year available	1978	1976	1977	1977	1977	1977	1976	1977	1977

*Midpoint of $25 to $33 million range.   **Midpoint of $6.20 to $8.80 million per year range.

***Includes by-product Btus.   †For corn at $2.50 per bu

Source: DOE HCP/ET-2854

Table 2.2: Summary of Financial Parameters Used

	Schaffer*	Schaffer/ Battelle	. . . . . . Battelle				Mitre	. . . . SRI.		
			cane juice, debt		corn, debt					
			high	low	high	low				
Feedstock	sugarcane	molasses	cane juice, debt high	cane juice, debt low	corn, debt high	corn, debt low	wood	wheat straw	molasses	mission analysis (wheat straw, sugarcane, algae)
Analytical approach	fixed selling price	fixed selling price	revenue requirements		revenue requirements		revenue requirements	revenue requirements	discounted cash flow	revenue requirements
Capital structure, %										
Debt	60	100	60	30	60	30	53	0	0	65
Equity	40	0	40	70	40	70	com eqty 35 pref eqty 12	100	100	35
Interest on debt, %	9	9	8.75	8.5	8.75	8.5	8	—	—	9
Return on equity, %	(15)	—	14	15	14	15	common 25 preferred 15	15	(15)	15
Economic life of facility, yr	(30)	(30)	20	20	20	20	30	15	(30)	30
Tax life of facility, yr	18	18	11	11	11	11	22	10	(20)	20
Income tax										
Federal and state, %	48	48	50	50	50	50	50	48	(52)	52
Other taxes, %										
Gross receipts	—	—	—	—	—	—	2	—	—	—
Property	—	—	—	—	—	—	1.4	—	—	—
Investment tax credit, %	—	—	24	24	24	24	4	—	—	—
Facility construction time, yr	(2)	(1)	(2)	(2)	(3)	(3)	2.5	(2)	(1)	2
Working capital required, as % of total capital costs**	10***	9.4*** † 8.2*** ††	32.8	32.8	19.4	19.4	9.1	13.2	none specified	none specified
By-product credits, as % of annual operating cost	—	—	60.4	60.4	80.8	80.8	—	—	—	12.8 (wheat straw only)

(continued)

Table 2.2: (continued)

	Scheller			Lawrence Berkeley Labs	DP&RA††††	USDA (ESCS)
Feedstock	corn and distressed grain	grain sorghum	milo	molasses	corn	corn
Analytical approach	fixed selling price	fixed selling price	fixed selling price	fixed selling price	revenue requirements	revenue requirements
Capital structure, %						
Debt	70	70	70	60	60	60
Equity	30	30	30	40	40	40
Interest on debt, %	10	(10)	(10)	(9)	(9)	(9)
Return on equity, %	(15)	(15)	(15)	(15)	(15)	(15)
Economic life of facility, yr	20	(20)	(20)	10	(30)	(30)
Tax life of facility, yr	(13)	(13)	(13)	10	(20)	(20)
Income tax						
Federal and state, %	50	(50)	(50)	(50)	(50)	(50)
Other taxes, %	—	—	—	—	—	—
Investment tax credit, %	—	—	—	—	—	—
Facility construction time, yr	(3)	(2)	(3)	(2)	(3)	(3)
Working capital required, as % of total capital costs**	none specified	11.4	14.8	none specified	none specified	none specified
By-product credits, as % of annual operating cost	143.3	64.7	153.3	41.7–84.3§	94.1	78.6–111.7§§

Note: All numbers were taken from the studies except those enclosed by parentheses. These figures are assumed by Metrek.
      Criteria for assuming these parameters are discussed in the text.
 *Subcontract to Battelle.
 **Interest during construction omitted.
***Where working capital not explicitly provided, used summation
   of costs for spare parts, contractor's fee, contingency and
   miscellaneous (no inventory feedstock assumed in studies).

†Half-year operation.
††Full-year operation
†††Parameters are provided in study, but not yet available to public.
§By-product credit varies based on fermentation technology.
§§By-product credit varies based on price of corn.

Source: DOE HCP/ET-2854

- Feedstock costs: These costs vary widely, depending on the feedstock used, its availability and the extent of preprocessing needed. Costs are given in terms of dollars per year and dollars per million Btu of feedstock used.
- Process efficiency: This is a measure of the overall conversion efficiency of the facilities studied and is defined as the ratio of Btu produced to Btu of feedstock required. The Btu produced include all salable by-products, notably ddg and other stillage. It should be noted that when operating and capital costs are given in terms of dollars per million Btu and dollars per million Btu per hour, respectively, the Btu include those of salable by-products as well as ethanol. This is so because the Metrek economic assessment model computes a life cycle cost which is a weighted average of all salable products. Since the capital, operating and feedstock components of product costs cannot be allocated among the products of the ethanol facility, the Metrek model divides the sum of these costs, i.e., the total revenue requirements, by the total Btu production to obtain an average product price in dollars per million Btu.
- Ethanol production: This is given in terms of million gallons per year and billion Btu per year. The conversion from gallons to Btu is based on a lower heating value for 95% ethanol of 75,670 Btu/gal. If a higher concentration ethanol (example: 99%) is produced, the lower heating value is adjusted accordingly.
- By-product production: For each by-product produced, the annual quantity involved is given in Btu and based upon the information provided in the reports reviewed. The market value of each by-product is given in dollars per year. This information is used to adjust the average life cycle cost of all products provided by the Metrek model in order to obtain a life cycle cost for ethanol alone.
- Operating availability: This is defined as that portion of a year during which the ethanol facility is operational.
- Year dollars used: This is self-explanatory. If, for example, one of the reports reviewed presented cost data in 1976 dollars, then 1976 is listed.
- Year available: This is the first year that the ethanol facility is expected to be in commercial operation.

Table 2.2 presents the financial parameters which were assumed for each of the ethanol facility configurations. Comparing Table 2.1 with 2.2, three basic distinctions should be made regarding the preparation and ultimate use of each table in this study:

(1)   In Table 2.1, all of the performance and process economics data were directly derived or inferred from the data in each study. In Table 2.2, however, many of the financial parameters were not provided, as the studies varied widely in the levels of financial sophistication employed. For parameters essential to the running of Metrek's life cycle cost model, assumptions were made where necessary, as indicated by values in parentheses.
(2)   While Table 2.1 presents 28 different facility configurations, Table 2.2 presents only 8. This consolidation of studies in the second table is based on the fact that many of the process configurations

by the same author, or configurations in a given study, used a single set of financial parameters (except in the Battelle studies which used separate high debt and low debt financial parameters).

(3) Finally, and perhaps most important, the financial parameters presented in Table 2.2 are more discretionary and unrelated to process technology and feedstocks than are the technology performance and economic data provided in Table 2.1. Therefore, a standardization of financial parameters will help in making a more accurate comparison of system economics without changing the technology or related performance assumptions.

For each study or group of process configurations, the following financial parameters were obtained or (when in parentheses in Table 2.2) assumed:

- Analytical approach: This simply describes the manner in which the economics of ethanol production were dealt with in each study. Revenue requirements is a generic approach (usually employed by regulated utilities) in which the total revenue required to meet all expenses over the facility's economic lifetime is calculated, including a desired return on equity and the assumed cost of capital (debt). Discounted cash flow (DCF), and specifically the net present value method as employed in one of the studies, may be considered a variation of the revenue requirements approach. Here, all future cash flows (revenues and expenses) are simply discounted at a required rate of return on investment, which implicitly includes a return on equity and interest on borrowed capital. Finally, fixed selling price is a method whereby ethanol is assumed to be sold at current market value rather than being related in any way to system costs. The rate of return on investment is solved for, rather than specified in advance as in the case of the revenue requirements approach. Thus, process configurations in a given study using the fixed selling price approach may yield vastly different returns on investment, depending on variations in operating costs, feedstock costs, and other expenses.
- Capital structure: The relative percentages of the initial capital investment which are obtained through various portions of debt and equity financing.
- Interest on debt: The rate of compound interest paid in equal annual installments on the debt portion of capital financing. For use in Metrek's life cycle cost model, all loans are assumed to be amortized over the economic life of the ethanol facility.
- Return on equity: The desired rate of return to company shareholders expressed as a percentage of net annual profit after taxes over the equity portion of the initial investment.
- Economic life of facility: The number of years the ethanol facility is expected to operate. This is the basis over which all expenses and revenues are spread.
- Tax life of facility: For tax purposes, many companies assume a tax life of about two-thirds of the potential economic life of the facility. Over this period, the initial value of the facility, as an asset, is deductible from gross earnings as an operating expense, thus lowering the effective tax rate. The Metrek life cycle cost model assumes use of the sum-of-the-years'-digits method of depreciation, which ac-

celerates the depreciation and, thus, lowers taxes in the early years
of an asset's life, a common industry practice.

- Income tax: The assumed rate of taxation covering federal and state
  income taxes.
- Other taxes: Property taxes and gross receipts (or sales) tax, the most
  common other taxes imposed on industrial production sites.
- Investment tax credit: A federal tax credit applicable to certain types
  of assets that reduces directly, by a percentage, the amount of corporate or individual income taxes which must be paid during the
  year the asset is placed in service. The maximum credit presently
  available is 10%, but this has been varied in the past and is considered a flexible tool of fiscal policy to stimulate certain investments.
- Facility construction time: The number of years during which the
  plant is under construction, and used by the model to compute the
  cost of interest during construction.
- Working capital required: Generally considered to be the necessary
  liquid assets which a company must have available to secure unin-
  terrupted operation, often expressed as a percentage of capital costs.
  In some studies this assumes a large inventory of biomass feedstocks
  while others assume only miscellaneous fees, contingencies and spare
  parts.
- By-product credits: A source of revenue calculated in many studies
  for salable nonethanol products such as fermentation stillage, yeast
  and other residual outputs. The studies reviewed range from conservative to optimistic in their economic assessment of such credits.

Table 2.2 shows that some of the studies have explicitly provided most of their
financial parameters while others were deficient in detailing this portion of their
analyses. For the first comparison of process economics for the 28 ethanol configurations, financial parameters were assumed only in those cases where some
value was necessary in order for the Metrek life cycle cost model to operate (or
unless omission would grossly distort comparative results). Thus, parameters
such as "other taxes" and "investment tax credit" were entered as "blank" where
not provided in a study, while "income tax burden" was assumed to be about
50%; the latter value was considered necessary since some equivalent federal and
state taxes would in reality be imposed in all cases, whether or not specified in
the study.

Where essential financial parameters were omitted in any study, two criteria were
employed by Metrek in assuming a necessary value. First, explicit financial parameters were drawn from other studies by the same author, if and where available.
For example, Scheller's corn study used a 20-year economic life. The same value
was therefore assumed in Scheller's grain sorghum and milo studies although not
explicitly stated. Second, in the absence of studies by the same author, necessary
parameters were assumed, or averaged, from other studies using the same feedstock.

Finally, a set of representative financial parameters has been developed to serve
two purposes. First, it represents probable values for each parameter based on
common financial experience for energy investments. These parameters are used
when guidance from the above criteria (e.g., studies carried out by the same
author or using the same feedstock) are either lacking or inconsistent. Second,

and most important, these representative parameters form the necessary common
basis for comparison of ethanol technologies without altering the original system
cost and performance characteristics. These parameters, therefore, were employed
for all of the ethanol configurations in the second comparison of process economics. In this comparison, the differences between technologies was based only
on explicit differences in feedstock types, prices, and conversion process parameters such as capital and operating costs. The representative (common) financial
parameters which were used included:

Capital structure	60% debt
	40% common equity
Interest on debt	9%
Return on equity	15%
Economic life of facility	30 years
Financial (tax) life	20 years
Income tax	50%
Other taxes	
Gross receipts	2%
Property	1.4%
Investment tax credit	10%
Facility construction time (de- pending on feedstock)	2 to 3 years

Cost/Performance Data Used: The basic cost/performance data elements used
to compute life cycle ethanol costs are capital, operating and feedstock costs,
process efficiency, operating availability, plant size and by-product credit information. The ranges of data inputs obtained from the studies reviewed for each
of these are discussed below.

Plant Size — The plant sizes given in the studies ranged from small facilities producing ethanol from molasses (Schaffer/Battelle: 3.3 MMgal/yr; SRI: 4.0
MMgal/yr) to large facilities proposed by Battelle in its five-volume study. These
facilities, using corn and cane juice, would produce over 68 MMgal/yr of ethanol.
The most common plant size used in the studies was in the 20 to 27 MMgal/yr
range. Sixteen of the 28 plant configurations evaluated ethanol plants of this
size.

Operating Availability — The operating availabilities of the plants are usually in
the 80 to 90% range indicating almost full-time operation, i.e., 7,008 to 7,884
hr/yr. When the feedstock is only available on a seasonal basis, however, operating availability may be considerably lower than this.

Capital Cost — Capital costs used vary widely and depend on both the plant size
and on the extent to which the feedstock used must be preprocessed. In absolute terms, these costs may vary from $5.2 million for a small molasses plant to
over $200 million for a large plant producing almost 50 MMgal/yr of ethanol and
using wood as a feedstock. The mean capital cost in the studies reviewed was
$56.04 million. In relative terms, capital costs can be expressed as a ratio of dollars
to annual ethanol production. When this is done, about half of the facilities
have capital costs that vary from $1.05 to $2.15 per gallon per year and about
one-third have capital costs ranging between $1.30 and $1.75 per gallon per year.
The large facilities using wood and wheat straw feedstocks have the highest capital costs, $4.40 to $5.20 per gallon per year, while those presented by Lawrence
Berkeley Laboratories for various fermentation systems have costs that range from

$0.19 to $0.65 per gallon per year. These latter costs are considerably lower than those obtained from the other studies reviewed.

Operating Cost — As with capital, these costs vary widely, from a low of $249,000 per year for a 3.3 MMgal/yr plant producing ethanol from molasses to a high of $64 million per year for a plant producing 25 MMgallons of ethanol per year from wheat straw. The mean annual operating cost was about $15 million per year although this figure has limited usefulness because of the variety of plant sizes and feedstocks involved. In relative terms, the operating costs can be expressed in terms of dollars per million Btu produced, including by-products. In these terms, a greater clustering of operating costs takes place, with two-thirds of the facilities having operating costs between $2.00 and $4.00 per MMBtu produced.

Feedstock Cost — As with the other cost elements, feedstock costs varied widely and to a great extent were a function of the amount of preprocessing needed. Feedstocks requiring substantial preprocessing such as wheat straw and wood were available for $0.60 to $1.00 per MMBtu. At the opposite extreme, the most easily usable feedstocks such as cane juice and molasses cost between $5.00 and $10.00 per MMBtu.

Process Efficiency — This has been defined as the ratio of Btu produced (including salable by-products) to Btu of feedstock used. Where salable by-products are produced, typical efficiencies vary from 70 to over 90%. For those feedstocks requiring more preprocessing, efficiencies are substantially lower. These feedstocks and the process efficiencies in parentheses are: wood (44%), wheat straw (24%), algae (32%) and sugarcane (26 to 31%).

By-Product Production — The major by-products produced by the plants studied are stillage and ddg. Both of these can be used as a cattle feed supplement. Significant quantities of these by-products are produced only for those facilities using corn, grain sorghum and milo as feedstocks. For these facilities, the annual by-product Btu production is one-half to two-thirds that of ethanol production. In one other facility, using wheat straw as a feedstock (SRI-Mission Analysis) suf ficient quantities of pentose sugar are produced so that the annual Btu production of this by-product slightly exceeds that of the ethanol. Note that in an earlier SRI study in which wheat straw was used as a feedstock, by-product credits were not taken for pentose sugar and thus the process efficiency was considerably lower than for the same facility given in the mission analysis.

Financial Parameters Used: The twelve financial parameters presented in Table 2.2 indicate both similarities and differences among the studies reviewed in this report. These parameters complement the basic cost and performance data presented in Table 2.1 in that both sets of data are required to compute life cycle ethanol costs in $/MMBtu produced. The range of financial inputs provided are discussed in this section, grouping several categories together where appropriate.

Analytical Approach — The studies were rather evenly split between the two dominant analytical approaches employed. Seven studies used revenue requirements, six used fixed selling price, and one used the net present value mode of the discounted cash flow method, which is essentially the same as revenue requirements. To a certain degree, this dichotomy represents a difference in relative emphasis among the studies. Those using the revenue requirements approach tended to have greater depth in terms of process economics and gave more detail about methods of project financing, while those using the fixed selling price approach tended to provide more detailed descriptions of the processes used.

Capital Structure, Interest and Return on Equity — A wide variety of capital structures was observed among the studies. It ranged from 100/0 to 0/100 in percentage of debt to equity financing. Few generalizations can be made about the distribution of debt and equity financing among the studies except that higher debt fractions are usually associated with higher interest rates.

The cost of debt used in the studies fell within the 8 to 10% range. Almost all of the studies assumed that the amortization period on the loan would coincide with the facility's economic life. The exceptions to this were the Schaffer studies, both of which assumed 10-year amortization periods.

The return on equity parameter, as with interest rates, did not vary greatly among the studies reviewed. This parameter is generally 14 to 15%. In the Mitre study, however, returns on common and preferred equity of 25 and 15%, respectively, were specified. In this study, a relatively high percentage (47%) of the capital investment was derived from stockholder equity. Therefore, this apparent difference is actually consistent with the observed financial trend of providing relatively higher returns on equity and debt (interest) based on their percentage contributions to the initial investment.

Economic and Tax Lifetimes of Facilities — Only about half of the studies explicitly provided economic and tax lifetimes for the ethanol facilities which they described. The range of economic lifetimes provided for these facilities is from 10 to 30 years. Although some authors felt such facilities could be made to last much longer with periodic component replacement, they felt that it would be more likely for technological improvement to require design changes before a facility actually becomes inoperative.

The tax lives of these facilities, where provided, ranged from 10 to 20 years, or approximately two-thirds of the economic life. This relationship is a common industry practice and usually falls within the flexible depreciation guidelines provided by the Internal Revenue Service.

Taxes — Of those studies which specified income tax rates, three used 48% (the Federal tax rate for corporations with annual earnings over $50,000), four used 50%, and one used 52%. Most studies which used higher than the 48% rate indicated that the difference was assumed to cover state and, in some cases, local taxes. Only the Mitre study explicitly provided for local taxes, in the form of a 1.4% property tax, as well as for a gross receipts or sales tax of 2%. Regarding the investment tax credit, only three studies explicitly included it. Mitre assumed a 4% tax credit in its study and Battelle assumed a 24% tax credit in both of its studies.

Facility Construction Time — Facility construction time was only explicitly provided in two studies. The Mitre study assumed two to three years for its wood-to-ethanol facility, and Stanford Research Institute assumed 2 years in its mission analysis of wheat straw, sugarcane, and algae.

Working Capital Required — Working capital was explicitly provided in six of the studies. This parameter showed wide variations ranging from about 9 to 33% of total capital costs. Much of this variation may be attributable to different assumptions regarding the amount of inventory feedstock that is needed as operations commence, with a maximum of a half year's supply of cane juice assumed by Battelle as part of its working capital. It was also possible to compute working capital requirements for the Schaffer studies, due to the very detailed breakdown of facility costs which were provided. In those studies, working capital ranged from 8 to 10% of total capital costs, based only on the costs of spare parts, contractor's fees, contingency and miscellaneous expenses, as no feedstock inventory was provided.

By-Product Credits — Nine of the fourteen assessments included by-product credits in some form, usually for fermentation stillage, distiller's dried grains and sometimes for yeast as well. The single treatment by-product credit relative to the size of operation was attributed to an ethanol plant using milo as a feedstock (by Scheller). For this plant, the by-product credit was estimated at 153% of annual operating costs. Among the remaining feedstocks, corn consistently provided the most valuable stillage by-product. The value of this by-product ranged from a low of 81% to a high of 143% of annual operating costs. When the price of corn was allowed to vary, as in the USDA-ESCS Report, the by-product cost also varied, from 79 to 112%. These ranges indicate the great uncertainty involved in estimating the value of by-products.

Specifically, they underscore the potential hazards of depending upon consistent by-product prices to assure the financial solvency of a given operation. Finally, lesser credits were claimed for by-products from cane juice, wheat straw, grain sorghum and molasses. These by-product credits, expressed as a percent of annual operating costs, ranged from 12% for wheat straw to 84% for molasses.

Comparison of Process Economics

The ethanol economic data base described provides a basis for comparing the results obtained in each of the studies reviewed. Using this data base, two sets of analyses were carried out to provide a comparative evaluation of ethanol production systems. The purpose of the first analysis was to transform a heterogeneous set of ethanol production costs, which was computed by differing means, into a set of life cycle costs which was obtained in the same manner. To achieve this, it was necessary to make specific assumptions and some modifications to include those values (such as facility construction time and economic lifetimes) which were not provided in certain studies but which are necessary to compute an initial set of life cycle costs. In essence, the first analysis provides a common basis for comparison of the 28 configurations while keeping the original study assumptions virtually intact.

The second analysis provides a comparison of ethanol production systems employing the common set of financial parameters derived. In this way, a common basis for comparing ethanol technologies was developed without altering original system cost and performance data. The purpose of the second analysis was not to achieve more uniform ethanol production cost estimates among the studies but to identify those factors such as capital and operating costs and by-product values which are most responsible for the remaining cost differences. The results of the second analysis were used as a basis for sensitivity analyses which are described in a later section.

Ethanol Life Cycle Costs Obtained Using the Metrek Model: Life cycle costs of ethanol production were computed using the Metrek Full Life Cycle cost (FLC) model. This model gives the life cycle levelized cost, in constant 1977 dollars, of ethanol production. The term "life cycle levelized cost" describes the ethanol sales price at the plant gate which is required to recover all costs incurred in producing the ethanol and to achieve the desired return on investment. Thus, the terms "life cycle levelized cost," "ethanol sales price" and "ethanol production cost" are used interchangeably in this report.

The life cycle costs provided by the Metrek model are presented in Table 2.3. The first column presents the average life cycle costs of all those cases in which more than one product is produced. The second column gives the ethanol life cycle cost alone, adjusted for by-products. The method used to adjust for by-products is described below.

Table 2.3: Ethanol Life Cycle Costs Based on Study Parameters and Assumptions

Study	Average All Products	Ethanol
	. . . . ($/MMBtu)	
Schaffer, cane		
330 days, low price	—	11.10
330 days, high price	—	13.92
180 days, low price	—	13.13
180 days, high price	—	15.94
Schaffer/Battelle, molasses		
150 days	—	8.53
330 days	—	9.72
Battelle, corn		
Low debt	13.28	15.72
High debt	12.73	14.87
Battelle, cane juice		
Low debt	19.13	17.58
High debt	18.28	16.69
Mitre, wood		
850 tpd, low price	—	27.91
850 tpd, high price	—	30.21
1,700 tpd, low price	—	24.05
1,700 tpd, high price	—	26.48
SRI, enzymatic hydrolysis, wheat straw	—	51.18
SRI, 10-site survey, molasses	—	17.09
SRI, mission analysis		
Algae	—	23.74
Wheat straw	23.47	42.82*
Sugarcane	—	25.46
Scheller		
Corn and distressed grain	10.51	12.04
Sorghum	13.24	15.52
Milo	9.01	9.91
LBL, molasses		
Batch	13.42	13.31
Continuous	12.56	12.22
Continuous with cell recycle	12.44	12.08
Vacuum with cell recycle	11.85	11.60
DP&RA, corn	12.18	15.96
USDA, corn	11.74	14.30

*Assumes a by-product credit taken. If no by-product credit assumed (as in SRI study), life cycle cost of ethanol produced is $47.05/MMBtu.

Source: DOE HCP/ET-2854

Table 2.3 indicates that a majority of the studies would have obtained ethanol prices in the $10 to $17 per MMBtu range if they had used a common methodology. Those studies providing the highest ethanol prices are for plants using feedstocks such as wood and wheat straw which require considerable preprocessing.

By-Product Adjustments — By-product credits were taken in several of the studies reviewed. These credits reduced the required selling price of the ethanol produced. Since the market value of the by-products is not a function of the cost of ethanol production, ethanol life cycle costs can be reduced by an arbitrary amount depending on the value assigned to by-products. Thus, those studies in which overly optimistic assumptions were made concerning by-product sales would give a disproportionate reduction in the required selling price of ethanol.

It was previously indicated that the life cycle cost model used by Metrek computes a product selling price that is a weighted sum of all products of the plant. This selling price, when multiplied by the annual Btu production, gives the total annual revenue requirements of the plant. This is the income needed to recover all costs and to realize the desired return on investment. In cases assuming by-product credits, the annual value of by-product production was subtracted from total revenue requirements to obtain the net annual ethanol revenue requirements.

In 10 of the 28 process configurations reviewed, significant by-product credits were taken. For these configurations, appropriate by-product adjustments were made to the weighted life cycle costs given by the model. These adjustments are shown in Table 2.4.

In almost all cases, the adjustment made for by-products did not significantly affect the weighted life cycle cost for all products. This was due to one of two reasons:

- Either small quantities of by-products are produced relative to ethanol production as in the case of Battelle cane juice where ethanol accounted for 95% of total Btu production; or
- The unit value of the by-product was similar to that of the weighted life cycle cost given by the model. For example, in the paper by Scheller on ethanol production from milo, the weighted life cycle cost was $9.00/MMBtu. The market value of the by-product was $9.90/MMBtu

The impact that by-products will have on the weighted cost will increase as the quantity of by-product produced and the difference between by-product and ethanol values increase. For example, in the SRI mission analysis of wheat straw, large quantities of a salable by-product (pentose sugar) are produced, 1.90×10^{12} Btu/yr vs 1.89×10^{12} Btu/yr of ethanol. In addition, the unit value of the by-product differs considerably from that of the weighted product price since the by-product can be sold for $4.20/MMBtu while the weighted product price is $23.50/MMBtu. As a result, the unit selling price of the ethanol alone is considerably higher, $42.82/MMBtu, than the weighted selling price given by the model.

It should be noted that these adjustments are not by-product credits in the sense that they represent downward changes in the required selling price of ethanol. Rather, they represent changes in a weighted selling price of ethanol and all coproducts as given by the model. The adjusted selling price or life cycle cost of the ethanol alone is approximately the same as would be obtained if a by-product credit were taken.

An Alternative Life Cycle Costing Approach — In order to compute levelized product costs, the Metrek model used here (FLC) assumed that the revenue stream in $/MMBtu is uniform in constant dollars and increasing in current dollars. An alternative approach is to assume that the revenue stream is uniform in current dollars (defined as constant dollars inflated at a nominal 5% per year) and thus is declining in constant dollar terms. Under this alternative approach, computed life cycle levelized costs are significantly higher than those obtained under the levelized constant dollar approach used in this evaluation.

In order to compare the two methodologies, life cycle costs for 11 of the facilities studied were computed using the calculator module of the Metrek SPURR ("A System for Projecting the Utilization of Renewable Resources: SPURR Methodology," The Mitre Corporation, September, 1977) model. This model computes life cycle ethanol costs using the levelized current dollar method and gives higher life cycle product costs. The results obtained using the two approaches were obtained using the assumptions and parameters given in the studies reviewed, and are summarized in Table 2.5.

In all cases, the results obtained from the life cycle cost model used in this study are lower than those obtained using the alternative levelizing approach. This is so because uninflated dollars are used in the methodology employed in this study. These differences are fairly consistent and FLC gives costs which are 60 to 70% of those obtained with SPURR.

What is important and should be noted is that while the model results differ, the rankings of the technologies are virtually identical. Thus, in any comparative assessment such as this one, technologies having the lowest/highest life cycle costs under FLC will also have the lowest/highest life cycle costs under SPURR. As analytical tools, either model is valid since both methods of levelizing costs have been used and accepted in industrial applications and in past studies. The FLC model was chosen for use in this study for two reasons. First, it gives life cycle costs which represent the lower end of a range of acceptable values, depending on the levelizing methodology chosen to compute revenue requirements. In effect, it presents marginally cost competitive technologies in a more favorable light than does SPURR. Second, in terms of the analysis carried out, the FLC file structure permitted more rapid data entry and manipulation than did SPURR and this facilitated the evaluation.

Life Cycle Cost Comparison Based upon Study Parameters: Ethanol life cycle costs computed with the Metrek model and based on the parameters given in the studies were presented in Table 2.3. In Figure 2.1, these costs are compared with ethanol prices reported in the studies. In more than half of the cases shown, the two costs given differ by less than 10%. This is not surprising, since the only parameters which were specified by Metrek are those which a study did not provide or implicitly use in its own calculations. Discrepancies among studies, on the other hand, tend to be much greater. Differences in system cost and performance assumptions are important reasons for this variation as are basic differences in analytical approach. Many of the studies which used the fixed selling price approach provided only a simple ethanol production cost estimate.

Table 2.4: Summary of By-Product Adjustments to Life Cycle Costs

	Battelle Vol. V Corn			SRI	Scheller				Battelle Vol. III Cane Juice		Lawrence Berkeley Laboratories Molasses			
	Low Debt	High Debt	USDA Corn	Wheat Straw	Corn	Sorghum	Milo	DP&RA Corn	Low Debt	High Debt	Batch	Contin-uous	Cell Recycle	Vacuum
Value of by-products given in study, $MM/yr	25.81	25.81	7.88	8.04	9.08	6.36	9.20	4.14	12.85	12.85	2.42	4.63	4.63	2.55
Total Btu production, 10^9 Btu/yr	8,078.7	8,078.7	2,513.4	3,795.2	2,597.4	2,254.4	2,684.4	1,332.0	5,414.0	5,414.0	2,253.3	2,402.0	2,402.0	2,262.0
Weighted life cycle cost from Metrek Model, $/MMBtu	13.28	12.73	11.74	23.47	10.51	13.24	9.01	12.18	19.13	18.28	13.42	12.56	12.44	11.85
Total revenue requirements, $MM/yr	107.29	102.87	29.52	85.05	27.29	29.84	24.19	16.22	103.58	98.98	30.24	30.17	29.88	28.80
Net ethanol revenue require-ments, $MM/yr	81.48	77.06	21.64	18.01	18.21	23.48	14.99	12.08	90.73	86.13	27.82	25.54	25.25	24.25
Annual ethanol production, 10^9 Btu/yr	5,183.4	5,183.4	1,513.4	1,392.0	1,513.4	1,513.4	1,513.4	756.7	5,160.3	5,160.3	2,089.7	2,089.7	2,089.7	2,089.7
Adjusted ethanol life cycle cost, $/MMBtu	15.72	14.87	14.30	42.82	12.04	15.52	9.91	15.96	17.58	16.69	13.31	12.22	12.08	11.60

Table 2.5: Life Cycle Costs Obtained Using Two Levelizing Methodologies

Study	Life Cycle Levelized Cost* ($/MMBtu, 1977 dollars) Full Life Cycle (FLC) Cost Model	SPURR Model	FLC/SPURR x 100 (%)
LBL – vacuum fermentation with cell recycle	11.85	18.32	64.7
SRI – enzymatic hydrolysis, wheat straw	51.17	70.00	73.1
Mitre – weak acid hydrolysis, wood, 850 tpd, low feedstock price	27.90	43.33	64.4
Mitre – weak acid hydrolysis, wood, 1,700 tpd, high feedstock price	25.48	41.80	61.0
Battelle – corn, low debt	13.28	18.80	70.6
Schaffer – sugarcane, 330 days/yr, high feedstock price	13.92	21.70	64.1
USDA – corn	11.74	18.60	63.1
SRI mission analysis – sugarcane	25.46	40.60	62.7
DP&RA – corn	12.18	19.90	61.2
Scheller – corn	10.51	15.00	70.1
Battelle – cane juice, low debt	19.13	26.80	71.4

*Unadjusted for by-products.

Source: DOE HCP/ET-2854

The prices presented in all the configurations of the Schaffer sugarcane, Schaffer/ Battelle molasses and the LBL molasses studies do not provide for returns on equity. In the SRI molasses and wheat straw cases, the 100% equity financing makes them more expensive than other studies using the same feedstock.

Key to Figures 2.1, 2.2 and 2.3

	Study	Feedstock	Configuration
a	Schaffer	sugarcane	330 days/yr, $10/ton
b	Schaffer	sugarcane	330 days/yr, $13.50/ton
c	Schaffer	sugarcane	180 days/yr, $10/ton
d	Schaffer	sugarcane	180 days/yr, $13.50/ton
e	Battelle	cane juice	high debt with by-product credits
f	Battelle	cane juice	low debt with by-product credits
g	SRI	sugarcane	65/35 (D/E) mission analysis
h	Schaffer/Battelle	molasses	150 days/yr, $0.18/gal
i	Schaffer/Battelle	molasses	330 days/yr, $0.18/gal in season $0.19/gal off season
j	SRI	molasses	0/100 (D/E), 10-site study
k	LBL	molasses	batch with by-product credits
l	LBL	molasses	continuous with by-product credits
m	LBL	molasses	cell recycle with by-product credits
n	LBL	molasses	vacuum with by-product credits
o	Battelle	corn	high debt with by-product credits
p	Battelle	corn	low debt with by-product credits
q	Scheller	corn and distressed grain	with by-product credits
r	DP&RA	corn	with by-product credits
s	USDA	corn	with by-product credits.
t	Mitre	wood	850 tpd, $1/MMBtu
u	Mitre	wood	850 tpd, $2/MMBtu
v	Mitre	wood	1,700 tpd, $1/MMBtu
w	Mitre	wood	1,700 tpd, $2/MMBtu
x	SRI	wheat straw	0/100 (D/E)
y	SRI	wheat straw	65/35 (D/E) mission analysis
z	Scheller	grain sorghum	with by-product credits
aa	Scheller	milo	with by-product credits
bb	SRI	algae	65/35 (D/E) with by-product mission analysis

To help explain some of the more substantial intra-configuration cost differences, it must first be noted that the economic and financial assumptions employed in the various studies differed greatly. Such differences include:

- Use of annualized fixed charges on capital vs levelized costs which cover capital, raw material and operation expenses in the Metrek model;
- Use of sinking funds for debt repayment vs conventional amortization schedule in the Metrek model;
- Loan amortization over short periods vs loan amortization over project's economic life in the Metrek model;
- Recovery of working capital at end of project vs assumed liquidation of working capital in the Metrek model; and
- Consideration of plant by-products as credits deducted from operating expenses vs consideration of plant by-products as joint output, with ethanol, valued on a Btu basis in the Metrek model.

Figure 2.1: Ethanol Costs Calculated Using Original Study Parameters

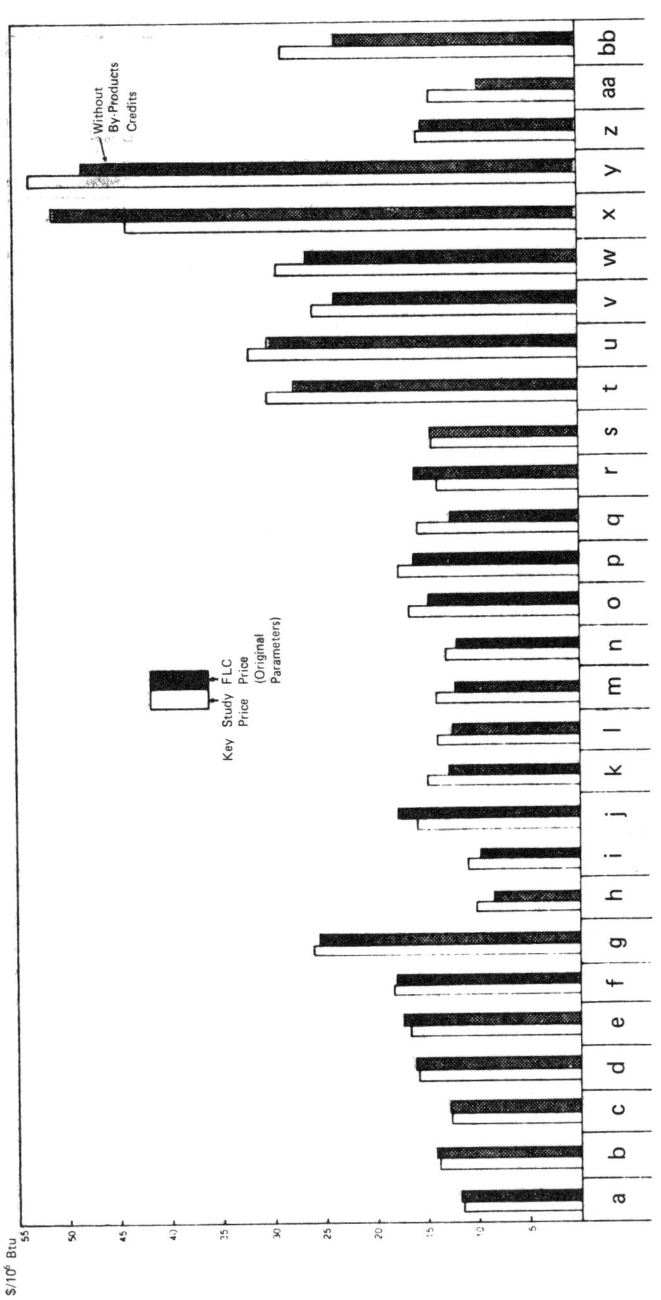

Source: DOE HCP/ET-2854

These and other differences in financial and accounting techniques underlie many of the differences between life cycle costs computed with the Metrek model and the individual studies. The method by which costs are computed is not important as long as a model uses sound financial assumptions and is designed to levelize costs in a technology-unbiased manner. The value of a model lies in its ability to provide a single analytical framework in which a cross-technology comparison can be made. A model also permits the use of a common set of financial parameters in a consistent fashion. This provided a more uniform comparison of ethanol production systems in which the only variables among configurations were the process technology and feedstock specifications presented in Table 2.2. The purpose of the second phase of the analysis was to provide such a comparison. This work is described below.

Life Cycle Cost Comparison Using Common Assumptions: For the 28 ethanol configurations studied, life cycle costs were computed using the Metrek model and a common set of financial parameters. In this part of the analysis, the only parameters which varied from study to study were the major cost/performance characteristics such as capital, operating, and fuel costs, and operating parameters such as thermal efficiency and capacity factor.

The objective of the analysis was to standardize both the financial parameters and assumptions used and the methodology employed in obtaining life cycle costs. Any cost differences remaining are those due to the differing cost/performance estimates presented in the studies. Such differences were expected, given the variety of processes and feedstocks. However, when more than one study used the same process/feedstock combination, differences in ethanol selling prices could be attributed to variations among cost/performance estimates.

In the remainder of this section, the life cycle ethanol costs obtained using common assumptions are presented and contrasted with those obtained using the assumptions given in the studies reviewed. The common assumptions are: common equity, 40%; debt, 60%; return on common equity, 15%; cost of debt, 9%; income taxes, 50%; gross receipts tax, 2%; property tax, 1.4%; investment tax credit, 10%; economic life, 30 years; and tax life, 20 years.

The resulting life cycle costs are given in Figure 2.2. The life cycle costs obtained using common financial parameters usually do not differ substantially from those obtained using the individual study parameters. In 22 of the 28 configurations, life cycle costs differed by less than 10%, indicating the relatively small impact that changes in financial parameters had on ethanol prices. In all but seven of the configurations, the use of common financial parameters increased life cycle costs by amounts no greater than 2.5%. The changes in life cycle cost for each study are summarized below.

- Schaffer—Sugarcane: For all four of the process configurations, ethanol costs increased by about 2%. Because several key parameters such as return on equity, and economic lifetime were already specified in order to perform the first comparison, only the gross receipts and property taxes remained to be specified. Modification of the tax life from 18 to 20 years resulted in a negligible decrease in costs. The relative cost differences among the four configurations were maintained in the second comparison.

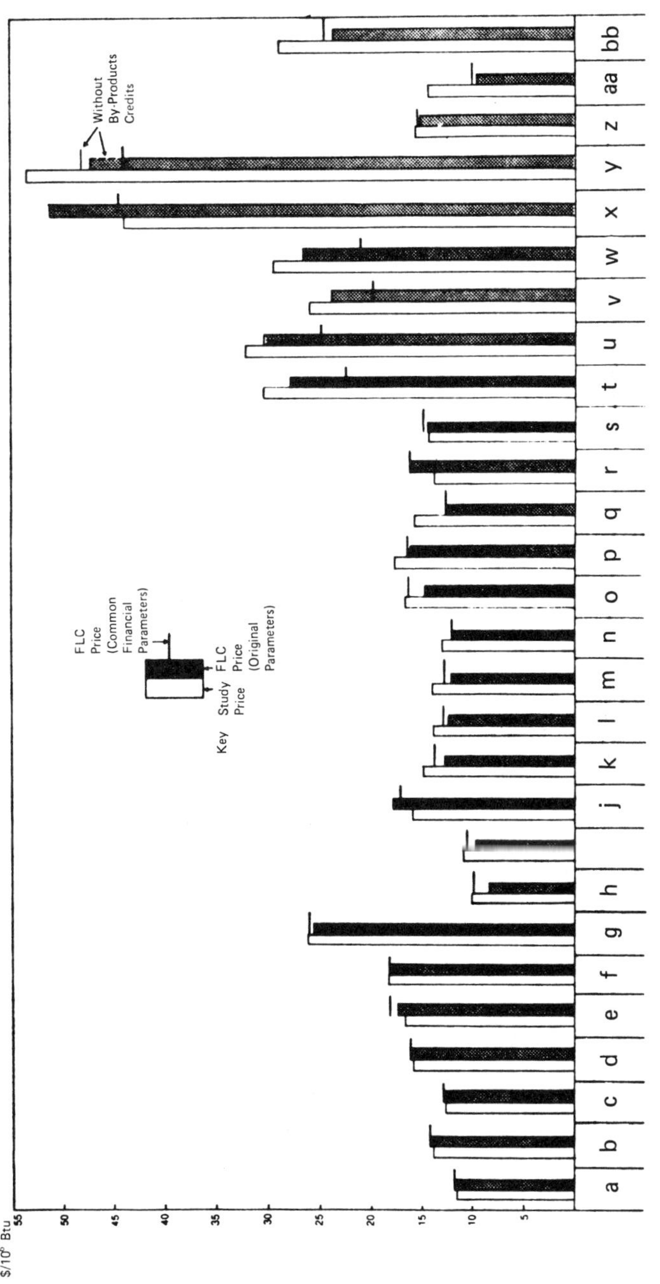

Figure 2.2:   Ethanol Costs Calculated Using Common Financial Parameters

Source: DOE HCP/ET-2854

- Schaffer/Battelle—Molasses: The half-year operating facility, self-sufficient in molasses, experienced an increase in its ethanol production cost of 13%, while the year-round facility which purchases molasses during the off-season had an increase of only 7%. Since both facilities were assumed to be 100% debt financed in the first comparison, the change in capital structure to 60% debt and 40% equity made ethanol from both facilities more expensive given the higher return on equity. The smaller increase in costs for the year-round facility was due mainly to an almost doubled ethanol output for its slightly larger capital cost, effectively diluting the effect of the increased cost. It is interesting to note that in the study, a much higher return on investment for the year-round facility is calculated based on an assumed fixed selling price. This occurred only because a smaller per gallon profit on the doubled sales volume yielded a greater return on investment than did the higher per gallon profit, or lower production cost, of the half-year facility, given its lower output.

- Battelle—Cane Juice: Ethanol costs increased by 7% in the high debt case, but only by 2% in the low debt case. Using common financial parameters, ethanol costs were the same for both cases. These costs increased for two reasons. First, the weighted cost of capital in the high debt case increased from 10.9% to the standard Metrek case of 11.4%, while the low debt case actually experienced a slight drop from its initial weighted cost of capital of 13.1%. Thus, the relative advantage of high debt (debt capital requiring a lower annual return than equity capital) was nullified when common financial parameters were used. And second, the high debt case started from a lower initial cost than the low debt case, and therefore, had to increase proportionately more to reach the same ultimate cost. Finally, the lower investment tax credit and assumption of gross receipts and property taxes in the common parameter case outweighed the effect of a lower economic lifetime and resulted in at least some net cost increase for both Battelle cases.

- Battelle—Corn: Ethanol costs increased by 7% in the high debt case but only by 2% in the low debt case. Again, using common financial parameters, ethanol costs were the same for both cases. The reasons for these relative increases are the same as those discussed above in the Battelle cane juice study. Similarly, the investment tax credit, gross receipts and property taxes, and economic lifetime were standardized, resulting in some net cost increase for both cases.

- Mitre—Wood: For the four configurations studied, 850 and 1,700 tpd of wood and wood prices at $1.00 and $2.00 per MMBtu, life cycle costs of ethanol decreased by about 20% when the common parameters were used. The major reasons for this were the use of a higher investment tax credit (10% vs 4% used in the Mitre study) and a lower weighted cost of capital (11.4% vs 14.8% in the Mitre study). Assumptions concerning tax rates and economic lifetimes were the same for both configurations.

- SRI—Wheat Straw: Costs decreased by 13% due to three parameter changes. First, the economic and tax lifetimes were increased from the studies' 15 and 10 years to 30 and 20 years, respectively. Second, the capital structure changed, from 100% equity financing to 40% equity/60% debt financing and consequently, the weighted cost of capital decreased. Finally, Metrek assumed an investment tax credit of 10% vs none in the study. Slight cost increases due to the higher tax rates assumed by Metrek were more than counteracted by these three parameter changes.

- SRI Molasses—Ten-Site Study: A slight cost increase of 2% occurred when the common parameters were substituted for those used by SRI. This was due to a slightly greater use of equity financing, 40% vs 35% and use of property and gross receipts taxes. Costs of debt and equity and economic/tax lifetimes were the same in both configurations. Two factors which reduced the cost increase were a somewhat lower income tax rate than used by SRI and a 10% investment tax credit (itc).
- SRI Mission Analysis—Algae, Wheat Straw and Sugarcane: Use of the common assumptions resulted in cost increases of about 2%. Lifetimes were the same in both sets of assumptions. Factors tending to reduce life cycle costs when the common set of assumptions was used included a slightly lower income tax rate of 50% vs 52% used by SRI and a 10% itc. Factors tending to increase costs were a 1.4% property tax and a 2% gross receipts tax used in the common assumptions and a capital structure that was slightly more equity-intensive, 40% vs 35% used by SRI. The net impact was that a slight cost increase takes place. It should be noted that while neither SRI wheat straw study employed by-product credits, SRI personnel involved in the mission analysis indicated that significant quantities of salable by-product, pentose sugar, are produced. An adjustment for this by-product was made for purposes of this analysis. Hence, when common parameters were employed, the use of a by-product adjustment resulted in lower cost ethanol from wheat straw in this study, outweighing the remaining advantage in the earlier study of about 40% lower feedstock costs.
- Scheller—Corn, Grain Sorghum and Milo: Slight increases in life cycle costs resulted from use of a more equity-intensive capital structure, 40% equity vs 30% equity used by Scheller, and slightly higher taxes. A longer economic life than that assumed by Scheller, 30 years vs 20 years, and a 10% investment tax credit, however, tended to reduce ethanol costs. The net result was an increase in ethanol costs of 1 to 2% when compared with Scheller's assumptions used in the life cycle cost model.
- Lawrence Berkeley Laboratory—Molasses: For the four configurations considered, cost changes increased by 2.0 to 2.4%. The only parameter changes made were a lengthening of the facilities' tax and economic lifetimes from 10/10 years to 20/30 years, respectively, and the imposition of the property tax, gross receipts tax and the itc. The higher taxes slightly outweighed the higher tax credit and longer economic lifetime and life cycle costs increased slightly using the common parameters.
- Development Planning and Research Associates—Corn: Life cycle costs increased by only 1% when the common parameters were used. This was due to the fact that almost all of the parameters were identical for both configurations. A life cycle cost reduction due to an itc of 10% in the common parameters was more than offset by the higher taxes assumed in the common parameter set.

- USDA–Corn: Ethanol costs increased by 2% due to the imposition of property and gross receipts taxes in the set of common assumptions. All other parameters were virtually identical.

Distribution of Life Cycle Costs: Most costs changed only slightly when common financial parameters were used. In part, this was due to the assumption of financial parameters for the first comparison when individual studies were not explicit. Since most of these parameters were retained in the second comparison, this minimized the effect of standardizing financial assumptions. Nevertheless, a basic question is raised by these results. That is, to what extent can changes in financial parameters influence the total cost of ethanol production, given the fact that different assumptions about plant financing influenced only the capital cost component of total production costs?

To answer this question, the influence of capital, operating and feedstock costs on total ethanol life cycle costs was investigated. The results of this work are presented in Figure 2.3. From observation alone, it can be shown that capital cost is in no case the largest life cycle cost component. In fact, for most cases, the capital cost is exceeded by both the operating and feedstock costs. The only cases where capital costs exceeded operating costs were the Schaffer 180-day sugarcane facility with low cost ($10/ton) cane, and the Schaffer/Battelle 150-day molasses facility which operates only on indigenously produced molasses. Both of these configurations inefficiently utilize their plants and equipment due to the short cane milling season in regions which they typify. Similarly, the only cases where capital costs exceeded feedstock costs were the Mitre wood facilities and the first SRI wheat straw facility which assumed 100% equity financing. The relatively inexpensive cellulosic feedstocks in these cases require hydrolysis prior to fermentation, and thus operating costs are by far the largest component of all wood and wheat straw configurations.

In Table 2.6, the distribution of life cycle costs among the three basic components are presented for each configuration. The distribution expresses life cycle costs both in $/MMBtu and in percent of total ethanol costs. Table 2.7 presents the average distribution of life cycle cost components by type of feedstock for cases where more than one configuration used that feedstock. From these two tables, the average life cycle costs by feedstock type may be compared. Average capital costs ranged from a low of 8.5% for molasses to a high of 29% of total costs for wood. Average operating costs ranged from a low of 17% for sugarcane and sugar juice to a high of 74% of total costs for wheat straw. Finally, average feedstock costs ranged from a low of 14% for wheat straw to a high of 72% for molasses.

This analysis of component costs indicates that capital costs are less important than operating and feedstock costs in influencing total production costs. Thus, a different capital structure may induce a 30% change in capital cost, but if capital only contributes 12% to total production costs, this will only change life cycle ethanol costs by about 4%. This point is significant in determining which types of federal incentives and additional research may have the greatest impact on reducing the projected price of ethanol from each biomass feedstock.

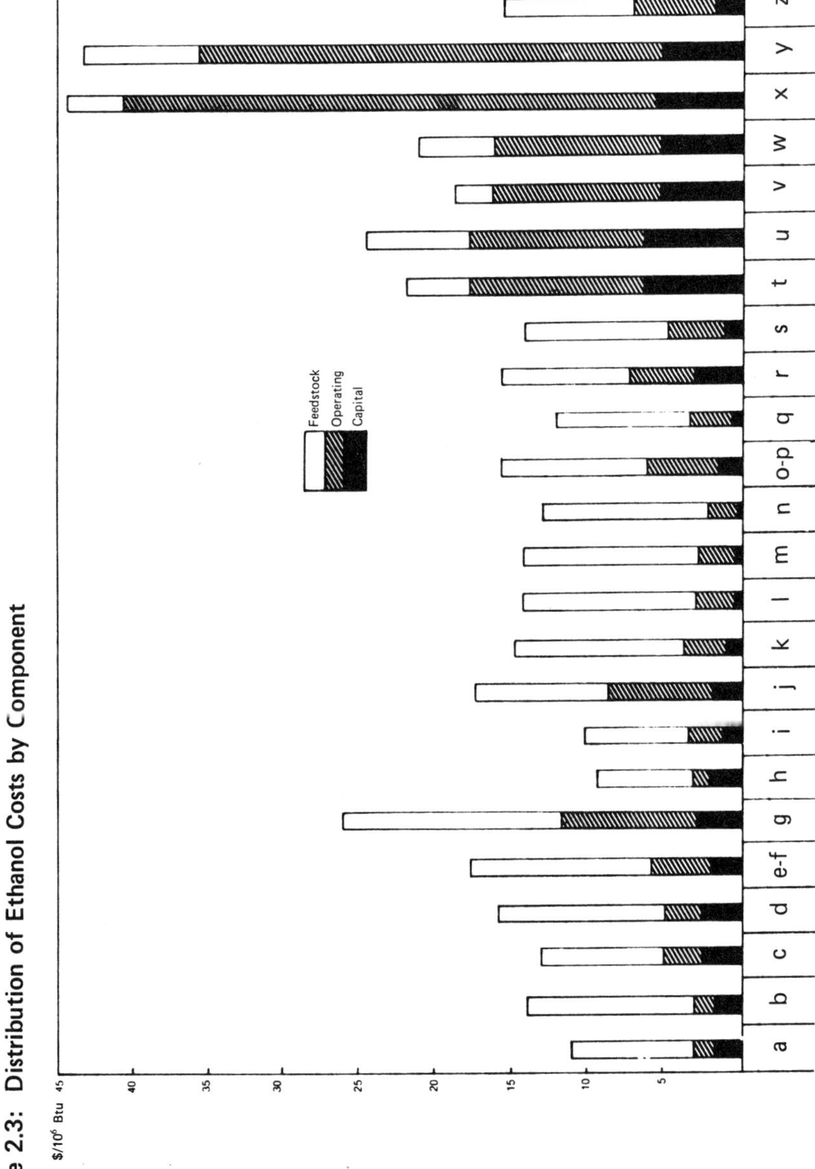

Figure 2.3: Distribution of Ethanol Costs by Component

Source: DOE HCP/ET-2854

Table 2.6: Distribution of Ethanol Life Cycle Cost Components

	Schaffer				Battelle	SRI
Feedstock data	sugarcane 330 days/yr $10/ton	sugarcane 330 days/yr $13.50/ton	sugarcane 180 days/yr $10/ton	sugarcane 180 days/yr $13.50/ton	cane juice high and low debt*	sugarcane 65/35 (D/E) mission analysis
Cost components $/10⁶ Btu						
Capital	1.50	1.50	2.73	2.74	1.95	2.68
Operating	1.56	1.56	2.37	2.37	3.95	9.29
Feedstock	8.23	11.10	8.24	11.10	11.97	14.03
Total	11.29	14.16	13.34	16.21	17.87	26.00
Percent						
Capital	13.3	10.6	20.5	16.9	10.9	10.3
Operating	13.8	11.0	17.7	14.6	22.1	25.5
Feedstock	72.9	78.4	61.8	68.5	67.0	64.2

	Schaffer/Battelle		SRI	LBL			
Feedstock data	molasses 150 days/yr $0.18/gal	molasses 330 days/yr $0.18/gal** $0.19/gal***	molasses 0/100 (D/E) 10-site study	molasses batch	continuous	cell recyc.	vacuum
				with by-product credits			
Cost components $/10⁶ Btu							
Capital	2.10	1.09	1.49	0.92	0.54	0.47	0.45
Operating	1.01	2.58	7.19	2.54	2.13	2.07	1.42
Feedstock	6.60	6.75	8.77	10.12	9.84	9.83	9.98
Total	9.71	10.42	17.45	13.58	12.51	12.37	11.85
Percent							
Capital	21.6	10.4	8.5	6.8	4.3	3.8	3.8
Operating	10.4	24.8	41.2	18.7	17.0	16.7	12.0
Feedstock	68.0	64.8	50.3	74.5	78.7	79.5	84.2

	Battelle	Scheller	DP&RA	USDA	Scheller	SRI
Feedstock data	corn high and low debt*	corn and distressed grain†	corn†	corn†	milo†	algae 65/35 (D/E) mission analysis*
Cost components $/10⁶ Btu						
Capital	1.60	0.91	3.20	1.33	1.04	2.62
Operating	4.79	2.86	4.42	3.73	2.53	6.59
Feedstock	9.63	8.46	8.53	9.58	6.45	15.03
Total	16.02	12.23	16.15	14.64	10.02	24.24
Percent						
Capital	10.0	7.4	19.8	9.1	10.4	10.8
Operating	30.0	23.4	27.4	25.5	25.2	27.2
Feedstock	60.0	69.2	52.8	65.4	64.4	62.0

	Mitre				SRI		Scheller
Feedstock data	wood 850 tpd $1/10⁶ Btu	wood 850 tpd $2/10⁶ Btu	wood 1,700 tpd $1/10⁶ Btu	wood 1,700 tpd $2/10⁶ Btu	wheat straw 0/100 (D/E)	wheat straw 65/35 (D/E) mission anal.	grain sor- ghum†
Cost components $/10⁶							
Capital	6.81	6.81	5.81	5.81	5.88	5.13	1.78
Operating	13.14	13.14	10.97	10.97	34.61	30.77	5.22
Feedstock	2.31	4.62	2.45	4.87	4.19	7.93	8.63
Total	22.26	24.57	19.23	21.65	44.68	43.83	15.63
Percent							
Capital	30.6	27.7	30.3	26.8	13.2	11.7	11.5
Operating	59.0	53.5	57.0	50.7	77.4	70.2	33.4
Feedstock	10.4	18.8	12.7	22.5	9.4	18.1	55.1

*With by-product adjustment   **In season   ***Off season   †With by-product credits

Source: DOE HCP/ET-2854

Table 2.7: Average Distribution of Life Cycle Cost Components by Feedstock

	Sugarcane	Molasses	Corn	Wood	Wheat Straw	Grain Sorghum and milo*
Cost Component						
Capital, $/10^6 Btu	2.18	1.01	1.76	6.31	5.55	1.41
Operating, $/10^6 Btu	3.52	2.71	3.95	12.06	32.69	3.83
Feedstock, $/10^6 Btu	10.78	8.84	9.05	3.56	6.06	7.54
Total, $/10^6 Btu	16.48	12.55	14.76	21.93	44.26	12.83
Capital, %	13.8	8.5	11.6	28.9	12.4	11.0
Operating, %	17.4	20.1	26.5	55.1	73.8	29.3
Feedstock, %	68.8	71.4	61.9	16.1	13.8	59.8
No. of configurations	6	7	4	4	2	2

*Milo is one type of grain sorghum.

Source: DOE HCP/ET-2854

Sensitivity Analysis

In the previous section, the costs of producing ethanol were computed using the Metrek life cycle cost model. In this section, the sensitivity of these costs to changes in several parameters is assessed. Those parameters which have been varied in order to assess their impacts on ethanol life cycle costs are: the value of by-products, the cost of feedstock, the capital structure (debt/equity ratio), the size of the investment tax credit allowed, and the cost of debt and common equity. By holding all of the input parameters constant and varying only those given above, the impacts of changing each parameter were computed.

Changes in By-Product Credits: Of the 28 ethanol plant configurations studied, 14 produced significant quantities of salable by-products. The market values of these by-products in $/MMBtu or million dollars per year were given by the authors of the reports reviewed and were used as the basis of this sensitivity analysis. These values were used to adjust the life cycle costs given by the model as described above. By using the same computation method, ethanol costs were adjusted for changes in by-product values. In this analysis, by-product values were adjusted upward and downward by 20% and the resulting changes in ethanol life cycle costs were computed. (The absolute quantity selected is not particularly significant, since the relationship between the two is linear; i.e., if a 20% increase in by-product value reduces ethanol costs by 6%, then a 10% increase will reduce these costs by 3%.)

The results of this analysis are presented in Table 2.8. By-product values increased/decreased by 20% are shown in the first column of the table. Life cycle costs associated with the by-product values given in the studies are presented in the second column of the table. Also given in the second column are the ethanol costs associated with the higher/lower by-product values. In the third column, the percentage change in ethanol cost is given.

Table 2.8: Sensitivity of Ethanol Costs to 20% Change in By-Product Value

Study	By-Product Value $/MMBtu	(±20%)	Ethanol Life Cycle Cost $/MMBtu	(Low/High)	Change in Ethanol Cost (±%)
Battelle, corn, low debt	8.91	(10.69) (7.13)	15.72	(14.73) (16.72)	6.4
Battelle, corn, high debt	8.91	(10.69) (7.13)	14.87	(13.88) (15.87)	6.7
USDA, corn	7.88	(9.46) (6.30)	14.30	(13.25) (15.34)	7.3
SRI, mission analysis, wheat straw	4.22	(5.06) (3.38)	42.82	(41.98) (43.67)	2.0
Scheller, corn	8.38	(10.06) (6.70)	12.04	(10.83) (13.24)	10.0
Scheller, grain sorghum	8.58	(10.30) (6.86)	15.52	(14.68) (16.37)	5.4
Scheller, milo	7.86	(4.43) (6.29)	9.90	(8.69) (11.11)	12.2
DP&RA, corn	7.20	(8.64) (5.76)	15.96	(14.87) (17.06)	6.8
Battelle, cane juice, low debt	50.65	(60.78) (40.52)	17.58	(17.08) (18.08)	2.8
high debt	50.65	(60.98) (40.52)	16.69	(16.19) (17.19)	3.0
LBL, molasses, batch	14.82	(17.78) (11.86)	13.31	(13.08) (13.54)	1.8
LBL, molasses, continuous	14.82	(17.78) (11.86)	12.22	(11.78) (12.67)	3.7
LBL, molasses, continuous with cell recycle	14.82	(17.78) (11.86)	12.09	(11.65) (12.53)	3.8
LBL, molasses, vacuum	14.82	(17.78) (11.86)	11.60	(11.36) (11.85)	2.1

Source: DOE HCP/ET-2854

The relationships between ethanol life cycle costs and by-product values are presented graphically in Figure 2.4. For a majority of the studies which assumed by-product credits, a 20% change in by-product unit value resulted in a less than 10% change in ethanol life cycle costs. Those studies in which changes in by-product values had the greatest impact on ethanol prices were conducted by Scheller using corn and milo as feedstocks. In both cases, relatively large quantities of ddg are produced (see Table 2.4), and the unit value of the by-product is high relative to that of ethanol. Consequently, for these studies the revenue requirements accounted for by the by-products are 33 and 38% of the total revenue requirements. These percentages are higher than for any other study. In addition to the greater sensitivity of ethanol costs to by-product values shown in Scheller's work, the life cycle costs of ethanol production are lower than for other studies in which grain crops are used as feedstocks.

Ethanol costs in the $10-16/MMBtu range were obtained for a majority of the studies reporting by-product credits. Three exceptions, however, are shown in Figure 2.4. The cost of producing ethanol from wheat straw, as reported by

SRI (Configuration 4), is high because of high capital and operating costs and a low process efficiency; and the reduction in ethanol costs for a 20% increase in by-product value is only 2.0% because of the value of the large quantity of by-product pentose sugar that is produced is relatively low ($4.22/MMBtu) and its contribution to total revenue requirements is also relatively low (9.0%).

Figure 2.4: Sensitivity of Ethanol Life Cycle Costs to By-Product Value

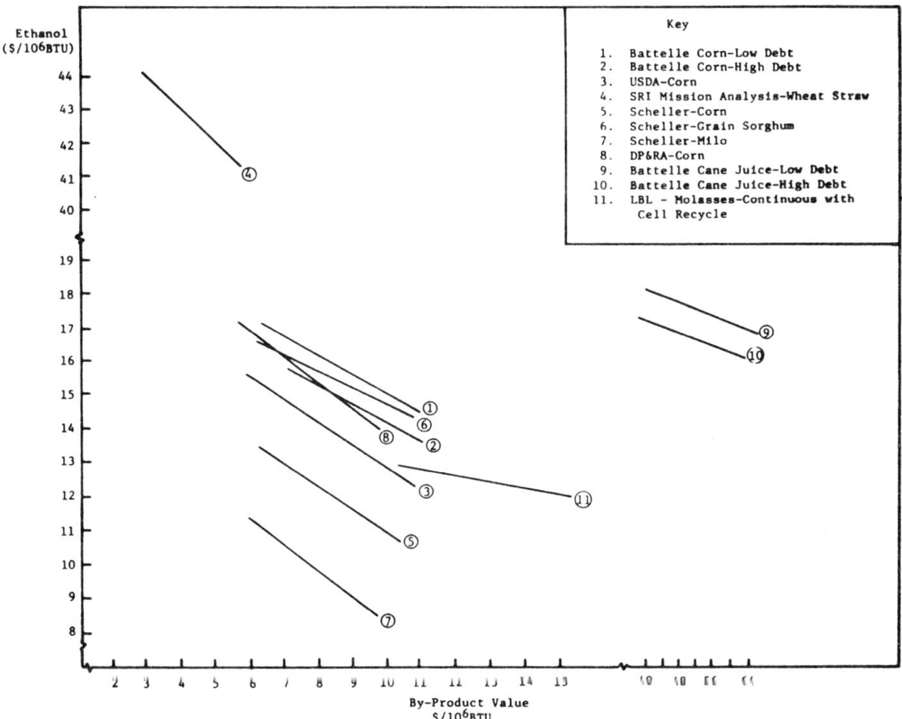

Source: DOE HCP/ET-2854

Configurations 9 and 10 represent low and high debt financial scenarios for a facility producing ethanol from cane juice reported by Battelle. In this case, while the value ($50/MMBtu) of the by-product is quite high, relatively small quantities are produced. As with SRI's wheat straw facility, the by-product contribution to total revenue requirements is low, 12 and 13%, compared to 20 to 35% for other plants using grains. As a result, the sensitivity of ethanol costs to changes in by-product value is also low with a 20% increase resulting in only about a 3% decrease in ethanol price.

Changes in Feedstock Price: For many of the ethanol configurations studied, the cost of the feedstock used represents the single most significant contribution to ethanol life cycle costs. For 17 of these configurations, feedstock accounts for over 60% of the total cost of producing ethanol. Consequently, for these configurations, any changes in feedstock costs could have significant impacts on ethanol costs.

The sensitivity of ethanol life cycle costs to changes in feedstock costs was treated in two different ways in this analysis. First, feedstock costs for all plants were increased and decreased by an arbitrary 25% and the impact of these changes on ethanol costs was completed. Second, for studies in which the same feedstock was used (e.g., corn by Scheller, Battelle, USDA and DP&RA), the average feedstock price was used and its impact on individual study results was assessed.

Uniform Percentage Change in Feedstock Cost — The percentage change in ethanol life cycle costs for a 25% change in feedstock cost is as shown in the first column of Table 2.9. The rank order of these changes in ethanol costs is given in the second column. For 14 of the 26 plants shown in this table, a 25% change in feedstock cost resulted in an ethanol cost change of between 13 and 17.5%. For all plants, the mean change in ethanol cost was about 14%. The percentage of the total production cost accounted for by feedstock is listed in the third column of Table 2.9 to emphasize that the sensitivity to changes in feedstock costs are closely related to the contribution of feedstock to total ethanol cost. The rank ordering of these percentage contributions is provided in the fourth column of the table.

Table 2.9: Sensitivity of Ethanol Life Cycle Costs to a 25% Change in Feedstock Costs

Ethanol Facility	Percent Change in Ethanol Cost*	Rank	Feedstock**	Rank
Schaffer				
Cane – 330 days, low price	18.2	6	72.9	6
Cane – 330 days, high price	19.6	4	78.4	4
Cane – 180 days, low price	15.4	16	61.8	16
Cane – 180 days, high price	17.1	8	68.5	8
Schaffer/Battelle				
Molasses – 150 days	16.9	9	67.8	9
Molasses – 330 days	16.0	14	64.8	12
Battelle				
Corn	14.7	17	60.0	17
Cane juice	16.8	10	67.0	10
Mitre				
Wood – 850 tpd, low price	2.6	25	10.4	25
Wood – 850 tpd, high price	4.7	22	18.8	22
Wood – 1,700 tpd, low price	3.2	24	12.7	24
Wood – 1,700 tpd, high price	5.6	21	22.5	21
SRI – wheat straw, enzymatic hydrolysis	2.4	26	9.4	26
SRI – molasses, 10-site survey	12.6	20	50.3	20
SRI mission analysis				
Algae	15.5	15	62.0	15
Wheat straw	4.5	23	18.1	23
Sugarcane	16.1	12.5	64.2	14

(continued)

Table 2.9: (continued)

Ethanol Facility	Percent Change in Ethanol Cost*	Rank	Feedstock**	Rank
Scheller				
Corn and distressed grain	17.3	7	69.2	7
Sorghum	13.8	18	55.1	18
Milo	16.1	12.5	64.4	13
LBL, molasses				
Batch	18.6	5	74.5	5
Continuous	19.8	3	78.7	3
Continuous with cell recycle	19.9	2	79.5	2
Vacuum with cell recycle	21.1	1	84.2	1
DP&RA, corn	13.2	19	52.8	19
USDA, corn	16.3	11	65.4	11

*Resulting from a 25% increase/decrease in feedstock cost.
**As a percent of total ethanol cost $/MMBtu.

Source: DOE HCP/ET-2854

Average Feedstock Costs — In addition to varying feedstock costs by a constant amount, a set of average feedstock costs was employed to test the sensitivity of relative ethanol costs using the same feedstock. Average feedstock costs were developed for sugarcane and sugar juice, molasses, wood, corn, grain sorghum and wheat straw. Average feedstock costs were calculated to be: sugarcane, $3.12/MMBtu; sugar juice, $9.18/MMBtu; molasses, $6.32/MMBtu; wood, $1.58/MMBtu; corn, $6.50/MMBtu; wheat straw, $0.80/MMBtu; and grain sorghum, $5.62/MMBtu.

Table 2.10 shows that considerable ethanol price fluctuation occurred even using average feedstock costs. The widest range occurred in the sugarcane and sugar juice configurations, followed by molasses, corn, wood, grain sorghum, and wheat straw.

Table 2.10: Impact of Average Feedstock Costs on Total Life Cycle Ethanol Costs

Configuration	Percent Decrease/Increase in Ethanol Costs Using Average Feedstock Costs
Schaffer, cane – 330 days, low price	+19.2
Schaffer, cane – 330 days, high price	–4.9
Schaffer, cane – 180 days, low price	+16.3
Schaffer, cane – 180 days, high price	–4.3
Schaffer/Battelle, molasses – 150 days	+17.4
Schaffer/Battelle, molasses – 330 days	+14.5
Battelle, corn	–5.9
Battelle, cane juice	–9.0
Mitre, wood – 850 tpd, low price	+6.0
Mitre, wood – 850 tpd, high price	–3.9
Mitre, wood – 1,700 tpd, low price	+6.2
Mitre, wood – 1,700 tpd, high price	–5.6
SRI, wheat straw – enzymatic hydrolysis	+3.1

(continued)

Table 2.10: (continued)

Configuration	Percent Decrease/Increase in Ethanol Costs Using Average Feedstock Costs
SRI 10-site survey, molasses	-6.5
SRI mission analysis, wheat straw	-4.0
Scheller, corn and distressed grain	+0.6
Scheller, grain sorghum	-3.0
Scheller, milo	+10.4
LBL, molasses – batch	-13.3
LBL, molasses – continuous	-17.0
LBL, molasses – continuous w/cell recycle	-19.0
LBL, molasses – vacuum w/cell recycle	-17.9
DP&RA, corn	+7.0
USDA, corn	-2.0

Source: DOE HCP/ET-2854

Perhaps more significant is a comparison of the range of ethanol prices before and after the use of average feedstock prices. Table 2.11 shows how the range of ethanol costs varied from the previous comparison which used standard financial parameters to the comparison which used both common financial parameters and average feedstock prices.

Table 2.11: Comparison of Cost Ranges Before and After Use of Average Feedstock Prices

	Sugarcane and Sugar Juice*	Molasses	Corn	Wood	Wheat Straw	Grain Sorghum**
Ethanol costs assuming only common financial parameters	11.29 to 17.87	9.67 to 17.45	12.23 to 16.15	19.23 to 24.57	43.83 to 44.69	10.02 to 15.63
Ethanol costs assuming average feedstock prices and common financial parameters	11.93 to 16.26	10.97 to 16.31	12.31 to 17.43	20.43 to 23.60	42.09 to 46.09	11.06 to 15.16

*SRI cane juice study omitted from both ranges.
**Includes milo.

Source: DOE HCP/ET-2854

Of the six feedstocks, the sugarcane, molasses, wood, and grain sorghum configurations showed a narrowing of the range of ethanol costs, while the corn and wheat straw configurations showed a widening range. This indicates that for studies using the first four feedstocks, ethanol cost differences may be more attributable to differing assumptions about feedstock prices. Similarly, factors such as by-product credit and basic technology cost and performance assumptions appear to be more important for the studies using corn and wheat straw.

In any event, ethanol costs within each feedstock category appear to be more sensitive to common feedstock prices than to common financial parameters.

Changes in the Investment Tax Credit:　Ethanol life cycle costs were computed under the common set of parameters described under "Life Cycle Cost Comparison Using Common Assumptions," assuming a 10% investment tax credit. The sensitivity of these costs to changes in this tax credit was computed for an itc of 0% and of 20%. Using the existing itc of 10% as a baseline, it was found that the change in itc had only a small impact on ethanol costs. This is shown in Table 2.12.

Table 2.12: Sensitivity of Ethanol Life Cycle Costs to Changes in the Investment Tax Credit

Configuration	Percent Change in Ethanol Cost for Increase in itc from 10 to 20% or Decrease in itc from 10 to 0%
Schaffer, cane – 330 days, low price	2.21
Schaffer, cane – 330 days, high price	1.77
Schaffer, cane – 180 days, low price	3.37
Schaffer cane – 180 days, high price	2.84
Schaffer/Battelle, molasses – 150 days	3.62
Schaffer/Battelle, molasses – 330 days	1.97
Battelle, corn	1.63
Battelle, cane juice	1.80
Mitre, wood – 850 tpd, low price	5.03
Mitre, wood – 850 tpd, high price	4.56
Mitre, wood – 1,700 tpd, low price	4.99
Mitre, wood – 1,700 tpd, high price	4.46
SRI, wheat straw – enzymatic hydrolysis	2.17
SRI 10-site survey, molasses	1.43
SRI mission analysis, algae	1.82
SRI mission analysis, wheat straw	1.92
SRI mission analysis, sugarcane	1.65
Scheller, corn and distressed grain	1.22
Scheller, sorghum	1.88
Scheller, milo	1.65
LBL, molasses – batch	0.94
LBL, molasses – continuous	0.41
LBL, molasses – continuous w/cell recycle	0.34
LBL, molasses – vacuum w/cell recycle	0.31
DP&RA, corn	3.25
USDA, corn	1.51

Source: DOE HCP/ET-2854

For 24 of the 28 plant configurations, increasing the itc from 10 to 20% only decreased ethanol costs by amounts of 0.3 to 3.68%. Decreasing the itc to 0% resulted in similar cost increases. This is due to the fact that the ethanol facilities are not capital intensive and that capital accounts for a relatively small percentage of life cycle costs. This was shown in Table 2.6. The only facilities for which changing the itc will have a more significant impact are those for which capital is a significant part of total costs. Those facilities using wood as a feedstock, have a capital cost component which is 27 to 31% of the total. For an itc of 0 or 20%, ethanol costs will be 4.5 to 5.0% higher or lower than costs associated with an itc of 10%.

Changes in Costs of Equity and Debt: For a majority of the studies reviewed and under the set of common assumptions used, the cost of equity capital was assumed to be 15%. For purposes of this analysis, ethanol costs were computed for a cost of equity of 10 and 20%. A lower cost of equity reduces the weighted cost of capital and income taxes which must be paid and thus, also reduces ethanol production costs. The impacts of the variation in cost of equity on ethanol costs are given in the first column of Table 2.13. Changes in the cost of equity financing changed the weighted cost of capital from 11.4% for the baseline assumptions to 9.4% for the 10% cost of equity and to 13.4% for the 20% cost of equity. The impact on ethanol costs was significant only for the more capital intensive facilities such as those using wood and sugarcane. As with the investment tax credit, there is a high degree of correlation between the extent to which a facility is capital intensive and the impact that changing the cost of equity has upon life cycle costs.

Table 2.13: Impact of Changes in Costs of Equity and of Debt on Ethanol Life Cycle Costs

	Percent Decrease/Increase in Ethanol Costs for . . . . .Changes in Cost of.	
	Equity from 15–10% and 15–20%	Debt from 9–3% and 9–15%
Configuration		
Schaffer		
Cane – 330 days, low price	3.54	3.72
Cane – 330 days, high price	2.82	2.97
Cane – 180 days, low price	5.55	5.85
Cane – 180 days, high price	4.57	4.81
Schaffer/Battelle		
Molasses – 150 days	5.48	5.48
Molasses – 330 days	2.40	2.40
Battelle		
Corn	2.75	2.90
Cane juice	2.94	3.09
Mitre		
Wood – 850 tpd, low price	8.40	8.98
Wood – 850 tpd, high price	7.61	8.14
Wood – 1,700 tpd, low price	8.11	8.84
Wood – 1,700 tpd, high price	7.34	7.85
SRI – wheat straw, enzymatic hydrolysis	3.54	3.74
SRI – molasses, 10-site survey	2.18	2.18
SRI mission analysis		
Algae	2.89	3.05
Wheat straw	3.17	3.34
Sugarcane	2.73	2.88
Scheller		
Corn and distressed grain	1.98	2.07
Sorghum	3.08	3.38
Milo	2.86	3.08

(continued)

Table 2.13: (continued)

	Percent Decrease/Increase in Ethanol Costs for . . . . .Changes in Cost of.	
Configuration	Equity from 15–10% and 15–20%	Debt from 9–3% and 9–15%
LBL, molasses –		
Batch	1.54	1.68
Continuous	0.27	0.34
Continuous with cell recycle	0.48	0.48
Vacuum with cell recycle	0.53	0.53
DP&RA, corn	5.45	5.86
USDA, corn	2.51	2.68

Source: DOE HCP/ET-2854

Table 2.13 also gives the relationship between ethanol costs and the cost of debt. This cost was changed to 3% and to 15% from the base case of 9%. The resulting life cycle cost differences are quite similar to those resulting from the variations in the cost of equity. Again, changes in the cost of debt have a greater impact on those plants which are capital intensive.

Changes in the Debt/Equity Ratio: The baseline capital structure used in the analysis with common financial assumptions was for a plant financed by 60% debt and 40% common equity. As the ratio of debt to equity financing changes, so does the weighted cost of capital and the life cycle cost of ethanol. The sensitivity of these changes to variations in the debt/equity ratio is illustrated in Figure 2.5 for ten of the ethanol configurations. No more than one case was selected from each of the studies reviewed. Taken together, these cases illustrate a broad range of cost variations. Figure 2.5 illustrates that those configurations which are most capital intensive will have ethanol costs that are most sensitive to changes in the debt/equity ratio. The weighted cost of capital will be 15% for a 100% equity financed plant; for a 100% debt financed plant, it will be 9% under the baseline assumptions. The impacts of this decrease in the cost of capital as the portion financed by debt goes up are shown.

Two conclusions may be drawn. All other parameters being equal, greater debt financing will reduce ethanol costs. Also, for a majority of the configurations, studied, capital is not a major component of ethanol costs and thus increasing the portion of capital financed by debt will not result in major decreases in ethanol costs.

Conclusions

The major conclusions resulting from this comparative economic assessment of the production of ethanol from biomass are summarized in the following points:

Ethanol process efficiencies and system designs vary significantly: Among the studies reviewed using the same feedstock, process efficiency, system capacity and equipment selection showed wide variations.

Figure 2.5: Sensitivity of Ethanol Costs to Changes in the Ratio of Debt to Equity Financing

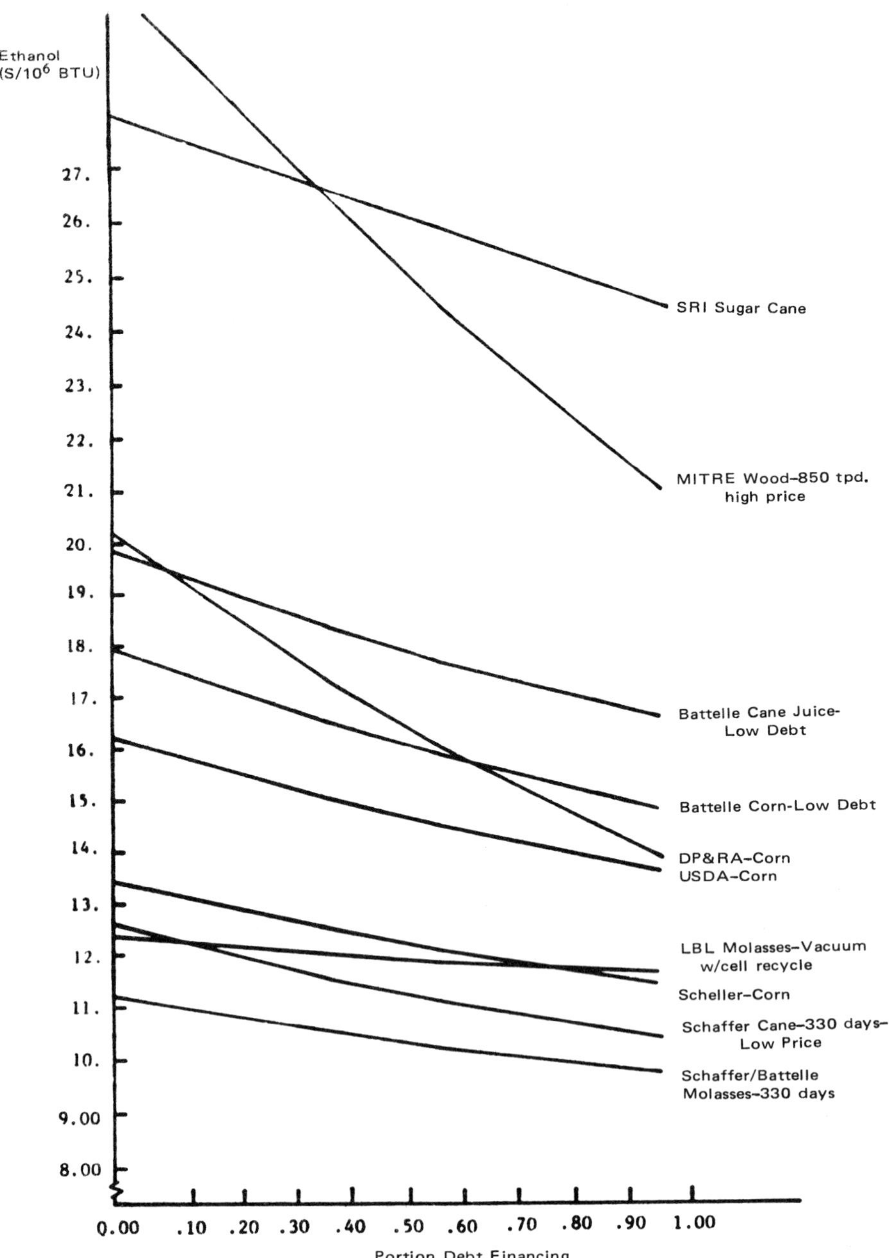

Source: DOE HCP/ET-2854

These variations contribute significantly more to the differences in ethanol costs than do the variations in financial assumptions originally made in the studies.

Different process efficiencies and system designs contribute to wide variations in capital and operating costs per gallon of ethanol produced: The differences include equipment and materials specifications, fermentation technologies, operating availabilities and other elements of system design. Among the 28 ethanol cases considered, the differences contribute to wide variations in capital and operating costs per unit of ethanol produced. Even among cases using the same feedstock, there are wide variations in capital and operating costs per unit of ethanol produced.

Capital costs have only a limited impact on total ethanol costs: Of the basic production cost components, capital costs are, on the average, the smallest component of total life cycle ethanol costs. In general, operating costs tend to be somewhat more significant, while feedstock costs tend to be the largest of the three components comprising life cycle ethanol costs.

Ethanol costs in general are more sensitive to feedstock prices than to common financial parameters: In more than half of the cases, there was less than a 3% change in life cycle ethanol costs when calculated using common financial parameters. By contrast, the calculation of ethanol costs using both common financial parameters and average feedstock prices showed greater than a 6% change in more than half of the cases. The different responses of ethanol costs to these changes reflect the lesser significance of capital costs (which are most influenced by financial parameters) compared to feedstock costs in determining total life cycle ethanol costs.

By-product credit values also vary widely among the studies and have a significant impact on relative ethanol production costs: By-product evaluations among the studies showed many differences. For molasses, cane juice and wheat straw, some studies claimed by-products, while others did not. Among those studies which did claim by-products, there was no standardization of key determinants of by-product value, such as Btu content, selling price, or quantity produced per unit of feedstock input or ethanol output. These differences are reflected in the magnitude of by-product credits claimed, which in some cases were assumed to be equivalent to about one and one-half times the annual operating cost of the entire facility.

The relative effect of using average costs for each feedstock on total ethanol production costs varies considerably among feedstocks: The range of ethanol costs was narrowed among configurations using sugarcane, molasses, wood, and grain sorghum when common financial parameters and average feedstock costs were used. Conversely, using the same assumptions, the range of ethanol costs became wider in those cases using corn and wheat straw. Thus, it appears that differences in feedstock costs were responsible for a larger range of ethanol costs than would otherwise have occurred in cases using the sugarcane, molasses, wood and sorghum. In cases using corn and wheat straw, other factors such as by-product credits and basic design and cost assumptions appeared to exert greater influence on the range of ethanol costs.

RAPHAEL KATZEN ASSOCIATES STUDY

> The information in this section is based on "Technical and
> Economic Assessment—Motor Fuel Alcohol from Grain and Other
> Biomass" by G.D. Moon, Jr., J.R. Messick, C.E. Easley and R.
> Katzen of Raphael Katzen Associates in the proceedings of the
> *Alcohol Fuels Technology Third International Symposium—Volume
> II*, Asilomar, California, May 1979 (hereinafter referred to as
> Asilomar Conference).

The primary objective of this study was to evaluate the technical and economical
feasibility for production of grain motor fuel ethyl alcohol from a grass roots
plant with a capacity of 50 million U.S. gal/yr. The base case evaluation includes a detailed process design using proven technology, a complete budget estimate of the plant investment including working capital, a detailed analysis of
the annual operating cost using corn as the primary feedstock, and a complete
financial analysis which establishes the alcohol selling price required to net to
the investor a 15% discounted cash flow-interest rate of return.

The study shows that motor fuel alcohol must have a selling price of $1.05 to
$1.13 per gallon (1978 dollars) to satisfy the requirements of the base case (the
lower selling price reflects the economics of a 20-year plant life whereas the
higher price reflects a 10-year plant life).

Variations in plant capacities cause the alcohol selling price to rise to $1.52/gal
for a 10 million gal/yr plant; and drop to $0.98/gal for a 100 million gal/yr
plant. Alternate feedstock materials cause the selling price to increase from
$1.05/gal for corn to $1.40/gal for sweet sorghum, and drop to a low of $1.02/gal
for milo (grain sorghum). Wheat requires an alcohol selling price of $1.31/gal.
Substitution of corn stover for coal as plant fuel causes a $0.04/gal increase in
the alcohol selling price. All of the above comparisons are based on a 20-year
production life for the plant.

The base case alcohol plant has a total capital requirement of $63.6 MM in
equivalent 1978 (year end) dollars. The fixed investment for the plant is $58.0
MM. The total production cost for the base case is $0.89/gal of alcohol.

Base Case Production of Alcohol

The alcohol plant is sized to produce 50 MM gal/yr of motor fuel grade alcohol
from corn. The alcohol product yield is 2.57 gallons per bushel of corn thereby
requiring 19.46 MM bushels of corn per year. In addition to the primary alcohol product, the plant produces 177,111 tons/yr of by-product Distillers' Dark
Grains (DDG), and 10,428 tons/yr of ammonium sulfate. The plant is assumed
to be located in Central Illinois, close to a source of Illinois #6 coal which is
to be used as fuel for the plant.

The alcohol plant, in general, uses existing process technology as currently employed in U.S. grain alcohol plants. The basic processing steps included in the
plant are grain or biomass feed preparation, starch removal and saccharification,
batch fermentation, alcohol distillation and purification, and by-product animal
feed recovery. The plant operates as a continuous flow process, except for the
fermentation and fungal amylase sections which are operated batchwise in order

to allow for frequent sterilization of the equipment. The distillation system employs a two-pressure concept which significantly improves steam economy. This process concept, along with other heat economy measures, results in a total plant steam usage of 31.7 lb/gal of alcohol product. The distillation system uses 21.4 lb/gal of which 2.8 lb/gal is obtained as flash vapor from the mash cooking operation.

All utility requirements, with the exception of electricity, are produced within the boundaries of the plant. Water is obtained from a well field located close to the plant. The boiler burns relatively low cost, high sulfur coal (approximately \$25/T compared to \$40/T for low sulfur coal). Boiler flue gas is used to dry the stillage residue in producing the Distillers' Dark Grains by-product. Wastewater is treated in a two-stage, activated sludge system. The sludge is dewatered and fed as supplemental fuel for the boiler. Cooling water is recycled from a two-cell cooling tower.

Approximately one-third of the plant power requirement is obtained from steam turbines. The coal fired boiler produces 600 psig/600°F steam which is used to drive turbines for the DDG evaporator recompression system. The turbines exhaust steam at 150 psig which is suitable for satisfying all other process steam requirements.

Overall Material and Energy Flows: The major input raw materials are:

Item	Daily Quantity	1978 Unit Value
Corn	58,955 bushels	\$2.30/bu
Coal	296.7 tons	\$24.50/ton
Yeast	1.2 tons	\$0.40/lb
Denaturant	1,500 gallons	\$0.60/gal
Ammonia	9.2 tons	\$120.00/ton

The plant products are:

Item	Daily Quantity	1978 Unit Value
Grain motor fuel alcohol	151,515 gal	\$1.05/gal
DDG	536.7 tons	\$110.00/ton
$(NH_4)_2SO_4$	31.6 tons	\$40.00/ton

The ammonium sulfate is a by-product of the flue gas desulfurization. The three major energy input items are corn, coal, and electricity. The two primary energy containing items are alcohol and DDG.

The plant thermal efficiency can be calculated by several different methods. One method is to consider only the fossil fuel input items. This method gives an energy efficiency of 154.4%. Another method for determining the thermal efficiency is to consider the energy required to produce the corn crop, with credit given for the energy required to produce DDG (equivalent corn basis) as an additional energy input item. In Illinois, the average energy usage is 95,200 Btu per bushel of corn (USDA 1974 Data Base) which includes energy for field preparation, harvesting, fertilizer supplied, transportation, and other miscellaneous usages. This method results in an efficiency of 105.4%. A third method of calculating the thermal efficiency is to include the total thermal energy contained in the corn and dried grains. This method yields an overall plant efficiency of 61.8%.

Economics for Base Case

Investment and Operating Cost: The investment required to build complete grass-roots facilities for a 50 MM gal/yr alcohol plant is $58.0 MM, last quarter 1978 cost basis. The investment includes all support facilities and fifty acres of land. Process steam is produced by a coal fired boiler. The boiler unit includes flue gas desulfurization equipment. The earliest start-up date, assuming a decision had been made to build the plant by January 1, 1979, is January 1983. Working capital is estimated at $7.9 MM for the first year of operation and is inflated at 7%/yr for the life of the plant.

The annual operating cost is $44.51 MM or $0.89/gal of alcohol, last quarter 1978 equivalent cost. This includes straight line depreciation over twenty years. The annual operating cost increases to $47.41 MM or $0.95/gal for a 10 year plant operating life. The operating cost is rather insensitive to plant investment since the capital investment items (fixed charges) are a small percentage of the total cost. The major operating cost item is the purchase price of corn.

Economic Analyses: The objective of the economic assessment is to establish a selling price for grain motor fuel alcohol which will cover the production cost and give a realistic return on company equity. The base case analysis uses a 15% discounted cash flow-interest rate of return (DCF-IROR). It is realized that a 15% DCF-IROR may not be large enough for a company to venture into production of grain motor fuel alcohol, consequently rates up to 20% DCF-IROR were also considered. (Since a cash flow analysis includes profits plus depreciation, a 15% DCF-IROR represents only about a 13% annual interest rate for a 20 year plant life.)

The procedure used to establish the 1978 selling price includes the complicating effects of an inflationary economy.

Base Case Selling Price: The base case selling price for grain motor fuel alcohol is $1.05/U.S. gal based on a 20 year plant life and a 15% DCF-IROR. The selling price increases to $1.13/U.S. gal when the plant is assumed to have an effective operating life of only 10 years. Many companies perform their venture analysis based on 10 years (although the plant has a life of 20 years) because the plant may become obsolete due to new technology.

Excursions on Plant Capacities

In addition to the base case production rate of 50 MM gallons per year, capacities of 10 MM and 100 MM gallons per year were evaluated. The 1978 equivalent alcohol selling prices are $1.55 and $0.98/gal respectively for the 10 MM and 100 MM gallons per year plants. Table 2.14 summarizes the investment, operating expense, and alcohol selling price for three production capacities.

Table 2.14: Summary of Investment, Operating Expense, and Alcohol Selling Price—20 Year Plant Life ($MM, 1978 Basis)

Plant Capacity, MM gal/yr	Investment		Operating Expense		15% DCF-IROR Alcohol Selling Price, $/gal
	Fixed	Working Capital	Total	$/gal	
10	25.2	2.7	12.2	1.24	1.55
50*	58.0	5.6	44.5	0.89	1.05*
100	100.0	10.9	83.6	0.84	0.98

*Base case.

Source: Asilomar Conference

Excursions on Feedstock Material

In considering the production of motor fuel grade alcohol, the most logical primary feedstock choice is corn, because it is produced voluminously in major growing areas of the United States. Producers of industrial grade alcohol prefer corn because of its economic advantage.

Alternate feedstock materials considered in this work were wheat, milo (grain sorghum), and sweet sorghum.

In addition to alternate feedstock materials, an evaluation was made of the effect of substituting corn stover biomass for Illinois #6 coal for the fuel requirement of the plant. The results of these excursions are summarized in Table 2.15.

Table 2.15: Summary of Excursions on Feedstock Materials–20 Year Operating Life

Excursion	Feedstock Variable	Deprec. Period yr	Investment*, $MM			Oper. Cost* Total,		Alcohol Selling Price* . . $/gal, DCF . .		
			Fixed	Work. Cap.	Total	$MM	$/gal	10%	15%	20%
Base case	corn	10	58.0	5.6	63.6	44.5	0.89	0.97	1.05	1.16
Wheat	wheat	10	58.0	7.9	65.9	57.1	1.14	1.23	1.31	1.44
Milo	milo	10	58.0	5.5	63.5	42.6	0.85	0.94	1.02	1.13
Sweet sorghum	sweet sorghum	10	91.6	6.4	98.0	58.8	1.18	1.28	1.40	1.59
Corn stover	corn with corn stover fuel	10	57.0	5.8	62.8	46.4	0.93	1.01	1.09	1.21

*1978

Source: Asilomar Conference

Wheat and milo can be processed in essentially the same equipment as the corn feed material. Sweet sorghum requires different front-end equipment, more steam generating capacity, and bagasse storage and handling facilities; consequently, the investment is considerably higher than for the base case (corn). The sweet sorghum excursion is designed for corn as the feed material during the dead season. It is assumed sweet sorghum can be harvested during only 165 days of the year. This becomes very expensive since a large portion of the plant equipment is idle during a large part of each year. In addition, sweet sorghum raw material is more expensive than corn; and the DDG by-product from sweet sorghum is less valuable than that from corn.

Another alternative would be to design the sweet sorghum plant to process double the sweet sorghum feed material during the active season, and concentrate approximately one-half the syrup for storage and use during the dead season. This alternative was not considered in detail because it became obvious that the investment would be higher than the combination corn/sorghum alternative due to the high cost of doubling the size of the front-end equipment, and energy use would increase.

The wheat excursion results in a relatively high alcohol selling price, whereas the milo excursion results in the lowest alcohol selling price. The wheat excursion is expensive for two reasons: (1) the raw material purchase price is high, $3.15 per bushel compared to corn at $2.30 per bushel, and (2) product conversion is less (2.5 gallons per bushel compared to corn at 2.6 gallons per bushel). The milo excursion results in the lowest selling price because the grain is less expensive than corn and product conversion is about the same. Milo production in this country is substantially less than corn. A major new market for milo could cause the milo market price to increase. This would reduce the advantage for production of alcohol from milo.

The corn stover (fuel) excursion results in a higher alcohol selling price because the corn stover fuel is approximately twice as expensive as Illinois #6 coal (on a net Btu basis). The selling price increases by $0.04/gal of alcohol. The plant investment is about $1.0 MM less than the base case because flue gas desulfurization is not required and the increase in investment for the boiler and fuel handling equipment is less than the cost of the flue gas desulfurization system.

Sensitivity to Financial Parameters

The alcohol selling price sensitivity to a number of financial parameters was evaluated for the twofold purpose of establishing the cost benefits of governmentally supported incentives, and determining the effect of variables which may differ from the values chosen from the base case.

All of the sensitivity analyses are compared to the base case with a 20 year plant life ($1.05/gal alcohol selling price). The purchase price of corn is the only variable that impacts a substantial effect on the alcohol selling price. The nonbudgetary governmentally supported incentives, such as loan guarantees, higher investment tax credit rates, and shorter depreciation schedules have only limited effect on reducing the base case selling price. The following general statements regarding the effects can be made.

Depreciation Schedule: Increasing the depreciation schedule (sum-of-years method) from 10 years to 20 years, increases the alcohol selling price by $0.02/ gal.

Reducing the depreciation schedule to 5 years does not reduce the selling price because the heavy losses which occur in the first five years of production are balanced by the investment tax credit.

Working Capital: Doubling the working capital from $5.6 MM to $11.2 MM causes a $0.03/gal increase in the alcohol selling price.

Purchase Price of Corn: The purchase price of corn is the most significant variable. For every 10% increase in corn purchase price, the alcohol selling price increases by about $0.12/gal assuming all other factors remain constant. As the purchase price of corn increases, the DDG by-product becomes more valuable; consequently, a 10% increase in the corn purchase price will increase the alcohol selling price by about $0.08/gal.

DDG By-Product: A 10% increase in the Distillers' Dark Grains by-product value effects a $0.04/gal decrease in the alcohol selling price.

Leveraged Capital: The leveraged capital situation is influenced by the high taxes on corporate profits. The high taxes foster a high debt to equity ratio since the interest on debt money is tax deductible, but the dividends on profits are not. A practical limit is reached, however, where the interest paid during the early years of the project, and especially the years prior to full production, becomes such a burden that an increase in debt beyond a certain level is not beneficial. At the 15% DCF-IROR rate and a 10% interest rate for a 20-year project life, financing 80% of the plant investment and 100% of the working capital results in a $0.10/gal decrease in the alcohol selling price (compared to 100% company equity).

The analyses herein are based on constant company equity throughout the life of the plant. Principal payments are subtracted from cash flow. This has the effect of minimizing company equity, minimizing total profits and increasing the risk of the venture. An alternative is to increase company equity by applying principal payments to equity, rather than subtracting the payments from cash flow. This should result in a higher alcohol selling price and reduce the risk of the venture. Highly leveraged capital situations affect investment tax credits which cannot be used during the 7 year time period allowed by the IRS; consequently, if the investment tax credit can be applied to other taxable profits, the leveraged capital situations would be marginally more favorable.

Investment Tax Credit: Increasing the investment tax credit rate from 10 to 50% reduces the alcohol selling price by $0.02/gal. The investment tax credit rate could be marginally more significant if the company producing the alcohol has profits from other ventures against which the investment tax credit could be applied immediately. The analyses herein are based on a single-venture company.

CALIFORNIA ENERGY COMMISSION STUDY—AGRICULTURAL WASTES

> The information in this section is based on *The Production of Ethanol from Agricultural Waste—An Economic Evaluation* (CAEC-28) by D. Paige and R. Boulton of the University of California, Davis, prepared for the California Energy Commission, January 1979.

In line with the current need to develop alternate energy sources, this report considers the economic feasibility of using certain agricultural waste materials for the production of ethanol, which in turn may be used as a substitute for hydrocarbon fuels. An attempt is made to integrate ethanol production considerations with existing problems in waste management.

The usual process for producing ethanol from biological materials involves the anaerobic fermentation of simple sugars by yeasts. These sugars may be present naturally or may be converted from other compounds such as cellulose or starch. Although some waste materials contain significant amounts of sugars, notably whole fruit culls, they are perishable and suffer from limited supply and length of season. Other feedstocks, such as straw or wood chips, require extensive pretreatment to produce the proper substrates for fermentation. Therefore, several degrees of process complexity are considered which, with increasing sophistication of treatment, will expand the scope of input feedstocks and increase the operating season. An economic evaluation is performed for each level of process complexity, and consideration is given to a range of operating parameters, e.g., plant size, ethanol value, and waste treatment credit.

Waste Materials

Any agricultural material which does not reach the market place as final product may be considered waste, and therefore, a candidate for use in ethanol production. However, to be economically practical, a material must be judged by certain criteria:

(1) Cost of collection;
(2) Content of fermentable substances, whether with or without conversion;
(3) Value for other uses, e.g., animal feed;
(4) Cost of present disposal methods;
(5) Perishability;
(6) Length of season available;
(7) Ability to be collected with existing agricultural practices and equipment; and
(8) Value of by-products or residue, e.g., for pyrolysis.

This report will focus on two types of waste material which, together, seem most attractive economically.

Vegetable and fruit wastes from processing plants and canneries constitute a considerable disposal problem because of their demand for oxygen while decomposing and the unpleasant odors and fly breeding which also develop. Present methods involve the dumping of this matter into shallow landfill sites, various forms of field application, use as animal feed, and treatment by conventional wastewater techniques. All of these methods are an economic liability and generally involve hauling the waste from the plant at an approximate average cost of $7/ton (8)(9), except if used as animal feed, where the hauling costs may be borne by the receiver. Table 2.16 lists the quantities and approximate composition of some of the most common cannery wastes in California.

It is apparent that this material, if fermented at the cannery site, satisfies most of the criteria advanced above, with the exceptions of perishability and length of season. There is even an economic waste disposal credit for any reduction in volume and biological activity due to the fermentation process.

According to an unpublished study by the Canner's League of California, the 52 canneries in the state generated about 750,000 tons of wet waste in 1977. 36 of these canneries, with a total output of 525,000 tons of wet waste, reported 29% disposed of as cattle feed, 47% by field application, and 24% by landfill. Hauling costs alone are therefore approximately $3.7 million per year statewide (533,000 tons not animal feed at $7/ton), and therefore an economic incentive exists to find a practical use for this material.

Tomatoes and peaches were chosen for consideration here because of the large quantities available of the former and the high sugar content of the latter.

Table 2.16: Cannery Wet Waste in California*

Material	Quantity (tons)	Total Solids (%)	Sugars (%)	Cellulose (%)
Total	750,000	—	—	—
Tomato, composite	500,000	14	2.2	2.2
Culls	150,000	6**	3.5***	1.0†
Peeling waste	100,000	12†	2.0***	2.0†
Pomace	250,000	20††	1.5***	3.0††
Peaches	50,000†	10†	7.5***	1.0†

   *Canners' League, unpublished report.
  **Reference (1).
***Analysis of samples collected from Sacramento Foods Cannery, Sacramento.
  †Estimated.
  ††Reference (2).

Source: CAEC-28

However, since wet waste is only available for approximately 100 days of the year, and cannot be stored due to its perishable nature, there is the major drawback of having the capital investment in the processing plant stand idle for the remainder of the year. This condition can be alleviated, for processes which include a hydrolysis step, by using a dry feedstock which can be stockpiled for continuous operation. Various crop residues fulfill the basic criteria, and some are listed in Table 2.17. These materials could be gathered and baled with conventional machinery and transported within a 50-mile radius of the field for around $25/ton (3). Many of these materials are presently burned in the field with consequent pollution problems, and there is strong pressure to find other disposal methods.

Table 2.17: California Crop Residues

Crop	Approximate Quantity per Year (million tons)	Cellulose (%)	Starch (%)
Wheat straw	1.0*	50**	5***
Rice straw	1.0*	40***	20†
Corn stalks	0.8*	35***	5***
Cotton stalks	2.0††	40***	—
Barley straw	1.5*	40***	5***
Oat straw	0.15*	40**	5***
Wood chips	—	65	—

  *Estimated from USDA Economics, Statistics, and Cooperative Service
    Bulletins, using a harvest index (grain wt ÷ grain + shoot wt) of 0.5.
  **From *Nutrient Requirements of Beef Cattle,* Nat'l. Acad. Sci., 1970.
***Estimated.
  †Reference (4).
  ††Estimate on acreage basis.

Source: CAEC-28

Rice straw was chosen for the purposes of this study because of a high reported starch concentration (4) and the farmer's desire to remove it from the fields for disease control and nitrogen management.

Process Alternatives

The present study considers a range of process technologies in the development of alternatives. These vary from the fermentation of natural sugars in the waste to the hydrolysis of cellulose and starch by state-of-the-art enzyme reactions. The former case uses dilute solutions and simple equipment while the latter employs solution concentration and continuous fermentation, which requires considerably more sophisticated units. Four basic processes are shown in Figure 2.6 and these are to be evaluated for their economic feasibility.

Figure 2.6: Alternative Processes for Producing Ethanol from Wastes

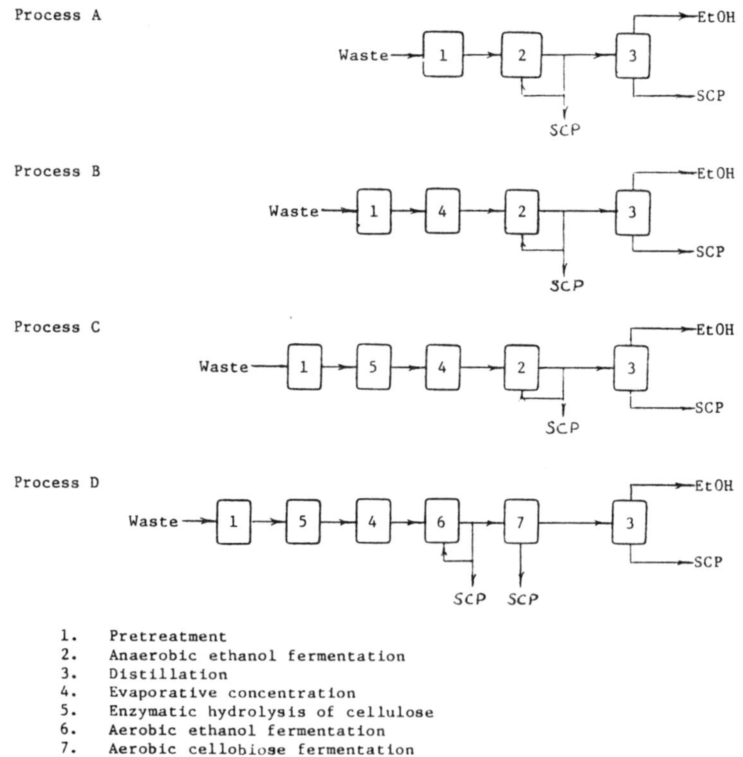

Source: CAEC-28

Process A — Pretreatment (grinding, extraction of sugars, SO_2 addition), fermentation with recycle and purge, distillation;
Process B — Pretreatment, concentration of dissolved materials, then the same as Process A;

Process C — Pretreatment, enzymatic hydrolysis, concentration, then the
same as Process A; and

Process D — Pretreatment, enzymatic hydrolysis, concentration, aerobic
ethanol fermentation, aerobic cellobiose fermentation, distillation.

The purpose of employing anaerobic fermentations is to minimize cell production and maximize ethanol production. The use of concentration stages is to
bring the sugar content up to approximately 10 wt % which leads to an optimum
fermentation rate. The cellobiose fermentation is to remove these sugars from
the final aqueous stream and to produce single cell protein (SCP) in the process.

Process Evaluation

The processes are evaluated using the discounted cash flow (or interest-rate-of-
return) method. In this method, the time value of money is taken into account
and the profits over the lifetime of the plant are expressed as an equivalent interest rate. This interest rate represents the rate at which the investment would
have been gathering interest to produce the same accumulated profits. Normally,
a rate of return of 10% pa on the investment would be considered acceptable in
many industries. Lower rates indicate poor investments; higher rates indicate
good investments.

The effects of installation capacity, dumping cost, and ethanol price on the
profitability of the investment are considered. The criterion is the rate of return
on the investment. Allowance has been made for labor and material costs, insurance and other overheads, and depreciation at 10% pa has been assumed. A
tax rate of 45% on profits has been used. The yeast produced by the fermentation is considered to be recovered and sold as single cell protein at $0.90/lb of
dry matter (6). This yields a considerable contribution to the profitability of
all these ventures with the amounts differing depending on the process used.
The 95% ethanol produced is high in quality and is assumed to be used in a
gasohol program or similar use giving it a value of $1/gal.

Tables 2.18 through 2.21 show the results for Processes A through D at several
input capacities. Peach waste is considered only for the simple fermentation
processes and rice straw only for the processes including hydrolysis. A 20% reduction in waste volume is assumed for simple fermentation, and 80% for hydrolysis plus fermentation, for the purpose of computing waste treatment credit.

Table 2.22 considers the use of tomato waste for a 100-day season and rice
straw for an additional 200-day period. (Since the rice straw is a concentrated
dry feedstock, the input tonnage is about one-tenth that of wet waste for the
same ethanol production.)

Table 2.23 shows the effect of increasing dumping costs of wet waste on the
profitability of Processes A and D. Similarly, Table 2.24 shows the influence of
increased ethanol price.

Results

Using the criterion of 10% as a minimum acceptable investment return, Tables
2.18 through 2.21 show that a number of conditions are satisfactorily profitable.

Table 2.18: Economic Evaluation of Process A

Plant Size (tons/day)	Plant Cost (10^3 $)	Labor ($/day)	Materials ($/day)	Return (%)	Ethanol (gpd)
. . . . Tomato Waste*					
100	248	412	172	7.3	256
200	376	488	344	11.5	512
400	570	576	688	13.5	1,024
. . . . Peach Waste**					
50	328	461	273	9.3	407
100	497	544	546	12.5	814
200	753	643	1,092	14.0	1,628
Pilot Plant					
4 tomato or 2 peach	36	200	—	—	10

*Composite (30% culls, 20% peeling waste, 50% pomace), 100 days/yr operation 20% reduction in waste volume, $1/gal ethanol, $7/ton waste removal.

**100 days/yr operation, 20% reduction in waste volume, $1/gal ethanol, $7/ton waste removal.

Source: CAEC-28

Table 2.19: Economic Evaluation of Process B

Plant Size (tons/day)	Plant Cost (10^3 $)	Labor ($/day)	Materials ($/day)	Return (%)	Ethanol (gpd)
. . . . Tomato Waste*					
50	178	466	86	loss	128
100	271	550	172	<1	256
200	411	650	344	10.0	512
400	623	768	688	12.8	1,024
. . . . Peach Waste**					
50	358	615	273	6.8	407
100	543	726	547	11.3	814
200	823	857	1,094	13.5	1,628

*Composite (30% culls, 20% peeling waste, 50% pomace), 100 days/yr operation, 20% reduction in waste volume, $1/gal ethanol, $7/ton waste removal.

**100 days/yr operation, 20% reduction in waste volume, $1/gal ethanol, $7/ton waste removal.

Source: CAEC-28

Table 2.20: Economic Evaluation of Process C

Plant Size (tons/day)	Plant Cost (10^3 $)	Labor ($/day)	Materials ($/day)	Return (%)	Ethanol (gpd)
. . . . Tomato Waste*					
50	644	597	31	loss	142
100	976	705	62	1.0	284
200	1,479	833	124	4.3	568
400	2,242	984	248	8.8	1,138
800	3,398	1,162	496	11.0	2,275

(continued)

Table 2.20: (continued)

Plant Size (tons/day)	Plant Cost (10^3 $)	Labor ($/day)	Materials ($/day)	Return (%)	Ethanol (gpd)
. . . . Rice Straw**.					
25	2,274	896	253	loss	771
50	3,446	1,058	507	loss	1,542
100	5,224	1,250	1,015	1.5	3,048

*Composite (30% culls, 20% peeling waste, 50% pomace), 100 days/yr operation, 20% reduction in waste volume, $1/gal ethanol, $7/ton waste removal.
**$25/ton collection cost, 100 days/yr operation, 80% reduction in volume, $1/gal ethanol, $7/ton waste removal.

Source: CAEC-28

Table 2.21: Economic Evaluation of Process D

Plant Size (tons/day)	Plant Cost (10^3 $)	Labor ($/day)	Materials ($/day)	Return (%)	Ethanol (gpd)
. . . . Tomato Waste*.					
50	805	716	31	loss	142
100	1,220	846	62	1.0	284
200	1,850	1,000	124	4.5	568
400	2,804	1,180	248	8.8	1,138
800	4,251	1,394	496	11.3	2,276
. . . . Rice Straw**.					
10	1,641	863	102	<1	308
25	2,844	1,075	253	7.8	771
50	4,312	1,270	507	11.3	1,542
100	6,534	1,500	1,020	13.5	3,080
. . . . Rice Straw***					
5	1,083	730	51	<1	154
10	1,641	863	102	3.5	308
20	2,488	1,019	204	8.5	616
40	3,771	1,204	408	10.5	1,232

*Composite (30% culls, 20% peeling waste, 50% pomace), 100 days/yr operation, 20% reduction in waste volume, $1/gal ethanol, $7/ton waste removal.
**300 days/yr operation.
***200 days/yr operation, no depreciation.

Source: CAEC-28

Table 2.22: Economic Evaluation of Combined Operation, Process D

Plant Size (tons/day)	Plant Cost (10^3 $)	Return (%)	Ethanol (gpd)
. . . . Combined Tomato and Rice Straw Wastes*.			
100 tomato	1,220	1.5	284
10 rice straw			

(continued)

Table 2.22: (continued)

Plant Size (tons/day)		Plant Cost (10^3 $)	Return (%)	Ethanol (gpd)
. . . . Combined Tomato and Rice Straw Wastes*				
200	tomato	1,850	5.5	568
20	rice straw			
400	tomato	2,804	11.0	1,138
40	rice straw			
600	tomato	3,900	14.0	1,706
60	rice straw			

*Tomato waste: composite, 100 days/yr operation. Rice straw: $25/ton cost, 200 days/yr operation. Overall: 80% reduction in volume, $1/gal ethanol, $7/ton waste removal.

Source: CAEC-28

Table 2.23: Economic Effect of Changing Waste Removal Cost

	. . . . Process A*	
Dump Cost ($/ton)	Return (Peaches) 50 tons/day (% pa)	Return (Tomatoes) 200 tons/day (% pa)
7	9.3	11.5
10	9.5	12.0
15	10.0	12.5
20	10.5	13.0

	. . . . Process D**		
Dump Cost ($/ton)	Return (Tomatoes) 200 tons/day, 100 days (% pa)	Return (Rice Straw) 20 tons/day, 200 days, No Depreciation (% pa)	Combined (additive) (% pa)
7	4.5	8.5	13
10	6.5	8.5	15
15	8.25	8.0	16
20	10.0	8.0	18

*$1/gal ethanol, 20% reduction in volume, 100 days/yr operation.
**$1/gal ethanol, 80% reduction in volume.

Source: CAEC-28

Table 2.24: Economic Effect of Changing Ethanol Price

	. . . . Process A*	
Ethanol Price ($/gal)	Return (Peaches) 50 tons/day (% pa)	Return (Tomatoes) 200 tons/day (% pa)
0.50	7.5	10.5
1.00	9.25	11.5
1.50	11.5	12.5

(continued)

Table 2.24: (continued)

	Return (Tomatoes)	Return (Rice Straw)	
Ethanol	200 tons/day,	20 tons/day,	Combined
Price	100 days	200 days, No	(additive)
($/gal)	(% pa)	Depreciation (% pa)	(% pa)
0.50	3.0	6.0	9.0
1.00	4.5	8.5	13.0
1.50	6.0	12.0	18.0

. . . . .Process D** . . . .

*$7/ton waste removal, 20% reduction in volume, 100 days/yr operation.
**$7/ton waste removal, 80% reduction in volume.

Source: CAEC-28

It is apparent that the larger plant sizes give higher returns due to higher production/capital cost ratios. Unfortunately, the processes involving cannery wastes are limited in size by the daily availability of feedstock. Exact amounts would depend upon the individual cannery and time of season, but an average upper limit would probably be on the order of 200 tpd for tomato waste and significantly less for peaches. This gives a daily ethanol production of about 500 gpd and yields approximately 10% return per annum for Process A under the assumed conditions. Where several canneries are located in close proximity, it might be possible for them to contribute feedstock to a common fermentation unit to gain the increased return of a larger size.

Rice straw, due to its localized production and ability to be stockpiled, can be sized for larger production capacities. Up to 100 tpd input is considered here, for an ethanol production of 3,000 gpd with Process D. Larger sizes would be more profitable but might involve longer hauling distances for the feedstock.

Table 2.22 shows the results of a combined operation where tomato waste is the feedstock for 100 days of the year and rice straw for 200 days. The quantities of each are adjusted to give the same ethanol production and therefore use the same size still. Profitability is low at the smaller, cannery-sized units, although larger sizes might be possible if rice straw was used to augment the tomato feedstock even during the canning season.

Table 2.23 summarizes the economic response to changes in dumping costs for the cannery waste. This constitutes a credit to the extent of the reduction in volume effected by the process. Process D for tomato waste benefits the most from increased dumping credit because of the 80% decrease in volume assumed for this process. There is a slight negative effect for rice straw because there is no waste treatment credit for the feedstock and a liability in disposing of the residue. Further increases in profitability might result from use of the residues in other waste utilization programs, such as pyrolysis.

Table 2.24 shows the economic effect of changing ethanol price. Note that, for Process A, peach waste is considered at a 50 tpd level while tomato waste is considered at 200 tpd, in line with assumption on the daily availability of the material at a large cannery. As expected, an increase in ethanol value to $1.50 per gallon increases the rate of return to a more respectable level, and, in the case of combined tomato waste and rice straw operation with Process D, the cannery-sized unit appears quite practical.

Discussion

The rates of return revealed by these analyses are generally marginal for the plant sizes considered, especially in view of the many assumptions made, including feedstock availability and composition process details, a stable market for single cell protein, and negligible distribution costs for the ethanol produced. Therefore, it would be unreasonable to recommend pursuing such an enterprise based on economic considerations alone. However, the waste treatment benefits, combined with the increasing need for petroleum fuel replacements, tip the balance in its favor. This would be especially true if federal or state agencies provided incentives in the form of low interest loans, tax credits or alternate fuel subsidies, and/or legislation against present waste disposal methods.

BATTELLE STUDY–CORN STOVER

> The information in this section is based on *Economics of Manufacturing Liquid Fuels from Corn Stover* (NTIS TID-29419), by D.M. Jenkins, T.S. Reddy and J.R. Harrington of Battelle Columbus Laboratories, October 1978.

A process and economic analysis was made on the manufacture of ethanol from agricultural residues, corn stover being selected as a typical residue. The economics were calculated for a conceptual process design based on advanced technological research done at Purdue University.

The conceptual process converts 1,427 tpd of corn stover (dry basis) into 209 tpd of anhydrous ethanol (plus 160 tpd of furfural). The estimated cost of manufacturing liquid fuels from corn stover is estimated at about $15/million Btu.

REFERENCES

(1) Stevens, M.A., "Relationships Between Components Contributing to Quality Variation Among Tomato Lines." *J. Amer. Soc. Hort. Sci.* 97:70-73, 1972.
(2) Hinman, N.H., Garrett, W.N., Dunbar, J.R., Swenerton, A.K., and East, N.E., "Utilization of Tomato Pomace by Ruminants."
(3) Dobie, J., Dept. Agric. Eng., Univ. of Calif., Davis, personal communication.
(4) Jones, D.B., Dept. Agron. & Range Science, Univ. of Calif. Davis, work in progress.
(5) Wilke, C.R., Yang, R.D., and Von Stockar, V., "Preliminary Cost Analyses for Enzymatic Hydrolysis of Newsprint," *Biotech & Bioeng. Symp. No. 6* 155-175, 1976.
(6) Cysewski, G.R. and Wilke, C.R., "Utilization of Cellulosic Materials through Enzymatic Hydrolysis. I. Fermentation of Hydrolysate to Ethanol and Single Cell Protein," *Biotech. & Bioeng.*, 18:1279-1313, 1976.
(7) Wilke, C.R., Cysewski, G.R., Yang, R.D., and Von Stockar, V., "Utilization of Cellulosic Materials through Enzymatic Hydrolysis. II. Preliminary Assessment of an Integrated Processing Scheme," *Biotech & Bioeng.*, 18:1315-1323, 1976.
(8) Hart, S., Davis Waste Removal Co., personal communication.
(9) Yates, E., Canners League of Calif., personal communication.

ETHANOL FROM MUNICIPAL WASTES

The information in this chapter is based on "Ethanol from Municipal Cellulosic Wastes," a paper by A.J. Parker, Jr. and T.J. Timbario of Mueller Associates and J.A. Mulloney, Jr. of Carling National Breweries, Inc., from *Alcohol Fuels Technology Third International Symposium—Volume I,* Asilomar, California, May 1979 (hereinafter referred to as Asilomar Conference).

This paper addresses the use of municipal cellulosic wastes as a feedstock for producing ethanol fuels, and describes the application of enzymatic hydrolysis technology for their production. The concept incorporates recent process technology developments within the framework of an existing industry familar with large-scale ethanol fermentation (the brewing industry). Preliminary indications are that the cost of producing ethanol via enzymatic hydrolysis in an existing plant with minimal facility modifications (low capital investment) can be significantly less than that of ethanol from grain fermentation.

Cellulose is one of the world's most abundant organic materials which can be used as a source of food, fuel and chemicals. The net worldwide production of cellulose is estimated at over one hundred billion tons per year (1). From one ton of waste paper, about one-half ton of glucose can be produced and subsequently fermented to 78 gallons of ethanol. Conversion of cellulose to glucose can be accomplished by either acid hydrolysis or by enzymatic hydrolysis.

Among the advantages of enzymatic hydrolysis is that the process takes place at moderate conditions so that the glucose yield is high and directly related to the weight of the cellulose used. Glucose syrups produced enzymatically are fairly pure and constant in composition.

The concept discussed in this paper describes a means for:

(1) Producing competitively priced ethanol from urban refuse,

(2) Placing a laboratory-level-proven ethanol production technology into commercial practice within the brewing industry, an industry

that is already familar with the technology of enzyme and large-volume ethanol production,

(3) Helping to solve a national problem of urban waste disposal,

(4) Incorporating an in-place, nationally dispersed network of plant facilities for ethanol production, and

(5) Taking advantage of the consolidation of brewing industry firms and production facilities which has led to an increase in idle, large-scale fermentation plants.

While a detailed engineering assessment has not been completed to date, preliminary findings and analyses (presented herein) indicate that the overall concept is feasible. Currently, the idle brewery alluded to in this paper has not undergone any modification or renovation to produce ethanol per the proposed concept.

The uniqueness of the concept described by this paper stems from:

(1) More broadly, the opportunity to utilize excess capacity in the nationally distributed brewing industry for the production of ethyl alcohol, and

(2) More specifically, the coupling of a readily available waste cellulosic feedstock (obtained as a product from a municipal waste resource reclamation facility) with minimal modifications to an idle brewery for ethanol production.

In 1978, the brewing industry sold 162.4 million barrels of beer and had a nameplate capacity of about 200 million barrels. Thus about 37.6 million barrels of capacity were not utilized (2). Additionally beer brewing is a seasonal industry with peak months and off-peak months. About 40 million barrels additional could be made available from off-peak production capacity for a total additional capacity of 77.6 million barrels (for the year 1978).

With a conversion factor of between 5 to 15 gallons of ethanol per barrel of beer, the total ethanol potential that could be supplied by the brewing industry, without impacting current beer production, is between 388 million to 1.16 billion gallons. This is 26 to 280% more than the entire capacity of the synthetic ethanol industry and equal to about 0.4 to 1.2% of current annual U.S. gasoline consumption (about 100 billion gallons per year). Viewed in another way, ethanol produced from excess brewing industry capacity could satisfy 4 to 12% of a national market for a 10% ethanol/gasoline blend.

The brewing industry is widespread and well distributed throughout the United States with operating breweries in 32 of the 50 states. Eleven breweries have become idled in recent years (thus exemplifying the current excess-capacity situation), mostly for reasons of economy. This is prompting an industry trend toward consolidation.

ENZYMATIC HYDROLYSIS OF CELLULOSIC WASTES

Perhaps the simplest way to describe the enzymatic hydrolysis process is to visualize a starch molecule as a long string of connected sugar molecules and the enzyme as a scissors that, starting from one end of the string, successively cuts off one sugar molecule at a time until the string is gone. Similarly, but in a more complex fashion and with different sugars, cellulose is a long-chain polysaccharide that can be enzymatically reduced to mostly glucose sugar via cellulase. (Cellulase is the enzyme which converts cellulose to sugars.) Once the simple sugars are produced, they can be fermented (via yeast) in much the same way as grains and fruits are fermented to produce alcoholic beverages. A distillation column is then used to separate high-purity ethanol from the fermentation liquid.

Over the last several years, significant technological advancements have been made in the enzymatic hydrolysis process. In particular, the U.S. Department of Energy's Fuels from Biomass Program and the National Science Foundation sponsor major projects in enzymatic process technology improvements. Notable accomplishments have developed from work conducted by the U.S. Army Natick R&D Center. Specifically, scientists at the Natick Laboratory have developed an enzymatic process which is based on the use of cellulase derived from mutant strains of the fungus *Trichoderma viride.* A prepilot-plant unit with a capacity of approximately 125 pounds per day is in operation.

Recent advances in enzyme production, saccharification (conversion to sugars) and materials pretreatment have brought this particular process nearly to the stage of commercial application. More recently, the laboratory is developing a process for production of large amounts of a cellulase enzyme from the fungus *Trichoderma reesei.*

Application to an Existing Brewery

In August 1978, for reasons of economy, the Highlandtown (Baltimore City, Maryland) Plant of Carling National Breweries was consolidated into Carling National's Beltway Plant in Baltimore County. The Highlandtown Plant still contains fermentation, packaging and warehousing facilities which have remained essentially undisturbed. With modifications basically that would require the installation of enzyme production and ethanol distillation apparatus, the plant could be converted to ethanol production at a fraction of the expense required for a completely new plant.

The remainder of this paper presents a preliminary evaluation of a concept which incorporates the use of enzymatic hydrolysis (on a cellulosic feedstock produced from municipal wastes generated in the metropolitan Baltimore area) in the idle Carling National plant to produce a fermentable product which can be distilled to anhydrous ethanol. The enzymatic process envisioned is based on the relatively new cellulase-producing fungus *Trichoderma reesei* developed by the U.S. Army Natick Laboratory.

Source and Characteristics of Cellulosic Feedstock

Feedstock for the envisioned concept would consist of the cellulosic fraction from municipal solid wastes. Currently about 4,200 tons of refuse is generated daily in the metropolitan Baltimore area. This amount is nearly ten times that needed to supply a converted brewery producing five million gallons per year of anhydrous ethyl alcohol.

A resource reclamation facility currently operating in suburban Baltimore County (near Texas, Maryland) has the capability of processing about 1,500 tons of refuse per day to about 1,125 tons of highly cellulosic material per day. At this facility, the incoming raw waste is shredded and mechanically separated into various fractions—ferrous metals, other metals, residue, and a highly cellulosic material commonly referred to as RDF (refuse-derived fuel). It is the cellulosic material or RDF that would serve as the ethanol feedstock for the proposed plant conversion concept. Characteristics of the RDF product are as follows (3) (4):

Method of production	Mechanically separated, air classified and trommeled
Fraction of RDF to waste input	70 to 75% (by weight)
Moisture content	10 to 30% (by weight)
Sulfur content	~0.2% (by weight)
Plastic and other inert materials	3.4% or less (by weight)
Ash content	6 to 12% (by weight)
Heating value (wet)	6,500 Btu per lb
Heating value (dry)	8,000 Btu per lb (calculated)

RDF could be supplied in the form of "fluff" (loose, unpressed material) or as pellets about one-inch diameter by two-to-three inches long. The fluff form is more amenable to hydrolysis. For several reasons, RDF from the Baltimore County facility appears to be a logical choice of feedstock:

(1) Transportation distance is minimal (about 10 to 12 miles) between the Baltimore County facility and the idle brewery in Baltimore City.

(2) The resource reclamation facility has the capability of supplying all the feedstock requirements for a 5 million gallon per year ethanol plant.

(3) The RDF output is a very suitable material for the envisioned process.

In addition to the above, it is contemplated that, in the near future, at least one additional resource reclamation facility may be built in the Baltimore metro area, thus providing an additional source of feedstock supply (if needed).

PLANT CONVERSION REQUIREMENTS AND PROCESS DESCRIPTION

Minimal modifications are required to convert an existing brewery to the production of ethanol from municipal cellulosic wastes. These modifications include the addition of enzyme production, cellulose pretreatment, hydrolysis processing, ethanol distillation and certain waste recovery equipment. Fermentation, materials handling, and storage equipment are essentially in place, as are portions of waste recovery equipment, all utility supplies (steam, electricity, water) and the physical housing structure. Labor is available and training requirements are minimal since brewing personnel are familiar with large-scale ethanol fermentation.

Figure 3.1 illustrates the envisioned process by which ethanol can be produced from municipal cellulosic wastes. Conversion of ethanol from cellulose begins with the production of the enzyme.

This is accomplished by obtaining a very small amount of the *Trichoderma* fungus and growing it in a culture medium containing shredded cellulose (most probably from the same source of cellulosic material that will be hydrolyzed) and various nutrients.

Figure 3.1: Simplified Process Flow

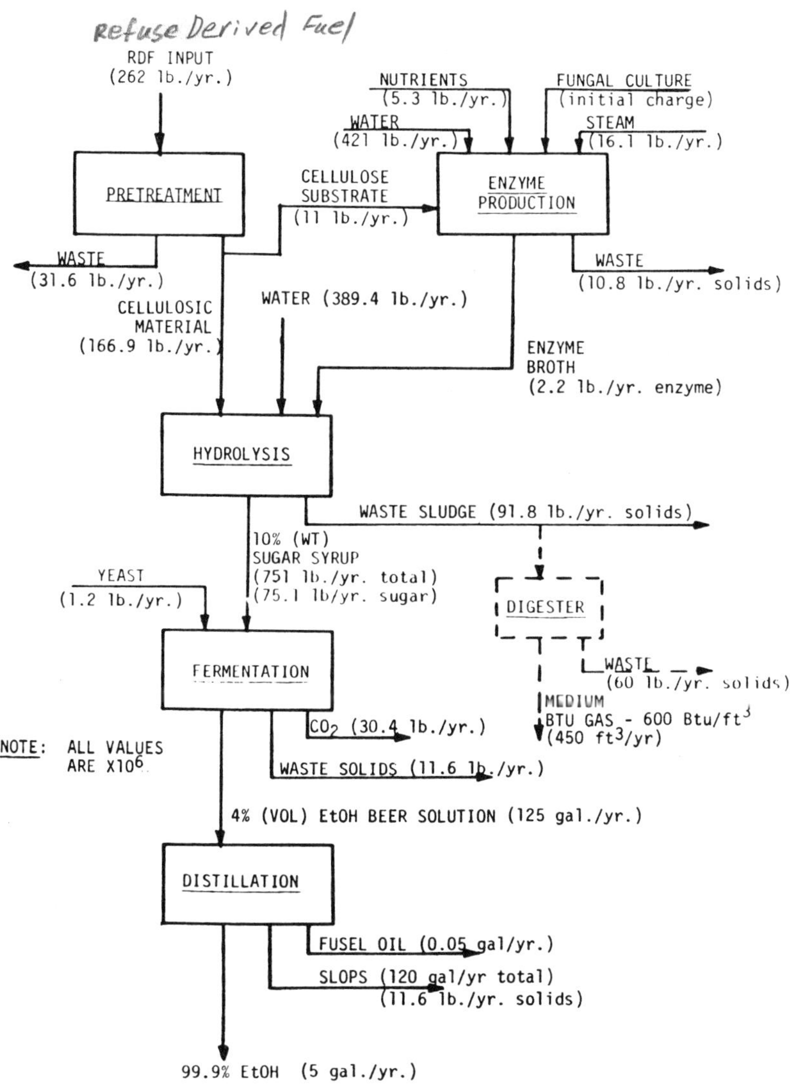

Source: Asilomar Conference

As the fungus grows, it secretes cellulase enzyme into the medium. Following its growth, the culture is filtered to remove the filamentlike fungus (mycelium). The filtrate, or broth, contains the enzyme that is used in the saccharification (hydrolysis) reactor. After adjusting the acidity of the enzyme broth to the correct pH, pretreated cellulosic material is introduced into the enzyme solution and allowed to react with the cellulase to produce glucose. Saccharification takes place at atmospheric pressure and at a temperature of about 50°C.

The rate and extent of hydrolysis will depend on the particular composition of the cellulosic material and the type of pretreatment employed. For the cellulosic material (RDF) envisioned, pretreatment will entail removal of noncellulosic "contaminants" (plastics, ash, etc.) to the extent that it is practical, followed by the addition of water to form a mash.

A glucose syrup of about 10% concentration is the desired product of the hydrolysis process. In addition, by-products containing unreacted cellulose and enzyme are produced (sludge). These by-products could be recycled back to the reaction vessel to the extent possible, or could be used directly as fuel for process steam generation, but their fuel values will have to be established because they contain a large quantity of water.

Since the idle Carling brewery contains excess fermentation capacity (beyond that which is needed to produce 5 million gallons of ethanol per year), it is also possible to utilize this excess capacity for digesting the by-product sludge (via anaerobic methanogenic bacteria) to produce a medium Btu gas which in turn can be used for generating process steam.

After the crude glucose syrup (from the hydrolysis process), is filtered, yeast is added so that fermentation can take place. This part of the process yields carbon dioxide, spent yeast, and a fermented liquid (called beer) which contains about 4% (by volume) ethyl alcohol. Carbon dioxide can be removed, compressed and sold as a by-product through existing market channels. A portion of the spent yeast product can be recovered and recycled. The nonrecoverable portion contains protein which is valuable as an animal feed ingredient.

The 4% alcohol product is fed to a beer still for the first distillation step in the concentration of the alcohol. The overhead, containing about 50 to 60% ethanol in water with other volatiles is condensed while the bottom residue, known as stillage or slops, can be utilized as livestock feed as it contains proteins, residual sugars and vitamins.

In the Baltimore area, a market exists for the wet stillage product, thereby eliminating the costly and energy-consuming process of stillage drying. It should be noted, however, that in the case of sugar solutions produced via enzymatic hydrolysis, there may be some constituents that make this stillage product unacceptable for feed; therefore, no facilities for the production of distiller's feed from the stillage have been included in the concept design.

The dilute alcohol solution passes to another distillation column, termed an aldehyde or purifying column, where aldehydes and other low-boiling impurities are removed as overhead products. The column bottoms are primarily fusel oil (mainly amyl alcohols). The aldehydes and fusel oil may be purified and sold for industrial use but it may be more economical to burn them in the plant boilers.

The effluent from the middle of the aldehyde column is sent to a rectifying column where the alcohol solution is concentrated to about 95% (volume) ethanol. The low-boiling overheads contain some residual aldehydes in alcohol and are returned to the aldehyde column. Some fusel oil is removed further down the column and combined with the fusel oil from the aldehyde column. Water is discharged from the bottom.

The usual method for producing anhydrous ethanol from a 95% (volume) solution is to add a third component which forms an azeotrope with one or both of the original constituents. Benzene has generally been used for dehydrating ethanol in this manner but other compounds (e.g., ethers, hexane) can be used as well. Although only one additional distillation column is needed to obtain the dehydrated ethanol product, two or more columns plus auxiliary equipment are required to recycle and recover the benzene and aqueous alcohol (5).

Since the anhydrous ethanol product will not be used for beverage purposes, it must be denatured in accordance with procedures/formulas acceptable to the U.S. Treasury Department's Bureau of Alcohol, Tobacco and Firearms. While the BATF currently authorizes six formulas which may be used for ethanol fuel purposes (6), it is important to note that the selection of denaturants must properly consider the effect of the denaturant on fuel properties and combustion products. The DOE's Alternative Fuels Utilization Program is currently sponsoring research that will investigate the most appropriate denaturants for fuel-grade ethanol (7). Accordingly, results of this work would serve as the basis for selecting denaturants to be used in a fuel-grade ethanol product.

Material Flows, Energy Use and Economics

Figure 3.1 also presents the approximate input/output material flows required to produce five million gallons per year of anhydrous ethanol in a batch-type process operating at about 7,920 hours per year. The input feedstock requirement is 262 million pounds of RDF per year, of which 6.6% is needed for enzyme production. Enzyme production is estimated to occur at a rate of 100 IU (International Units) per liter per hour. Hydrolysis conversion efficiency is taken as 45% (i.e., 45% of the cellulose is converted to fermentable sugars). Fermentation of sugars to ethanol is assumed to occur at an efficiency of 40%.

All yearly flow values noted in Figure 3.1 are approximate and those for the enzyme production and hydrolysis steps were developed from several sources, principally, references (8), (9) and (10).

At this point, it is worthwhile to mention several comments about process design characteristics. For example, the hydrolysis step is based on an input cellulose/water slurry containing 30% cellulose solids. This, combined with the noted cellulase enzyme input is expected to permit cellulose hydrolysis in a cycle period of 24 hours. The energy requirements for pumping a 30% solids slurry are a function of several factors including solids concentration and pumping distance.

While pumping/transfer equipment exists for this kind of high-solid content material (e.g., wood pulp and coal slurry applications), a detailed engineering analysis of equipment sizing, energy use and layout is required.

Several studies have examined the potential for utilizing the hydrolysis sludge effluent directly as a fuel for process heat. Process research to date has shown that this effluent contains a minimum of 50% water; thus, its energy value (for direct combustion) is probably on the order of 5,000 Btu/lb at best. Separation of the water from the sludge would be needed prior to any use as a fuel.

Note that the process described in Figure 3.1 illustrates the application of anaerobic digestion (of the hydrolysis sludge effluent) and subsequent generation of medium Btu gas. This gas could supply about 30% of the fuel needed for process steam. It should be noted, however, that residence times for digestion are on the order of 6 to 10 days (as opposed to a 24-hour hydrolysis cycle); thus materials flow considerations and any need for sludge "storage" may outweigh the practicality of this concept. More detailed engineering analysis is required.

Although several studies have identified the potential cost-saving opportunities offered by incorporating recycle of process materials (enzyme, process water, etc.), no recycle steps have been shown on the process flow chart. The reason for this is that the trade-off between the reduction in materials cost and the increase in process complexity and equipment costs has not been fully assessed. Analysis of recycle systems is an area in need of hardware-oriented R&D.

Estimated process utility requirements are summarized below:

Steam	76,000 lb/hr
Cooling water	3,411 gpm
Refrigeration	620 tons
Process water	741 gpm
Electricity	4,982 kW

From the above, the net process thermal efficiency is estimated to be:

$$100 \times \frac{5 \times 10^6 \text{ gal/yr} \times 84,200 \text{ Btu/gal}}{(262 \times 10^6 \text{ lb/yr} \times 6,500 \text{ Btu/lb}) + 600 \times 10^9 \text{ Btu/yr} + 135 \times 10^9 \text{ Btu/yr}} = 17\%$$

Gross (output ethanol to input RDF only) process thermal efficiency is estimated to be about 25%. Including the generation of medium Btu gas increases the net process efficiency to about 19%.

Estimated capital, operating and ethanol production costs are summarized below. Profit is not included.

Fixed costs	
Capital investment (equipment)	$7,000,000
Working capital	500,000
Total capital	$7,500,000
Annual costs	
RDF feedstock ($15/ton, delivered)	$1,965,000 (39¢/gal)
Plant operating expenses and other	
materials costs*	$3,200,000 (64¢/gal)
Sales, general and administrative	300,000 (6¢/gal)
Total annual cost	$5,465,000 ($1.09/gal)

*Includes all utility, maintenance, labor, materials (other than feedstock), and equipment amortization/depreciation costs.

From Carling National's experience, it has been found that the cost of producing salable by-products (CO_2, distillers' feed, etc.) is equal to or less than the income generated by their sale. Thus, the above cost estimate has conservatively estimated by-product production costs to equal sales revenues.

SUMMARY AND CONCLUSIONS

This paper has presented a preliminary analysis of the use of municipal cellulosic wastes in conjunction with the application of enzymatic hydrolysis technology for the production of ethanol. The envisioned concept appears feasible, but more thorough engineering analysis is required before firm process details can be established. The cost of producing ethanol in a converted idle brewery was shown to be competitive with synthetically derived ethanol from petroleum. The per-gallon-of-ethanol feedstock cost of 39¢ is significantly less than that of corn (about 90¢/gal) as used in several fermentation ethanol plants. The close proximity of cellulosic feedstock supply to the idle Carling brewery offers a significant cost advantage, as does the availability of the plant and associated fermentation equipment.

The nationally dispersed system of breweries (and its excess capacity) and their locations in urban/surburban waste generating areas may offer a means of developing a nationwide ethanol fuel supply.

REFERENCES

(1) "Enzymatic Hydrolysis of Cellulosic Wastes to Fermentable Sugars and the Production of Alcohol," Spano, L.A., U.S. Army Natick R&D Command, (January 1976).

(2) "Facing Future, Brewers Add Products, Capacity," *Beverage Industry,* (January 5, 1978).

(3) "Energy from Wastes," Maryland Environmental Service, (June 1978).

(4) Personal Communication with Mr. Donald Bowman, Maryland Environmental Service, Annapolis, MD, (March 1, 1979).

(5) "The Potential Role of the Distilling Industry in Supplying Ethanol Fuel," The Aerospace Corporation, Report No. ATR-79(7756)-1, (January 1979).

(6) *Denaturants for Ethanol/Gasoline Blends,* Mueller Associates, Inc., prepared for the U.S. DOE, HCP/M2098-01, (April 1978).

(7) "Modification of Ethanol Fuels for Highway Vehicle Use" (contract title), Union Oil Company of California, Contract No. EY-76-C-04-3683, U.S. DOE.

(8) "Reassessment of Economics of Cellulase Process Technology: For Production of Ethanol from Cellulose," Spano, L.A., et al, presented at The Second Annual Symposium on Fuels from Biomass, Troy, NY, (June 20-22, 1978).

(9) "Production of Sugars and Ethanol Based on the Enzymatic Hydrolysis of Cellulose," Wilke, C.R., University of California, Berkeley, presented at the Fuels from Biomass Symposium, University of Illinois, (April 18-19, 1977).

(10) *Mission Analysis for the Federal Fuels from Biomass Program, Volume V: Biochemical Conversion of Biomass to Fuels and Chemicals,* SRI International, Menlo Park, CA, (December 1978).

AVAILABILITY OF BIOMASS
SUGAR CROPS

The information in this chapter is based on *Fuels from Sugar Crops* (NTIS TID-22781) edited by R.A. Nathan of Battelle Columbus Laboratories for the U.S. Department of Energy, July 1978.

Sugarcane appears to be the most promising sugar crop for conversion to fuels and chemicals in the short and intermediate term. The costs of sugarcane juice and bagasse are lower than those for the corresponding sugar beet products. Agricultural and processing improvements in the offing can increase yields and reduce costs further. Of the U.S. sugar crops, sweet sorghum has the greatest long-range appeal because the crop can grow over a much wider geographical area than sugarcane. It is an experimental crop, however, and much development is needed before it can make a significant impact.

SUGARCANE

Since the concept of fuels from biomass was conceived, sugarcane has been a leading contender due to its composition, high yield, low unit costs, and concentration of large quantities of residues at a few central processing points. Drawbacks are high opportunity cost of land for sustaining high sugarcane yields, high opportunity cost of sucrose, limited availability of land with the needed water and solar insolation, and poor storage stability of freshly cut cane.

Four states produce sugarcane. Estimated 1975 cane production (including cane for seed) was 25.9 million tonnes vs an average of 19.5 million tonnes produced from 1960 through 1965 and 22.13 million tonnes from 1965 through 1970. The 312,900 total ha harvested in 1975 compares with an average of 222,695 ha from 1960 through 1965 and 242,212 ha from 1965 through 1970.

Prospects for using sugarcane products as raw materials for fuels vary substantially from one region to another. The discussion of production methods for each region is followed by a discussion of pests, potential area, yields, composition, costs, energy budgets, and mass balances.

Production Methods

Methods used to obtain high yields of sugarcane have evolved through an exemplary cooperation among farmers, the sugar industry, the U.S. Department of Agriculture, universities, trade associations, and suppliers of equipment and chemicals. The following sections describe technology and geographical areas in use and the yields obtained.

Florida: Sugarcane is grown on about 112,000 ha around the periphery of Lake Okeechobee in southern Florida. Of this area, 90% is muck soils (50% organic matter) and 10% shallow fibrous peat soils (10% organics) (Herbert and Dire, 1972). Although mill capacity (a 5-month milling season) seems to be the major factor limiting cane production, seasonal temperature is also important. Freezing is a serious problem; except for 30,000 to 35,000 ha of "warm" land adjacent to Lake Okeechobee, freezing temperatures can be expected in 8 out of 10 years.

The Florida planting-harvesting regimen results in one plant-cane harvest and two to four (average of three) stubble harvests per planting; for any given 5-year period, four crops of cane are harvested. The first crop (plant cane) is harvested on a 15-month growing cycle. Sugarcane is planted in August, when soil temperatures are optimum (greater than 26°C) for rapid germination. During the first fall and winter the stool (root system) is established, and foliar and stem growth occurs when air temperatures reach 15.5°C or higher.

Yields of cane with the ratoon system (5 years, four harvests) average approximately 105 tonnes/ha wet weight per year. (This value includes tops and trash, which average approximately 30% of the net cane weight.) If the ratoon system were to be abandoned and seed cane planted each year, approximately 240 tonnes/ha wet weight of whole cane could be produced in 15 months. However, costs would be higher and more area would be fallow each year.

Experimental data at the USDA Agricultural Research Service Laboratory, Canal Point, Florida, indicate yields as high as 294 tonnes/ha of wet weight sugarcane biomass in a 15-month plant-cane crop. If emphasis is on biomass, data indicate that average commercial yields within the next 10 years (ratoon system) could be increased to 147 tonnes/ha. The increase could be supplied by four experimental varieties now considered (larger stalks and more stalks per acre). The U.S. Sugar Corporation expects to increase yield by 7 to 14 tonnes/ha within 3 years.

Substantial land in Florida could produce sugarcane for conversion to fuels. Depending on the economics, cane could be grown as far north as Tallahassee. Skillful farmers in Louisiana, where similar growing conditions exist, average 110 tonnes/ha wet weight of millable stalks per year.

Increasing production, however, would mean that more sand lands would be planted. Fertilizer recommendations for current sand-land production are 166 kg of nitrogen, 27 kg of P_2O_5, and 333 kg of K_2O per hectare. Irrigation water may not be available for all the sand-land area. In fact, water control probably would be the major limiting factor to substantially increased cane hectarage. Current irrigation practices in Florida pertain to water-table control. Land is

drained by pumping water from drainage systems around the cane fields and irrigated by stopping the pumps. Although cane needs 1 m of water-free soil
above a permanent water table to grow best, high yields have been achieved
in parts of Florida, Louisiana, and Hawaii where water-free soil depths are
much less.

Louisiana: Approximately 136,000 ha (U.S. Department of Agriculture, 1974b)
of sugarcane is being grown in central and south-central Louisiana; however, almost half of it is produced in Assumption, Aberia, St. Mary, and Terrebonne
Parishes. Louisiana has little room for expansion in cane-growing areas; however, expansion into the northern cotton and soybean delta areas is considered
a possibility. In addition, lands could be available along the southern Louisiana
coast if it were drained.

The major factor limiting cane production farther north in Louisiana is soil
temperature. Seed cane must be able to overwinter in Louisiana (optimum
planting depth is around 7 to 8 cm). If the soil freezes to this depth for just
a few hours, no germination occurs. Since surface storage of seed cane is not
technically or economically feasible, seed cane must overwinter in the ground.
In addition, large mature stalks are used as seed stock; immature stalks tend to
rot at the ends.

Attempts to protect the seed stock with fungicides and pesticides are effective
for only a month or two. Soil temperatures must reach 21°C before the seed
will germinate; in Louisiana a soil temperature of 21°C marks the start of the
growing season. Soil temperature-growth relationships are reviewed by Humbert
(1968). Research at Louisiana State University to develop cane varieties more
tolerant to frost conditions shows experimental varieties tolerating down to –3°C.

The second most important factor limiting cane production farther north is
water availability. On the northern and northeastern portions of the central
districts of Louisiana, summer rainfall can be insufficient for cane production.
This situation is directly opposite to that in the south-central and southeast regions where drainage is the major problem. However, drip irrigation technology
being developed in Hawaii may prove applicable for some Louisiana conditions.
The growing period in Louisiana is about 5 months, and the milling season,
about 70 days. Most of the growth occurs during July, August, and September
(0.4 to 0.6 cm/day).

Cane yields vary considerably with the location, precipitation, management, and
size of the cane. The average plant crop yields approximately 74 tonnes/ha wet
weight with a second ratoon crop of 50 tonnes/ha wet weight. Skillful commercial growers produce 90 to 100 tonnes/ha wet weight, and with some of the
more highly developed varieties and top-notch crop management, 150 tonnes/ha
have been produced. Since tops and leaves are not taken into account in all
these values, some Louisiana growers are producing as much as 100 to 195
tonnes of wet biomass per hectare annually. Since cane is harvested mechanically
in Louisiana, the tops and leaves percentage may be higher because (in a good
growing year) the cane grows above the optimum height for the mechanical
harvester. As a result, up to 0.6 m of top may be left in the field.

J.E. Irvine, of the Agricultural Research Service Sugar Station at Houma, La.,
is conducting biomass trials by manipulating row spacing. His data indicate

that (1) cane yields can be tripled by broadcast planting; (2) with 1-m spaces between rows, yields can be doubled; and (3) with ⅓-m row spacings, 245 tonnes/ha wet weight can be produced. Dr. Irvine has stated that the aforementioned results were entirely experimental. No information is yet available on second- and third-year yields in ratooned fields. Questions of adequate nutrients, percent infestation by stem borers, etc., were not addressed. An additional benefit of close-row spacing is that weed control is less of a problem.

Texas: In 1974, 115 valley growers formed a cooperative, the Rio Grande Valley Sugar Growers Inc., and constructed a mill. The cooperative applied for and obtained designation as a mainland sugar-producing area under the Sugar Act (Anderson, 1973). The annual quota limit was 90,720 tonnes of sugar, and the area allotment was 10,409 ha.

Production in 1975 yielded an average of 90.8 tonnes of millable cane per hectare with a sugar content of 83 to 100 kg per tonne of cane. On repeal of the Sugar Act, the 1975 cane production area was increased by 4,050 ha. The maximum area in the Rio Grande Valley which could be devoted to cane production, under current management and water availability, is estimated at 40,500 ha.

The Rio Grande Valley is subtropical, semiarid, with short, mild winters and long, hot summers. The Valley has an annual freeze-free growing season of approximately 330 days.

Annual rainfall ranges between 76 and 51 cm on the coast to 160 cm inland. Valley agriculture is irrigated because cane requires from 127 to 152 cm of water per year. Most irrigation water comes from the Rio Grande River, but supplemental wells are used. The major problem with both well water and river water is salinity. River water averages around 900 ppm salt (during low-flow periods, as high as 1500 ppm); well water ranges from 900 to 2500 ppm salt. These high salt levels necessitate reliance on rainfall to leach the salts from the soil to prevent salt buildup to levels toxic to cane. Many growers install field drains to increase the movement of water through the soil. Drainage tiles 1.5 m deep are sufficient for subsurface drainage of the salty irrigation water.

In Texas, sugarcane is planted in the fall and harvested the following fall. Growers generally plan on four ratoon crops per planting of seed cane. Although the current average population of millable cane is 90 tonnes/ha, a few growers average 180 tonnes/ha. Production could average 134 tonnes of millable cane per hectare in 10 to 20 years.

Nutrient requirements of sugarcane are met by the soils of the Valley. The soils generate up to 134 kg of nitrogen per hectare annually from mineralization. Sugarcane depletes the soil of nitrogen at a rate of only 0.8 to 1 kg per tonne of cane. At the average yield of 90 tonnes/ha, nitrogen needs of sugarcane are adequately met. Increased yields may necessitate the addition of nitrogen to the soil.

Sugarcane production in the Rio Grande Valley has been substituted for some cotton and grain production. However, 1976 plantings of cotton were expected to exceed those of previous years. Thus sugarcane has not replaced other crops.

Considerable testing and developing of new sugarcane varieties to find those most suitable have been conducted in Texas. A major criterion is cold tolerance. Cane will normally withstand a light freeze of 0° to -1.7°C; however, temperatures below -3.3°C may kill the plant or just the terminal bud, depending on duration. Even if the entire stalk is killed, freezes above -6.7°C do not usually damage the root system seriously. In Weslaco, Texas, cane would have been injured from low temperatures in only 7 of 31 years. This can be compared to 17 of 30 years for the cane areas of Florida. Losses vary extremely owing to variation in duration and severity of freeze. In 37 years of recording, the lowest temperature recorded in Weslaco, Texas, was -6.7°C (U.S. Department of Commerce, 1965). The risk, therefore, of loss to low temperature is less in this area of Texas than in Florida.

Hawaii: About 46,000 ha of sugarcane is harvested annually in Hawaii. Since the cane-growing cycle is approximately 24 months, the total area under cultivation is about 92,000 ha. Of the total, roughly 44% is being grown on the island of Hawaii; the remainder is almost equally distributed among the islands of Oahu, Kauai, and Maui. The potential for substantially increased sugarcane production is limited. Only 4% of Hawaii's land is potentially arable; and most of it is under cultivation (sugarcane and pineapples).

The Hawaiian Sugar Planters' Association (Cushing and Heinz, 1975) estimates that 10,000 to 20,000 ha would be the maximum amount of additional land that could be planted to sugarcane. Although 612,000 ha is classified as grazing lands in Hawaii, shallow soils, steep slopes, and lack of available water limit the use of most of these lands for sugarcane production. Urbanization has removed some land from pineapple cultivation on the island of Oahu. However, more land has been removed from pineapple production because of competition from the Taiwanese.

Water availability seems to be the major limiting factor to increasing sugarcane plantings on lands of suitable slopes and soils. One side of the islands receives large amounts of moisture, and the other side, almost none. For example, on the Hilo side of the island of Hawaii, annual precipitation is approximately 381 cm at sea level, 635 cm at a 900 m elevation, and 101 cm at a 1500 m elevation. On the dry side of the island, annual precipitation averages around 50 cm. R. Fox, University of Hawaii, believes that the excess moisture on the island could be collected and transported to the dry side. Estimates from the Pepeekeo area indicate that as much as 1 million gallons of water is lost by runoff each day. On Oahu, elaborate schemes are used to collect runoff to irrigate the sugarcane fields. Since there is such a large demand for moisture control in the cane fields, much drip irrigation research and development is under way.

Soils of the sugarcane-producing area in Hawaii range from coarse sands to clays (U.S. Department of Agriculture, 1972; 1973). Soil management in Hawaii's sugarcane agriculture is complex. The sugar industry maintains laboratories where samples of plant tissue and soils are analyzed regularly. Nutrient deficiencies are identified and corrected.

Hawaiian sugarcane is produced on a 24-month cycle. With the favorable climate, fields are planted and harvested 10 to 11 months per year, and the milling season is essentially continuous. The entire process is mechanized except for seed-cane production.

Yields vary widely, depending on the location of the field, the quality of management, and the varieties planted. On the island of Hawaii, the maximum elevation where cane is grown is approximately 900 m; above this elevation growing conditions are poor because of low temperatures and reduced solar insolation. Cane doubles its growth rate when air temperatures rise from 26.7° to 32.2°C. Since Hawaii has air temperatures in the 26 to 28°C range, optimum growth conditions do not occur. However, average yields of about 261 tonnes per hectare are obtained. Unless otherwise specified, Hawaii yields are expressed for the full 2-year growing cycle, e.g., 261 tonnes/ha is 130.5 tonnes/ha/year.

In addition, row-spacing experiments with increased fertilizer and seed stock resulted in 12-month yields of 250 tonnes/ha wet weight of cane stalks (Bull and Glasziou, 1975). If we assume 30% solids and a trash dry weight of 12 tonnes per hectare, 87 tonnes of dry material per hectare could be available for conversion. Average availability would be much lower than this experimental quantity.

Puerto Rico: Although total hectarage is small, sugarcane is still considered a valuable industry in Puerto Rico. Sugarcane is grown in a belt approximately 13 km wide around the island's coast. Owing to the trade winds, there is considerable difference between precipitation rates on the north side of the island and those on the south side. On the north side, precipitation ranges from 110 to 220 cm, whereas the drier southern side averages 50 to 150 cm. In the south, sugarcane is gravity-irrigated from government-owned water supplies (Barnes, 1974). Soils in the sugarcane areas are similar in ambient potash and nitrogen levels. In the humid north, however, soils are richer in organics and slightly more acid.

Sugarcane is grown year-round in Puerto Rico. There are two general planting seasons, July to November and January to June. The former planting is harvested in 16 to 18 months and the latter in 12 to 13 months. In the humid areas up to seven ratoon crops have been grown; in the south, only three ratoon crops are produced. All ratoons are harvested after 12 to 13 months (Barnes, 1974).

The area devoted to sugarcane cultivation has steadily declined, and in recent years the sugar content has also decreased. For example, between 1959 and 1963, yields averaged around 10% sugar content, but, since 1968, they have steadily declined to approximately 7% (U.S. Department of Agriculture, 1974d). This decline has been attributed to the government takeover of sugar manufacturing and the resulting disruption of large sugarcane-producing companies and corporations (Barnes, 1974). In addition, labor shortages and slow mechanization of all aspects of sugarcane production, compounded with alleged shortcomings in management, resulted in losses in productivity.

In 1974, 3.6 million tonnes of cane were harvested, an increase in cane volume. Crystalline sugar recovered, however, was still only 7.2%, or 260,610 tonnes (Universidad de Puerto Rico, 1975). In 1975, yields averaged approximately 67.2 tonnes of millable stalks per hectare, with a sugar content of 8%. With proper management and the selection of appropriate varieties, yields of 112 to 134 tonnes/ha (with 10% sugar content) could be achieved.

Drainage is a major limiting factor in many areas. Research on the effectiveness of drain tiling has shown increased production of over 50% of cane and a similar increase in the yield of sugar.

Approximately 56,700 ha of sugarcane is under cultivation, all on land suited to mechanical management. The maximum area on which machines can be used is approximately 81,000 ha. All sugarcane production is now mechanized. Whether significant increases in sugarcane hectarage could occur is problematical since the government has developed a commonwealth policy oriented toward land use for food production. Thus any competing crop contributing to food self-sufficiency would receive considerable attention. Another factor is that Puerto Rico consumes approximately one-half its refined sugar output; the rest is exported. All molasses is used in the local rum industry. Therefore, Puerto Rico reflects the basic question: What are the priorities (food or fuel) and at what level (local or national)?

Pests and Diseases

Sugarcane cultivation involves more than supplying adequate water and nutrients; it involves a rigorous effort to minimize losses from pests and diseases. Sugarcane is attacked by a few mammals, numerous insects, nematodes, weeds, and several diseases. Control techniques for the numerous pests and diseases can be grouped into three major categories: biological, chemical, and crop management.

Biological methods most commonly use an organism in some way parasitic to the pest organism. Continuous research is attempting to develop new varieties of sugarcane more resistant to particular diseases or pests. However, it takes about 10 years to develop a new variety since sugarcane is a very complex heterozygous plant that complicates breeding efforts. Usually 40,000 seed clones are selectively reduced to 25 for use in further studies through 3 years of replicate trials.

Chemical controls generally eliminate a problem-causing organism. Poisoned baits are used against rats, insecticides against various insects, and herbicides against weeds. Management practices are directed at minimizing losses through the efficient use of all available methods of controlling pests and diseases. The cane grower must be cautious of disease symptoms, and must know what control method is most effective. Pest and disease control is greatly enhanced by management practices, such as planting disease- and pest-free seed cane, eliminating disease-infested plants through various harvest techniques, and controlling field-edge weeds and other plants that provide a habitat for pests and vectors for disease.

Mammals: Cane growers in the United States have only one mammal to deal with, the rat (*Rattus* spp.). A few different species have been shown to occur in all cane-growing areas of the United States. Rat populations in Florida have been shown to range from 75 to 250 individuals per hectare. Losses attributed to rats are both direct, from destruction of part of the stalk, and indirect, from the entrance of other organisms and diseases. Rats have also been found to be carriers of several cane diseases.

In one study in Florida, rats damaged 35% of the test plots (Humbert, 1968). Their high reproductive rate, mobility, and quick reaction to unfavorable conditions make rats difficult to control. Control measures cannot be limited strictly to the cane field; weedy or shrubby areas around the cultivated fields provide cover and nesting places.

Three forms of control have been attempted: biological, trapping, and poisoning. Biological control methods have included introduction of virus (unsuccessful owing to a loss of virulence of the disease and the development of immunity by the rats) and some introduction of new predators. However, predator introduction has not been pursued. Trapping is too costly to be effective in large-scale control. Poisoning has proved to be the most effective means of control, but the system must be designed to eliminate the possibility of adding toxic substances to the ultimate sucrose product.

Once an effective bait has been developed, several methods of distribution are applied. Prebaiting with nonpoisonous bait for a short time and then substituting poisoned food results in a fairly successful eradication. However, this method requires substation time and manpower. Poisoned bait is commonly dispersed from an airplane with good success as well as being economical.

The hazards involved in any type of pest control are concerned, not with the target organism, but with nontarget individuals, such as man or other animals or plants. Biological controls may introduce diseases that are toxic not only to rats but also to other species. New rat predators may reduce populations of birds or other organisms. Controlling the effect of trapping and poisoning presents similar problems. Poisoned bait attractive to rats may also attract birds, other rodents, and domestic animals. Hawaiian researchers (Cushing and Heinz, 1975) have developed a rodenticide (zinc phosphide) that has been approved for field use in crops.

Insects: Insects damage sugarcane by boring into or feeding on various parts of the plant. Virtually all parts of the plant are susceptible to attack; stalks are bored into and tunneled, leaves are chewed, roots are pitted and bored, and the seed cane is attacked. Some species also indirectly damage sugarcane by transmitting diseases. The kinds and importance of insects that feed on, attack, or affect cane in any way vary greatly among countries. For example, the leafhopper, a vector for transmitting Fiji disease in Fiji, when accidentally introduced in Hawaii, developed rapidly and did extensive damage but did not transmit the Fiji disease. Insect species must be monitored constantly to detect known insect pests as well as to assess the threat to cane of insects new to a region.

In Florida and Louisiana the most destructive insects are species of the genera *Diatrea* spp., a moth borer. The larvae of this genus tunnel through the cane stalk, eating tissue and exposing the cane to other organisms and diseases. In 1953, sugarcane losses in the United States attributed to this borer were estimated at $5,000,000/year (Humbert, 1968). This borer has been effectively controlled both biologically and chemically. Insecticide has been most successful in Louisiana, where climatic conditions influence the life cycle of insects. As many as 90% of first-generation borers have been killed with insecticides.

However, biological control has been equally successful, especially in Florida. Parasitic insects, such as *Trichogramma minutum*, a native egg parasite, *Lixophaga diatraea*, the Cuban fly, and *Agathis stigmaterus*, a wasp, have become established in Florida and are responsible for effective control of the borer. In Louisiana,

because of its climate, only the *Trichogramma minutum* parasite can overwinter
and, therefore, be economically effective against the borer (Humbert, 1968).
Much biological control research is ongoing, and several new parasites are being
introduced. Research is directed toward developing more resistant varieties of
sugarcane. Management practices that reduce the borer population are to burn
trash or to plow it under so that it cannot be a site for borers to overwinter.

The cane weevil, or beetle borer (*Rhabdoscelus obscurus*), is the greatest insect
pest in Hawaii (Humbert, 1968). Better field management, however, has reduced
this pest to minor importance in some areas. Field sanitation by using weevil-
free planting cane, destroying crop residues, and preharvest burning has successfully reduced damage attributed to this insect.

Nematodes: Another group of organisms destructive to cane is nematodes.
They exist in the soil and primarily attack the root system. In experimental
tests in Hawaii, reduction in yields attributed to root-knot nematodes ranged
between 30 and 60% (Humbert, 1968). Nematodes have been controlled biologically and through crop management. Biological control, through parasitic
fungi, has reduced nematode populations substantially.

Diseases: Sugarcane is susceptible to numerous diseases. Most diseases are
caused by fungi; a few are caused by viruses and bacteria. Many cane diseases
are carried by rats and insects that feed on cane. The lesions caused by various
organisms while feeding provide a site for diseases to enter. All plant parts are
attacked by one disease or another. Symptoms of diseases manifest in various
ways; some cause color streaking, some mottled leaves, and others dead patches.
Commonly, vascular bundles involved in water and nutrient transport are affected.
Wilting, stunted growth, or death of the cane results.

One of the most important diseases in Louisiana has been the mosaic disease,
to which has been attributed losses up to 50% in affected cane stands. Numerous different strains of the mosaic disease virus compound the problems of control. Since the most effective method of control against this or any sugarcane
disease is to develop new, resistant strains of sugarcane, research must continue
to meet the challenge of the frequently changing disease strains.

Weeds: Weeds compete with cane for nutrients, water, and space. Winter weeds
in Louisiana are commonly broadleaf annuals, such as chickweed, henbit, and
speedwell. In the spring and summer, the major weed problems are perennial
and annual grasses, such as Johnson grass (*Sorghum helepense*), crabgrass (*Dig-
itaria sanguinalis*), jungle rice (*Echinochloa colonum*), and Bermuda grass (*Cyno-
don dactylon*) (Breaux et al., 1972). The most widespread and economically
important weed in the United States is Johnson grass. Effective control of
weeds has been accomplished only through the combined use of herbicides and
cultivation. Without proper weed control, new cane cannot effectively compete.

Hawaiian sugarcane growers have serious weed problems but with species different from those of the mainland. Oahu has major problems with two weeds,
Panicum repens and guinea grass. Guinea grass uses the same C_4 photosynthesis
mechanism as sugarcane and is not susceptible to known herbicides. It ratoons
as readily as the sugarcane and responds to the cane management. The only
control mechanism is multiple plowing (as many as 11 times). *Panicum repens*
responds to chemicals; however, care must be used because of sugarcane suscep-

tibility. Yields of cane and sugar have been doubled when *Panicum repens* was controlled (Humbert, 1968). The island of Hawaii does not have a problem with guinea grass; it has the Job's tears plant. This plant looks like corn and has an extremely hard seed coat. The thick seed coat resists normal application rates of herbicides.

Production Costs

Field investigations were made in Florida, Louisiana, Texas, and Hawaii to gather sugarcane production cost data. These investigations produced varying results in terms of data availability and level of detail.

Separate and very detailed studies of the various costs associated with sugarcane production have been made for Florida (Walker, 1972) and Louisiana (Campbell, 1968). Both studies delineate the quantities of physical inputs in sugarcane production, such as labor, chemicals, seed, and equipment. These studies present all major operations and individual tasks comprising sugarcane production from land preparation through sugarcane harvesting. From the quantity and price data associated with each specific production input, many anticipated alternatives and their impacts on costs can be evaluated.

Similar detailed studies of sugarcane production costs are not available for Texas and Hawaii. Sugarcane is a relatively new crop in Texas, and detailed economic analyses for that area have not been conducted; they are planned. Even though the Hawaiian sugarcane industry is quite mature, the concentration of production among relatively few large plantations has resulted in a lack of published data on production costs. However, Hawaiian cost data have been assembled on the basis of information provided by the Hawaiian Agronomics Company.

Estimation Procedure: As a result of the varying quantity and quality of production cost data among producing regions, the basic information (except for Hawaii) for estimating 1976 sugarcane production costs has been supplied by the USDA's Firm Enterprise Data System (FEDS).

The production costs represent current average technology since the major anticipated use of the USDA budgets is in dealing with aggregate supply questions in response to government supply management policies and programs. Data to develop the production costs come from a variety of sources. The USDA's Statistical Reporting Service (SRS) reports information on yields, acreages, use of some agricultural inputs, and some production practices. The SRS also supplies input and product prices.

Surveys are being conducted by the Economic Research Service (ERS) branch of the USDA to estimate costs of producing various agricultural products as required by the 1973 Agricultural and Consumer Protection (ACP) Act. These survey data indicate machinery sizes and types along with operations performed in order to assess fixed and variable costs for crop production. Other miscellaneous data come from state experiment station and extension service staff (Krenz, 1975).

This review does not rely solely on USDA budgets in the estimation of current sugarcane production costs. Some of the data were aggregated to eliminate unnecessary detail. Also, since the USDA budgets were based on 1975 data, they were updated to 1976 price levels. Various price indicators found in *Agricultural Prices* (U.S. Department of Agriculture, 1974d; 1975a; 1976a) were used along with primary data collected during field interviews.

Some primary data do not coincide with numbers in the USDA budgets. For example, land charges for sugarcane production as stipulated in the USDA budget were overstated; consequently these charges were reduced to reflect information acquired during interviews with the University of Florida sugarcane economists.

Mainland United States: Estimates of 1976 sugarcane production costs for the U.S. mainland are shown in Table 4.1. Estimates are provided for the Texas Rio Grande Valley, Louisiana, and three Florida soil types (muck, peat, and sandy). Yields of millable cane per unit of area are differentiated according to expected productivity for each region with the use of current average technology for that region. This breakdown shows the decreased productivity and increased production costs resulting from use of less fertile soils as compared with highly fertile soils.

For example, yields on Florida muck soils are estimated at 94 tonnes/ha, compared with 75 tonnes/ha on peat soils and 70 tonnes/ha on sandy soils. Since Florida muck soil is limited, significant increases in cane hectarage will be on peat and sandy soils. Thus the implementation of new technologies will be required to obtain the high yields desired for biomass production.

The differences in productivity among regions make it necessary to compare costs on a dollars-per-tonne basis rather than on dollars per hectare. For example, estimated production costs on Florida muck soils are $1,584/ha compared with $1,083/ha in Louisiana. However, on a per-tonne-of-millable-cane basis, the Florida muck soil costs are $16.81/tonne vs $19.32/tonne in Louisiana. Current estimated costs per tonne of millable cane range from $15.61/tonne in the Texas Rio Grande Valley to $21.49/tonne on Florida sandy soils.

Although items in Table 4.1 are self-explanatory, machinery ownership, land charges, and management charges should be clarified. Machinery ownership costs are the fixed costs associated with ownership of planting and harvesting equipment, including depreciation, interest, insurance, and taxes. Where irrigation is necessary, ownership costs on irrigation equipment are also charged.

Land charges have traditionally been estimated by agricultural economists on the basis of net share rent, cash rent, or an annual interest charge on some specified value of the land. Land allocations deserve special attention because of their current magnitude relative to total costs, the diversity of methods for estimating land-use allocations, and the sensitivity of agricultural land values to product prices, government support programs, and potential urban development.

No single procedure is appropriate for determining land cost allocations for all purposes. Often land allocation is omitted from the cost of producing a commodity. The annual decisions of farmers as to what to produce may be made without considering an allocation for land they own. However, for long-run analysis pertinent to this study, land allocation is an appropriate part of the total cost of production.

Table 4.1: Estimated 1976 Sugarcane Production Costs for Mainland United States*

			Cost, ($/tonne)		
	Texas	Louisiana	Florida (muck soil)	Florida (peat soil)	Florida (sandy soil)
Preharvest Costs					
Seed	0.35	0.73	0.44	0.53	0.60
Fertilizer	1.20	1.59	0.77	1.09	3.23
Chemicals	1.20	1.16	0.84	0.89	1.02
Labor	1.12	2.35	1.19	1.69	1.97
Fuel and lubricants	0.19	0.46	0.32	0.37	0.48
Repairs	0.29	0.78	0.36	0.44	0.54
Interest on operating capital	0.34	0.31	0.19	0.24	0.44
Miscellaneous	0.50	0.32	0.36	0.43	0.58
Subtotal	5.19	7.70	4.47	5.68	8.86
Harvest Costs					
Labor	1.00	1.52	3.03	3.18	2.28
Hauling	2.20	1.32	1.32	1.32	1.32
Fuel and lubricants	0.60	0.41	0.45	0.54	0.61
Repairs	1.11	1.37	0.68	0.83	0.94
Interest on operating capital	0.10	0.03	0.28	0.29	0.30
Miscellaneous	–	–	0.37	0.45	0.51
Subtotal	5.01	4.65	6.13	6.61	5.96
Machinery ownership costs	1.66	2.39	1.25	1.50	1.74
Land charge	2.61	3.31	3.67	3.63	3.74
Management charge at 7% of gross receipts	1.14	1.27	1.29	1.29	1.29
Total Costs	15.61	19.32	16.81	18.71	21.59
Millable cane per hectare (tonnes)	85	56	94	78	70

*Estimates based on unofficial U.S. Dept. of Agriculture crop budgets and Texas A&M University crop budgets (adjusted to 1976 price levels where necessary).

Source: NTIS TID-22781

The 1973 ACP Act specified that a "return on fixed cost equal to the existing rate charged by the Federal Land Banks" be included in the land cost estimates. This implies the use of an "opportunity cost" allocation, or what could be earned if the "value" of the land were invested at the current federal land bank interest rates. For example, if land were valued at $2,471/ha and the current federal land bank interest rate were 9%, the allocated land cost would be $222/ha. However, the Act contains no such specific definition of value as current market price or purchase price (U.S. Department of Agriculture, 1976c).

Estimated land charges (shown in Table 4.1) in this review are based on either cash rent per hectare or the opportunity cost allocation method, depending on the availability of information. For example, cash rents of up to $346/ha/year are reportedly being paid for prime sugarcane-producing lands on the Florida muck soils. This is roughly equivalent to the opportunity cost concept where the land value would be $3,954 to $4,200/ha at a federal land bank interest rate of 8.5%.

In the pricing of management charges, there is no theoretical guide since management's input is usually rewarded by profit. Management input is not commonly

separated in farm management work, and there are no standard fee schedules
on which to base charges.

The 1973 ACP Act specified that a "return for management comparable to the
normal management fees charged by other comparable industries" be charged.
In this review the "comparable return" was based on fees charged by farm man-
agers working for banks, insurance companies, and farm management companies
which provide overall planning assistance but ordinarily not day-to-day management decisions.

These service charges vary widely and are calculated in a number of ways. A
charge of 7% of total gross receipts was used. With the use of this method, the
management charge will vary according to the level of product prices and crop
productivity levels. For the purposes of estimating the management charge, on
the basis of approximate current price levels, sugarcane prices were placed at
$16.25/tonne in Texas, $18.18/tonne in Louisiana, and $18.40/tonne in Florida.

Cost Component Percentage: A percentage breakdown of the major components
of sugarcane production costs by region shows that with few exceptions these
costs are comparable among the five regions. Fertilizer costs are higher on Florida
sandy soils owing to the high nitrogen requirements. Labor costs are relatively
low in Texas owing to the combination of a high degree of mechanization and
low wage rates. Labor costs are proportionally higher in Florida owing to the
high use of offshore workers. Hauling costs are proportionally greater in Texas
owing to the combination of longer hauling distances and lower total production
costs per hectare.

The breakdown of production costs by item among regions is, at best, an estimate rather than a precise measurement. The fact that one cost component is
higher or lower in a given region may be due to variations in the accounting
system. Even though the FEDS is intended to provide uniform data across
regions, it has been instituted only recently, and data refinements are certain
to be implemented. Also, in addition to accounting deficiencies, local supply
and demand conditions for purchased inputs in a given region can create significant variations in production costs.

Hawaii: Total sugarcane production costs in Hawaii in 1976 are estimated at
$17.57/tonne of cane on the basis of cost per hectare of $4,020 and a yield of
229 tonnes/ha. This includes a somewhat arbitrary allocation of $3.28/tonne
on the basis of one-half of total general and administrative expenses plus services.
The source of these data presented information on costs per tonne of raw sugar,
which included manufacturing and processing costs. In the allocation of these
costs of raw sugar to costs per tonne of cane, one-half was attributed to cane
cultivation and harvesting and one-half to cane processing.

The comparison of individual cane-production cost categories for Hawaii with
those for the mainland states is somewhat tenuous owing to the differences in
accounting and reporting format. However, the overall production costs should
be reasonably indicative of true costs and, therefore, should provide a basis for
comparing potential raw-material costs of fuels derived from sugarcane.

Puerto Rico: Present-day costs incurred in Puerto Rico sugar production are
generally higher than those in mainland U.S. sugarcane operations. These high

costs result from factors that are not relevant to the ultimate productivity of sugarcane in Puerto Rico. In some fields many ratoon crops are collected to avoid the plant-cane costs. The varieties grown in Puerto Rico are not optimal. Labor productivity is said to be low. Rather than present a dismal cost picture for Puerto Rico, detailed estimates are omitted from this review. Prospects for a future Puerto Rican sugar operation are discussed in another publication (Battelle Memorial Institute, 1977a).

Biomass Costs: In the production of sugarcane for fuel and/or chemicals production, the assumption is that all the plant's biomass will be used. Therefore, tops and leaves will be collected along with millable cane. Currently, tops and leaves are left on the field. Table 4.2 summarizes estimated 1976 sugarcane and biomass production costs; tops and leaves are included.

The quantity of tops and leaves per hectare was obtained by multiplying the millable cane production by 30%. It had been reported that tops and leaves ranged from 20 to 40% on hand-stripped cane, depending on the variety. Actually, reliable statistics do not exist on the quantity of sugarcane tops and leaves relative to millable cane. Since tops and leaves have had no real economic value, there has been no need to measure their weight.

Table 4.2: Estimated 1976 Sugarcane and Biomass Production Costs

State	Millable Cane, tonnes per ha	Tops and Leaves, tonnes per ha**	Delivered Costs of Sugarcane, $/tonne* . . . . Millable Cane		Delivered Costs of Biomass, $/tonne	
				Plus Tops and Leaves		Plus Tops and Leaves
Hawaii	229	68	17.57	14.52***	65	54
Texas	85	25	15.61	13.28	57	50
Louisiana	56	18	19.32	15.69	72	58
Florida						
Muck soil	94	29	16.81	14.26	63	53
Peat soil	78	25	18.71	15.75	69	58
Sandy soil	70	20	21.59	18.17	80	67

*Fresh weight
**Assumes that dry-weight content of aerial part of mature sugarcane is 27%.
***Assumes a 30% increase in total harvesting cost for collecting tops and leaves.

Source: NTIS TID-22781

Another assumption noted in Table 4.2 is that the dry-weight content of the aerial part of mature sugarcane is 27% (National Academy of Sciences, 1971). The dry-weight content will vary according to the time of harvest (mature cane will have a higher dry-weight percentage than green cane) and according to the different components of the sugarcane (dry weight of fresh sugarcane leaves is indicated to be 35%, and that for fresh leaves and tops combined is 25% or less) (National Academy of Sciences, 1971).

In the estimation of delivered costs, including tops and leaves, total harvesting costs were assumed to also increase by 30% over current costs. Since tops and leaves have a lower density than millable cane, this is probably a conservative estimate. Under these assumptions the delivered costs per tonne of biomass, including tops and leaves, range from $50/tonne for biomass from Texas to $67/tonne for biomass grown on Florida's sandy soils. This translates into raw-material costs per kilogram of biomass ranging from $0.051 to $0.068.

Fermentable Solids and Combustible Organics Costs: Since the total biomass of sugar crops consists of fermentable solids plus combustible organic material (fiber), production costs must be allocated between these two components owing to the different values and amounts of fuel available from each; i.e., the fermentable solids portion has a higher fuel value than does combustible organic material. The desired costs of fermentable solids, on the basis of their value in producing fuel, might be in the range of approximately $0.066/kg ($66.12/tonne). The desired combustible organic-material costs approximate $0.014/kg ($14.33/tonne). The basis for these desired costs is discussed in another publication (Battelle Memorial Institute, 1976b).

The steps followed in estimating costs of fuel raw materials from sugarcane are shown in Figure 4.1. The first step in allocating total sugarcane production costs between fermentable solids and combustible organic material was to iden-tify the composition of the sugarcane plant in terms of its dry weight and dry-weight components. The typical composition of sugarcane shown in Figure 4.2 is based on data provided by the National Academy of Sciences (1971) and through conversations with Battelle's chemical economists. The composition of the dry-matter portion of the plant is based on that reported for the aerial part of fresh, mature sugarcane.

Figure 4.1: Steps in Calculating Raw-Material Costs from Sugarcane

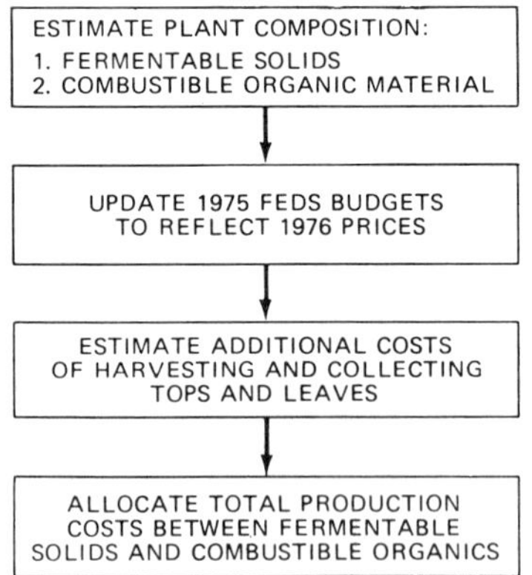

Source: NTIS TID-22781

The assumptions were that 80% of the ash, 100% of the ether extract and waxes, 100% of the nitrogen products, and 95% of the unattached solubles would remain in the soluble solids portion of the dry matter; that, conversely, 20% of the ash, 100% of the fiber, and 5% of the attached solubles would be in the insoluble portion of the dry matter; that 90% of the unattached solubles would consist of fermentable solids usable for fuel; and that all the insoluble portion, with the exception of the ash content, would be usable as combustible organic material.

These analyses approximate average quantities and will vary by region and variety of sugarcane. So that regional variations could be accounted for, the percentage of fermentable solids and combustible organics for each of the four U.S. cane growing regions was estimated. The total percentage dry-weight sugarcane, however, was assumed to remain the same among all regions, i.e., 27.2%. On the basis of USDA's statistics on raw sugar and molasses production by region, the fermentable solids were estimated to be 12.8% in Hawaii, 12.7% in Florida, 11% in Louisiana, and 10.6% in Texas. The relatively low percentage of fermentable solids in Texas cane may be due to adverse weather conditions and to the newness of Texas cane-milling operations.

It is important to note that the term "fermentable solids" is not synonymous with raw sugar production. For example, the typical composition of sugarcane shown in Figure 4.2 is approximately 144.5 kg of fermentable solids per ton of millable sugarcane, whereas the average raw sugar recovery per ton of millable cane in the United States is only about 95 kg. The quantity of fermentable solids is based on the reported nitrogen-free extract content equal to 52.7% of the sugarcane dry weight. This material contains other fermentable materials (e.g., starches and carbohydrates) in addition to raw sugar.

Figure 4.2: Typical Composition of Sugarcane

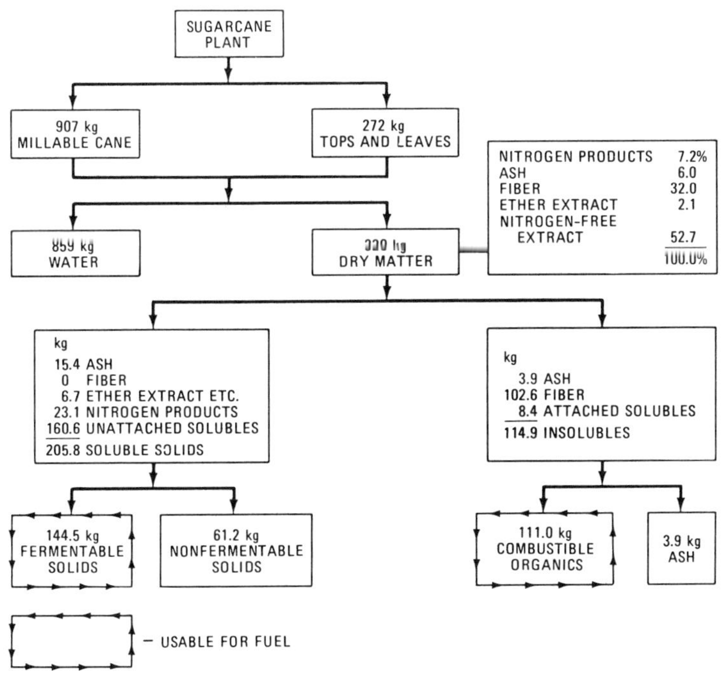

Source: NTIS TID-22781

Also to be included in the fermentable solids is material found in molasses plus material from the tops and leaves, which naturally is not included in raw sugar produced from millable cane. The "nonfermentable solids" in Figure 4.2 include the ash, ether extract, nitrogen products, and 10% of the unattached solubles contained in the total soluble solids (205.8 kg) portion of the dry matter.

The sum of fermentable solids and combustible organics was assumed to be constant for all sugarcane-growing regions (i.e., 144.5 kg plus 111.0 kg = 255.5 kg in Figure 4.2). In other words, the regions producing lower quantities of fermentable solids are assumed to produce higher quantities of combustible organic material and vice versa.

Total sugarcane biomass production and delivery costs shown in Table 4.2 provided the basis for allocating these costs between fermentable solids and combustible organic material. The cost allocation was based on the ratio of desired prices for fermentable solids and combustible organics. The desired costs were assumed to be $0.066/kg for fermentable solids and $0.0143/kg for combustible organic material. Thus a price ratio (4.62 to 1) was used in assigning the proportion of total costs to fermentable solids and combustible organic material (total costs = X + 4.62X, where X is the assigned costs of combustible organic material).

Estimated 1976 raw-materials costs of fermentable sugars and combustible organic material from sugarcane can be based on the preceding methods of calculation (see Table 4.3). Yields per hectare are based on average reported sugarcane yields for each growing region plus 30% additional biomass for tops and leaves. Based on these calculations, the raw-material cost per kilogram of fermentable solids ranges from $0.092 in the muck soils of Florida to $0.12 in Louisiana. The cost per kilogram of combustible organic material ranges from $0.02 in Texas to $0.035 in the Florida sandy soils. As milling operations are optimized in Texas, the cost per kilogram of fermentable solids should decline with higher recovery rates. At the same time, a slightly higher cost for combustible organic material will accompany its lower quantity.

Table 4.3:   Estimated 1976 Raw-Materials Costs of Fermentable Sugars
and Combustible Organics from Sugarcane

| | Production Costs, $/ha | | Yield, kg/ha | | Costs, $/kg | |
State	Fermentable Sugars	Combustible Organics	Fermentable Sugars	Combustible Organics	Fermentable Sugars	Combustible Organics
Texas	1,196	259	11,749	12,296	0.101	0.020
Louisiana	953	207	8,050	7,769	0.119	0.026
Florida						
Muck soils	1,445	314	15,570	10,996	0.092	0.029
Peat soils	1,346	289	12,975	9,163	0.103	0.031
Sandy soils	1,329	289	11,492	8,116	0.117	0.035
Hawaii*	1,773	383	18,682	12,946	0.095	0.029

*Annualized data are presented; over Hawaii's normal 24-month growing seasons, the costs per hectare and yields per hectare would be double the amounts indicated. The costs per kilogram would be unchanged.

Source: NTIS TID-22781

Current vs Desired Yields: Table 4.4 compares current and desired yields of fermentable solids and combustible organic matter from sugarcane in various sugarcane-growing regions. The desired yields are based on 1976 production costs and a cost of $0.066/kg for fermentable solids and $0.0143/kg for combustible organic material.

Table 4.4: Yields of Fermentable Solids and Combustible Organic Material from Sugarcane at 1976 Production Costs*

| | . . . . . Fermentable Solids . . . . . . | | | . . . . Combustible Organics . . . . | | |
State	Estimated Current Yield, kg/ha	Desired Yield, kg/ha	Desired Increase, %	Estimated Current Yield, kg/ha	Desired Yield, kg/ha	Desired Increase, %
Texas	11,749	18,123	54	12,296	18,108	47
Louisiana	8,050	14,424	79	7,769	14,487	87
Florida						
Muck soils	15,570	21,860	41	10,996	21,902	99
Peat soils	12,975	20,178	56	9,163	20,178	121
Sandy soils	11,492	20,103	75	8,116	20,178	149
Hawaii	18,682	26,870	44	12,946	26,731	107

*Desired raw-materials costs are assumed to be $0.066/kg for fermentable solids and $0.0143/kg for combustible organic material.

Source: NTIS TID-22781

For these levels of costs to be achieved, the yield of fermentable solids must increase from 41% (Florida muck soils) to 79% (Louisiana) over current levels. The yield of combustible organic material would be required to increase from 47% (Texas) to 149% (Florida sandy soils).

Large increases in fermentable solids and combustible organic-material yields are unlikely to be achieved simultaneously in the same sugarcane plant. Yield increases will result from higher green-cane yields per unit of area and/or higher concentration of fermentable solids or combustible organic material in the green cane.

With regard to the second alternative, if it is assumed that total plant biomass (dry weight) per tonne of cane remains fixed, an increased concentration of fermentable solids will result in a lower concentration of combustible organic material and vice versa. Past sugarcane breeding and culture has been directed to achieving higher yields of sucrose per tonne of cane with a minimum fiber content. Therefore, a redirection of effort will be required if higher fiber sugarcane is desired, i.e., to yield a greater volume of combustible organic material for use in fuel production.

SWEET SORGHUM

The name sweet sorghum is specifically applied to varieties of a species of sorghum, *Sorghum bicolor.* Sweet sorghum is grown for syrup and forage; other sorghums are grown for grain. New varieties are being developed specifically for sugar production also.

Sweet sorghum is not a new crop to U.S. agriculture. It has been used for many years as a source of table syrup. Sweet sorghum is adapted to climatic and soil conditions ranging from those of Alabama to those of Minnesota. In 1964, sweet sorghum was grown for syrup production in 19 states (Freeman, Broadhead, and Zummo, 1973).

States in which more than 12 ha of sweet sorghum were raised in 1964 include California, Texas, Wisconsin, Iowa, Missouri, Arkansas, Louisiana, Indiana, Kentucky, Ohio, Tennessee, Mississippi, Alabama, Georgia, Florida, West Virginia, Virginia, North Carolina, and South Carolina (Coleman, 1970). Few producers cultivate more than 20 ha of sweet sorghum, and the majority of producers grow less than 0.4 ha (Freeman, Broadhead, and Zummo, 1973). Since its peak in 1946, sweet sorghum production has generally declined. The primary cause has been a decrease in the demand for sorghum syrup. Until recently raw sugar could not be economically refined from sweet sorghum because of starch in the sorghum juice. Extensive experimentation is directed specifically at sweet sorghum as an additional sugar crop.

Production Methods

Production and cultural methods for sweet sorghum are summarized in Table 4.5. Note that both syrup and sugar varieties of sweet sorghum are grown. Syrup varieties are grown primarily in the southeastern United States, whereas sugar varieties are tentatively planned for the lower Rio Grande Valley.

**Table 4.5: Summary of Sweet Sorghum Production and Culture in
the United States**

Where grown	Southeastern United States (eight states) produces about 90% of total production, but production has occurred in 19 states.
Area harvested, 1973–1975, United States	Probably less than 4,000 ha (no official production statistics available).
Yield	34 to 85 tonnes unstripped stalk/ha, 27 to 58 tonnes stripped stalk/ha in 1975 research trials for selected syrup varieties. Biomass yields of sugar varieties about 30% less than syrup varieties.
Typical farm size	Experimental and traditional. Average traditional producer grows less than 0.4 ha for syrup.
Planting date(s)	April–June in southeastern United States; March–August in Texas Rio Grande Valley.
Harvest season	August–October in southeastern United States; July–December in Texas Rio Grande Valley; must be harvested before first autumn frost.
Length of growing season	110 to 140 days.
Rotation	Normally rotated with cotton, corn, or soybeans.
Soil requirements	Many different soil types can be used; loam and sandy soils preferred; organic matter improves soil's water-holding capacity.

(continued)

Table 4.5: (continued)

Fertilizer application (actual fertilizer application based on local soil requirements)	Nitrogen, 45 kg/ha; phosphate, 45 kg/ha; potash, 35 to 45 kg/ha. Sandy soils usually will require more fertilizer than heavy soils. Lime should be applied to acid soils.
Water managment	Similar to sugarcane (3 cm per tonne of stalks); ample moisture required during growing season. Most moisture obtained from upper 0.3 m of soil layer.
Disease problems and control	Downy mildew, anthracnose and red rot, mosaic, and gray leaf spot are major diseases of economic importance. Primary control measures include planting resistant varieties, crop rotation and clean cultivation, and chemical acid treatment.
Pest problems and control	Insect damage usually slight but may be severe particularly during dry periods. Lesser cornstalk borer, sorghum midge, sugarcane borer, corn leaf aphid, fall armyworm, corn earworm, and wireworms may cause damage. Control by planting early maturing varieties, insecticides, and proper cultural practices.
Weed problems and control	Weeds controlled by good planting methods ensuring adequate stands. Cultivation important weed-control measure with hoeing and thinning sometimes used for heavy infestations. Preemergence application of 2.4 to 3.6 kg propazine per hectare controls many small-seed annual weeds.
Harvesting procedure	Current harvesting methods vary considerably depending on acres involved and local customs. Can be harvested by hand or machine, but there is a lack of suitable mechanical equipment. Corn binders may be used; also, silage harvesters have been modified to harvest sweet sorghum. Leaves may be stripped from standing stalks.
Prospects for additional production	Adapted to diverse climatic and soil conditions found from Alabama to Minnesota.

Source: NTIS TID-22781

At this point it is not clear whether energy research on sweet sorghum should be conducted on syrup or sugar varieties. Total production of biomass per unit of area is greater for syrup varieties; however, the total soluble solids (Brix) content of sugar varieties is higher. Preliminary composition and cost estimates for both varieties are reviewed here.

Sweet sorghum will grow on a variety of soils, but it grows best on loams and sandy loams. Adequate soil moisture and good drainage are important to good yields. Optimum planting time for satisfactory germination occurs when the soil reaches 21°C and sufficient moisture is available for germination. Germination occurs in 3 to 5 days. Cultivation is necessary to minimize weed competition until canopy closure. Maturity at harvest varies with varieties and regions. Clarification of the juice depends on the maturity of stalks (Coleman, 1970).

In the southern areas the planting period is usually from mid-April to mid-May. In the northern areas the best planting season may be limited to 10 to 15 days during mid-May to achieve maturity before the first killing frost. Specific planting times are dependent on local climatic features and on the sweet sorghum variety planted. For example, in southern Texas Rio Grande Valley sweet sorghum can be planted from March until August. This variety also matures in 100 to 120 days; thus harvesting is possible from July until December. Consequently, two or even three sorghum crops per year are possible where a long growing season exists (Cowley and Smith, 1972).

Because the harvest seasons for sugarcane and sweet sorghum for the various production regions are not coincident, the production of sweet sorghum in conjunction with sugarcane would enable southeastern mills to begin operating in August, as opposed to the present early to mid-October. Texas mills could begin processing sweet sorghum in July and sugarcane in mid-October. Longer mill operation would allow greater coverage of fixed costs and would also help to alleviate seasonal labor problems.

Yields vary with variety and location. Average yields have been reported to range from 22.4 to 44.8 tonnes of millable stalks per hectare (Freeman, Broadhead, and Zummo, 1973). The most recent data from Texas indicate that yields (wet weight) ranging from 44.8 to 112 tonnes/ha are possible during a 130- to 150-day growing season. Irrigated sorghums in Tucson, Arizona, produced 90 tonnes of wet biomass (stalk, leaves, and seeds) per hectare (Freeman, Broadhead, and Zummo, 1975). Sorghum yields have been correlated to day length and solar radiation (Cowley and Smith, 1972).

In experimental plantings of the Rio Grande variety in Texas, yields of 35.8 to 44.8 tonnes of millable stalks (wet weight) per hectare were obtained from May plantings. Plantings before and after May yielded less tonnage. Sucrose content is not directly related to yields but generally averages 15% of yields (Cowley and Smith, 1972). If 100% extraction of sugars is assumed, 0.91 tonne of sweet sorghum yields approximately 147 kg of solids, which is approximately 80% total sugars.

Sweet sorghum is subject to insect and disease problems. A few of the major diseases are red rot, one of the most destructive in the Southeast; downy mildew; mosaic; gray leaf spot; bacterial stripe; leaf blight; smut; and seedling blight. Disease control measures include the planting of resistant varieties, crop rotation, and chemical treatments. Insect damage to sweet sorghum is generally slight but occasionally is severe. The most common sorghum insect pests include lesser cornstalk borer, sorghum midge, sugarcane borer, aphids, armyworms, and wireworms. Seeds are also attacked by the grain moth and weevil, which can cause losses, especially in stored seed.

The biological tolerance of sweet sorghum with respect to soils combined with the capability to reach maturity in 90 to 150 days, depending on variety, give sweet sorghum considerable potential. It has the potential to grow anywhere cotton, corn, grain sorghum, sugar beets, or sugarcane can grow. Its rapid maturity makes more than one crop per year possible.

Composition

Figures 4.3 and 4.4 give estimates on the composition of sweet sorghum. The term estimates is emphasized owing to the relative lack of hard data on sweet sorghum as compared with that for sugarcane and sugar beets. Typical yields are restated in terms of salient composition parameters in Table 4.6.

Figure 4.3: Estimated Composition of Sweet Sorghum, Syrup Varieties

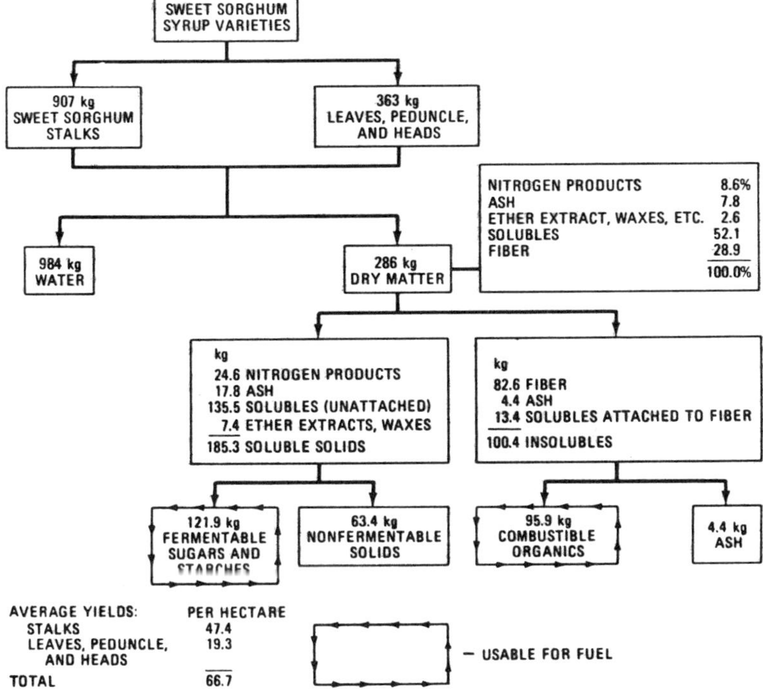

Source: NTIS TID-22781

Syrup Varieties: Based on 1975 growing trials, the average gross yield per hectare of sweet sorghum syrup varieties from nine locations was 66 tonnes/ha. This yield includes all parts of the plant—stalk, leaves, peduncle, and head. The average yield per hectare of stripped stalks was 43 tonnes for the nine locations. The average time to crop maturity was 121 days, with a range from 97 to 135 days.

The ratio of gross yield to stripped-stalk yield was used as a basis for estimating the quantity of leaves, peduncles, and heads per tonne of stalks. On that basis, the yield per acre is increased by approximately 40% when the leaves, peduncles, and heads are added to stripped stalks. This compares with an estimated average of 30% add-on for tops and leaves of sugarcane. The dry-matter content of sweet sorghum has been estimated at between 20 and 25%. The midpoint, 22.5%, was used as the basis for the estimated 286 kg of dry matter as shown in Figure 4.3. This estimate may be conservative. Substantive data are unavailable, and the dry-matter content of some sorghum species ranges upward of 30% of total fresh weight (National Academy of Sciences, 1971).

The dry-matter composition of syrup sweet sorghum is estimated in terms of solubles, fiber, nitrogen products, ash, ether extractables, and other products. These percentages are based on averages for eight different species of sorghums (National Academy of Sciences, 1971).

In the allocation of the dry matter between soluble and insoluble solids, the assumption was that none of the fiber, 80% of the ash, 91% of the solubles, 100% of ether extractables, and 100% of the nitrogen products would go into the soluble solids portion of the dry matter. Conversely, all the fiber, 20% of the ash, and 9% of the solubles would be in the insoluble portion. A small amount of the soluble solids is assumed to remain attached to the fibrous portion of the dry matter, which accounts for the apparent misnomer "solubles" in the "insoluble" portion.

Figure 4.4: Estimated Composition of Sweet Sorghum, Sugar Varieties

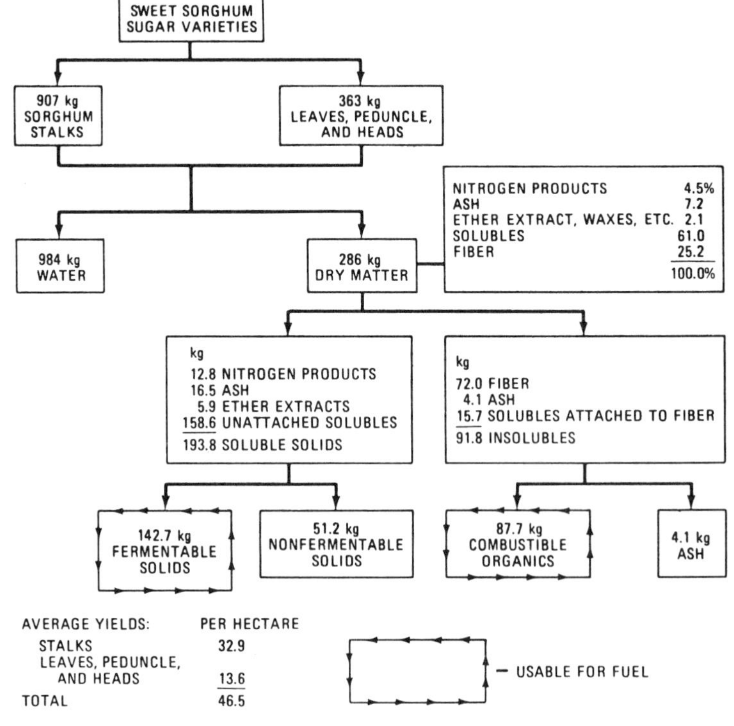

Source: NTIS TID-22781

It was arbitrarily assumed that 90% of the unattached solubles would consist of fermentable sugars and starches and would be usable for fuel or chemicals manufacture. All but the ash portion of the insoluble dry matter is assumed to have fuel value and is labeled combustible organics.

With this estimation procedure, approximately 122 kg of fermentable sugars and starches plus 96 kg of combustible organics are obtained from 907 kg of sweet sorghum stalks and 363 kg of attached material (leaves, peduncle, and heads).

Sugar Varieties: The basis for estimating the composition of sugar varieties of sweet sorghum was similar to that for syrup varieties. The major difference was in the breakdown of the dry-matter portion shown in the middle of Figure 4.4. The average Brix percentage of eight sugar varieties of sweet sorghum was approximately 18% higher than the Brix content of nine syrup varieties. On this basis the solubles percentage of the dry matter was estimated to be 61%, as compared with 54.1% in syrup varieties. The content of nitrogen products, ash, ether extractables, and fiber was assumed to be essentially equal to the Darso variety of sorghum, which also has a solubles content of 61% (National Academy of Sciences, 1971).

Table 4.6: Estimated Sweet Sorghum Product Yields, Sugar and Syrup Varieties

Variety	Fresh Biomass	Total Dry Matter	Fermentable Solids	Combustible Organics
. . . . Average Experimental tonnes/ha				
Syrup	60.5	13.7	5.8	4.5
Sugar	42.6	9.4	4.7	2.9
. . . . High Experimental tonnes/ha				
Syrup	81.4	18.4	7.8	6.1
Sugar	61.2	13.7	6.9	4.3

Source: NTIS TID-22781

Approximately 143 kg of fermentable solids plus 88 kg of combustible organic material are estimated to be derived from 907 kg of sweet sorghum stalks plus 363 kg of attached plant material.

Based on 1975 sweet sorghum growing results reported by the USDA, the average yield per hectare of sugar varieties was only about 43 tonnes of total plant material per hectare vs 61 tonnes of material for syrup varieties. These data indicate that the total biomass yields of sugar varieties are only about 70% of that of syrup varieties. This is a question that needs to be examined more closely in future research.

Production Costs

Table 4.7 gives a breakdown of estimated sweet sorghum production costs for 1976. Crop yields are based on the average yields achieved in the 1975 growing trials at various locations, as mentioned previously. Variable preharvest and harvest costs are based on information provided by K.C. Freeman of the U.S. Sugar Crops Field Station, U.S. Department of Agriculture, Meridian, Miss. So that these estimates could be compared with those of sugarcane and sugar beets, costs were added for hauling, miscellaneous expenses, interest on operating capital, machinery ownership, land charge, and a management charge. Interest, miscellaneous expenses, and machinery ownership costs are based on averages for production of corn silage and grain and forage sorghums in Oklahoma and Texas. Hauling costs are estimated at $2.20/tonne, which is comparable to the estimates used for sugarcane and sugar beets. The basis for deriving a management charge is explained in the footnote in Table 4.7. Except for hauling and the management charge, the production costs for syrup and sugar varieties of sweet sorghum are assumed to be the same.

Table 4.7: Estimated 1976 Sweet Sorghum Yields and Production Costs*

	Syrup Varieties	Sugar Varieties
Average yield (tonnes/ha)		
Stripped stalks	42.6	30.3
Leaves, peduncle, and heads	17.9	12.3
Total yield	60.5	42.6
Costs ($/ha)		
Preharvest costs		
Seed	18.53	18.53
Fertilizer	88.96	88.96
Chemicals	7.41	7.41
Labor	25.95	25.95
Machinery operating expense	25.95	25.95
Interest on operating capital	9.88	9.88
Miscellaneous (crop insurance)	11.86	11.86
Custom spraying	4.94	4.94
Harvest costs		
Labor	8.65	8.65
Machinery operating expense	8.65	8.65
Handling at $2.20/tonne	133.43	93.90
Miscellaneous	6.18	6.18
Machinery ownership costs	32.12	32.12
Land charge (1976 $/ha x 8%)	158.14	158.14
Management charge at 7% of gross receipts	55.10**	42.99***
Total costs	595.75	544.11

*Based on information provided by K. Freeman, Agricultural Research Service, U.S. Department of Agriculture, Meridian, Miss.

**Gross receipts based on fermentable solids prices of $0.11/kg and 2341 kg fermentable solids per hectare ($258) plus combustible organic-material prices of $0.033/kg and 1825 kg of combustible organics per hectare ($61).

***Based on fermentable solids prices of $0.11/kg and 1901 kg fermentable solids per hectare ($210) plus combustible organic-material prices of $0.033/kg and 1166 kg of combustible organics per hectare ($39).

Source: NTIS TID-22781

Note that production costs for sweet sorghum are considerably lower than those for sugarcane. This is due to a number of factors, with lower costs for chemicals, labor, machinery ownership, and land, the most significant being land. Sweet sorghum costs are comparable to production costs for other types of sorghums, as indicated by various production costs budgets. Also, an experimental crop may appear to have a low production cost since cost factors increase with experience, e.g., pesticides to battle insects.

Sweet sorghum production costs can be allocated between fermentable solids and combustible organics. The desired cost of fermentable solids was estimated at $0.066/kg and of combustible organic material at $0.014/kg. This price ratio (4.62 to 1) can be used in assigning the proportion of total costs to fermentable solids and combustible organic material. These estimates combined with the yield data result in costs of fermentable solids at $0.0849/kg for syrup varieties and $0.095/kg for sugar varieties. The cost of combustible organic material is estimated at $0.0232/kg for syrup varieties and $0.033/kg for sugar varieties. Many arbitrary assumptions, both technical and accounting, were used in arriving at these cost estimates. Because of the numerous uncertainties concerning the estimating basis, these costs must be regarded as crude at this point.

Potentially Limiting Factors: Sweet sorghum has a quite small seed and a longer germination period than corn. A clean seed bed is required, and care must be exercised with regard to both preemergence and postemergence herbicides. Drought greatly inhibits the growth and development of sweet sorghum, as it does most plants. However, sweet sorghum is more tolerant than corn with regard to moisture stress. Soil temperature must reach about 21°C for seed germination. Once the seed has germinated, sweet sorghum is relatively frost resistant. Since it will continue to grow well until the first frost, sweet sorghum can be grown in many areas of the United States, as far north as Ohio and Michigan. Sweet sorghum is a short-day plant, and reduced sugar yields of current sugar varieties will occur if seed set is inhibited.

Budworms, sugarcane borers, lesser corn borers, and wireworms are typical pests. Varieties have been developed with good resistance to downy mildew, rust, and other microbial diseases. Birds can be a serious problem in seed-producing fields. The limiting factors in large-scale cultivation of sweet sorghum for sucrose and biomass-fuel production are unknown at this time.

Possible Improvements: The present and anticipated near-term future yields of sweet sorghum are not impressive in terms of the quantity of energy produced per hectare. However, there are several factors that keep this crop in the running. (1) The development efforts have not been aimed at maximizing biomass production. (2) Sweet sorghum has responded well to a relatively low level of genetic improvement effort. The ingenuity and funding used to make the dramatic improvements in yields of corn, sugar beets, and sugarcane have not been applied to sweet sorghum. (3) The opportunities to double crop sweet sorghum have not been explored. (4) Harvesting methods for sweet sorghum are directed specifically at sucrose maximization and are at a primitive development stage compared with other major U.S. crops.

One of the more promising opportunities is double cropping with wheat; the wheat can be harvested in many areas before the time comes for planting sweet sorghum. In some areas of Texas and Puerto Rico, it may even be possible to obtain two crops of sweet sorghum. Virtually nothing is known about growing sweet sorghum for harvesting via mowing operations of the type that are standard for alfalfa.

SUGAR BEETS

Production Methods

The sugar beet is a biennial plant of the goosefoot family (Chenopodiaceae).
It is vital to man as a high-energy source and accounts for 45 to 50% of all raw
sugar produced in the United States (Hills and Johnson, 1973). Unlike sugarcane
and sweet sorghum, sugar beets use the C_3 photosynthetic pathway, which is
well suited to temperate climates (Loomis and Gerakis, 1975).

The sugar beet is tough and adaptable. Seedlings can withstand mechanical abuse,
cold, heat, drought, and defoliation. It is successfully cultivated in four fairly
distinct climatic and geographic areas in the United States. These areas include
(1) the humid zone in the North-Central States; (2) the prairie soils of the Great
Plains and the Red River Valley; (3) 2,100-m altitudes in sedimentary, arid
Mountain States; and (4) the semitropic below-sea-level California Imperial Valley
(Johnson et al, 1971). The last two areas are irrigated. The plant is compatible
with modern agricultural technology and adapts well to complete mechanization
for production. Its deep-rooted nature, to 1.83 m or more, and weed-free culture make it well suited to crop-rotation practices with crops such as corn, soybeans, and small grains.

From 1973 through 1975 sugar beets were grown in 17 states. Sugar beet acreage represents 0.7 to 0.9% of the total principal crop area harvested in these 17
states during the same period. States having the highest average area harvested
per year for the 3-year period (in thousands of hectares) were California, 112;
Minnesota, 69; Colorado, 53; Idaho, 53; and North Dakota, 47.

A review of yields and total production of sugar beets by state from 1973
through 1975 indicates that average yields ranged from 30.6 tonnes/ha in North
Dakota up to 57.7 tonnes/ha in Washington. The U.S. average yields were 43
tonnes/ha during this period. California is the leading producer, averaging 6.3
million tonnes, 27% of the total U.S. sugar beet production. All yields and
production are reported on a fresh-weight basis. Dry weight of sugar beet roots
equals 21 to 22% of fresh weight.

Sugar content and purity is highest in the roots. Between 70 and 75% of the
dry matter is sucrose. This represents a relatively small percentage of the total
root weight since dry weight of the roots is approximately 21 to 22% of the
fresh weight. Sugar content varies slightly in the root, with amounts and purity
being somewhat lower in the center and near the skin. Generally, the larger the
root, the lower is the percentage of sugar (Coe and Menser, 1976).

During the peak growing season, the greater part of the sugar made by the leaves
may be used by the plant. However, when the vegetative growth declines in late
season owing to lower temperatures and nitrogen deficiencies, a large part of
the sugar is stored in the roots. Commercial crops are harvested at or near the
end of the period of the first year's growth when the roots are large and contain
the most sugar. However, harvesting schedules necessary to permit an orderly
flow of beets to the processor may not always coincide with the maximum sucrose yield.

Sugar beets require at least 5 months of good growing weather to produce a profitable crop. Seedlings usually emerge within 4 to 5 days after planting (Loomis, Ulrich, and Terry, 1971). In the presence of light, the newly formed leaves expand rapidly and are usually horizontal to the soil surface. When the first leaf is fully developed, the taproot may reach 30.48 cm or more. Damage to the main root at this time may later result in the formation of spangled or forked storage roots.

The rate of storage root growth, which is fairly constant in most climates, varies with the supply of surplus sugar. Leaf growth is cyclic, but the amount of living leaves remains fairly constant whereas total dry matter of the tops, living and dead, increases at a relatively uniform rate.

Sugar Beet Management: In recent years sugar beet culture has become increasingly mechanized. Hand labor and low-capacity implements have been replaced by gang plows, combination implements for seedbed preparation, multiple-row cultivators, stand-reducing implements, and mechanical harvesters (Lill, 1964). Fertilizer and pest control are often combined with seedbed preparation, planting, and cultivating operations to enhance the efficiency of sugar beet culture.

Planting — Planting time depends on the weather and the proposed harvest date. This varies in different parts of the United States where sugar beets are grown. Sugar beets are planted somewhere in the United States each month except July and August.

Row spacings vary in different localities from 70 to 76 cm. Reliable ground rules for the per hectare populations, which are essential before maximum yields can be achieved, are difficult to establish. Generally, plants should be no closer than 12.7 to 15.24 cm apart in the row for current sucrose and beet-size objectives. In some instances, in-row spacing can be as great as 40.64 cm without reducing root yield and corresponding sugar percentage, but 20.32 to 30.48 cm is usually considered optimum. This spacing is obtained by thinning with machines or hoes or by precision planting.

Harvesting — Sugar beets are generally harvested when the sucrose content reaches a maximum. This degree of maturity occurs when the lower leaves turn brown and the upper leaves turn yellow. Harvest time depends on climatic conditions. In northern areas (such as Colorado and Nebraska), harvesting must be completed before the ground freezes. Sugar beets grown in California (except northern California) can be harvested any time of the year. Because high temperatures reduce the sucrose content in the roots, harvesting is usually completed by the end of June in hot climates. Three fundamental steps are involved in sugar beet harvesting:

    (1)   Remove the foliage (tops) from the root,

    (2)   Lift the roots from the ground, and

    (3)   Remove most of the soil and trash from the two crops

Two types of harvest systems are followed. The first involves an integrated harvester that lifts the beet plants from the soil and then conveys them to a set of rotating disks that separate the tops from the roots. The second, and more popular, system requires two pieces of equipment, a top-recovery machine and a root harvester. The top-recovery machine slices the tops from the roots and then gathers the tops in windrows. The largest of these machines can top up to six rows at a time. The windrows are picked up later by hay-gathering equipment. After the plants have been topped, root harvesters lift and shake the roots free of soil and convey them to a hopper. Root harvesters generally have a two- or three-row capacity.

Today's systems can harvest 8 to 10 ha per 24 hr day under favorable circumstances. Most farmers own their own top-recovery machines but share-rent the root harvester with other farmers.

Effects of Climate — Climate affects the growth and development of sugar beets in three ways: (1) transitory effects, such as may be observed in photosynthesis, respiration, or transpiration; (2) additive effects as observed in total vegetation growth; and (3) developmental events as observed in the time-delayed reaction of flowering after an extended low-temperature period (Loomis, Ulrich, and Terry, 1971).

At harvest time low night temperature increases sucrose concentration, a transitory phenomenon, and decreases top and root size when adequate nutrition and moisture are available. Maximum sucrose production occurs at approximately 10° to 15°C night temperature (Loomis, Ulrich, and Terry, 1971). Sucrose yield fails to increase above temperatures of 30°C. Day temperatures have a similar effect on sucrose production.

Under ideal conditions, sucrose concentration and the size and shape of the leaves are determined by the climate prevailing just before harvest. Cold nights and nitrogen deficiency before the harvest will slow vegetative growth and increase the yield of sucrose (Hills and Johnson, 1973). Nearly 50% loss can be experienced when an early-season cold spell is followed by a late-season hot spell. An entire season of warm weather or a combination of early-season hot weather followed by late-season warm weather will result in best root growth. Only yield, not sucrose concentration, is affected by seasonal climate throughout the growing season. Concentration is affected at harvest time.

Cold climates favor top size all during the growing season, and hot weather reduces it (Loomis, Ulrich, and Terry, 1971). Life expectancy of sugar beet leaves is 44 days, 58 days, and 67 days in hot, warm, and cold climates, respectively.

Light Effects — Sugar beets are influenced by day length almost entirely through the effect of light on photosynthesis. Increased day length increases the photosynthates produced by the plant (Loomis, Ulrich, and Terry, 1971). The principal difference noted in laboratory experiments between 8-hr and 10- to 14-hr days was doubling of the root size, whereas top size and percentage of sucrose increased slightly. Thus supplies of sucrose not needed in top growth were used in root growth, and the concentration of sucrose remained nearly constant. A similar effect is observed when radiation is increased and day length held constant. This would indicate that as much sucrose would be produced under long days of moderate brightness as would be under short days of greater total radiation.

Day length also influences flowering and reproduction. Several weeks or more of temperatures at 4°C is sufficient to induce flowering. Long days then tend to accelerate seed-stalk production, or bolting.

Soil Requirements — Soil texture determines to a large extent the moisture-holding capacity, drainage, and aeration of a soil. A deep, well-drained loamy soil is generally the most satisfactory for sugar beets (Loomis, Ulrich, and Terry, 1971). Acid soils, of pH 5 to 5.5 or less, are generally not favorable for sugar beet growth. The greater part of the sugar beet production is grown on mineral-type soils (Lill, 1964). Although these soil types differ greatly in their physical properties, they should have good depth and adequate drainage and should be well aerated. Mineral-type soils should also have moderate organic content and moisture-holding capacity. Those which combine the necessary characteristics for sugar beet production are usually the dark, heavier types, such as loams, silt loams, clay loams, and clays.

Soils should be at least 0.91 m deep (Lill, 1964). Accumulated surface water can result in severe crop loss, and a high water table can cause short, ill-shaped roots. Although the sugar beet can withstand drought as well as, or better than, most other cultivated crops, too long a period results in lagging growth and reduced yield.

Fertilizer is almost always applied to growing sugar beets. Of the mineral elements essential for growth, nitrogen has by far the greatest influence on root quality and sucrose production (Hills and Ulrich, 1971). Although large amounts of nitrogen are required for plant growth and maximum root yield, the plants should be deficient in nitrogen before harvest to retard the use of sucrose for growth. For example, Fick, Loomis, and Williams (1975) cite one example where nitrogen deficiency increased sucrose concentration from 12.4 to 16.8% fresh weight and from 75 to 77.8% dry weight. This calls for an exacting nitrogen management program.

Although nitrogen can be absorbed by sugar beet roots as nitrate, ammonium ion, urea, water-soluble amino acids, and nucleic acids, by far the largest amounts are taken up as nitrate. Usually a single application at planting or thinning time produces results comparable to splitting the total amount required into two or three applications. Where sandy soil results in leaching, a second application may be made at or before midseason.

Crop Rotation — Crop rotation is a necessary practice in the culture of sugar beets because of pests. The cropping sequence may range from 3 to 8 years, depending on the severity of nematodes and other pests or diseases. A general guideline is one beet crop per 4-year period. Rotational crops should not be hosts for diseases and pests common to beets (Nichol, Burtch, and Traveller, 1971). In addition, care must be used in selecting chemicals to be used on rotation crops.

Weed, Disease, and Pest Control — Preemergence chemical weed killers are effective for weeds germinating prior to sugar beet planting. Postthinning weed control is usually by cultivation. A crop of weeds emerging after plants are too large for mechanical cultivation is sprayed with chemicals similar to those used in the preemergence spraying.

Nematodes are the major sugar beet pests, but insects, viruses, and fungi are also damaging. A number of plant diseases have also been identified. Included are fungus diseases described as damping-off or blackroot; foliage diseases, such as curly top, cercospora leaf spot, sugar beet mosaic, and downy mildew; and root diseases, including wet root rot, southern sclerotium rot, and Aphanomyces root rot.

Composition

A typical compositional analysis of sugar beets is indicated below in Figure 4.5. Although only sugar beet roots are used in producing sugar, the tops and crowns of the plant represent a significant additional source of biomass. If sugar beets are to be considered as a major energy source, the tops and crowns must be collected. Estimated production of fermentable solids from sugar beet roots is only about 40% of that derived from sugarcane per unit of growing area per year.

Figure 4.5: Typical Composition of Sugar Beets*

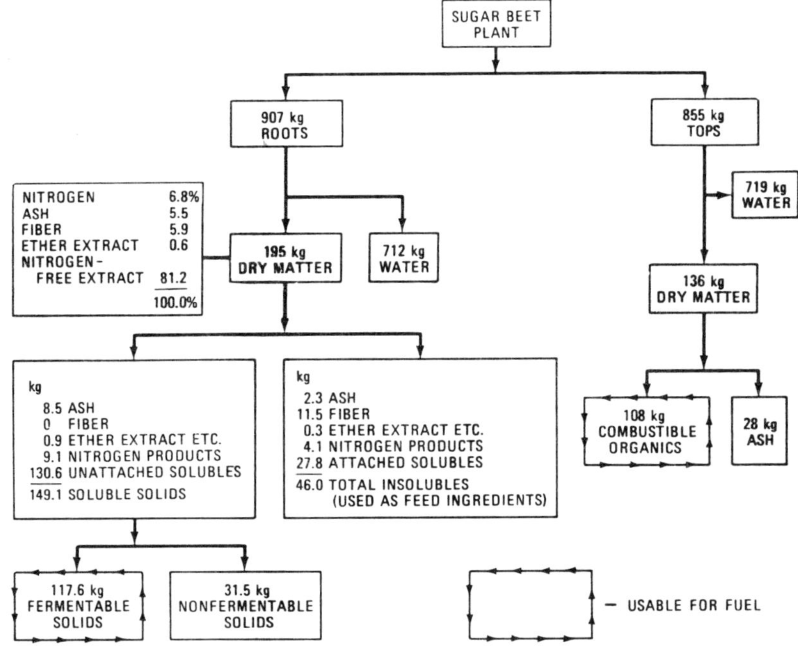

*Based on data from National Academy of Sciences (1971)

Source: NTIS TID-22781

Hills reports that 907 kg of fresh sugar beet roots are associated with about 855 kg of dry tops (Hills, Albaugh, and Pearl, 1955). Given a top dry-matter content of 15.9%, we can calculate that the fresh beet tops represent 136 kg of dry matter. Of this, 20.6%, or 28 kg, consists of ash. The remainder, or 108 kg, is combustible organic material having fuel value.

Estimation of the composition of sugar beet roots begins with an assumed 21.5% dry-matter content, which results in 215 kg per tonne of root dry matter. The recovery of beet pulp per tonne of roots is estimated to be approximately 5%

before the addition of molasses or Concentrated Steffen Filtrate (CSF) solids
(Norman, 1975). The constituents in the total insoluble portion also are identified in Figure 4.5 which is based on the reported composition of dried beet
pulp (National Academy of Sciences, 1971).

The 50.6 kg per tonne of total insolubles of beet roots is not considered a fuel
source since it is much more valuable as a cattle feed ingredient. The prevailing
price of molasses beet pulp at Pacific Coast markets was about $110/tonne as feed
or $142/tonne on a dry-weight basis (U.S. Department of Agriculture, 1975c).
This is a price of about $0.14 per kilogram of combustible organic material,
which is approximately 10 times the desired costs for this fuel raw material.

The quantity of soluble solids shown in Figure 4.5, along with the estimated
composition of these solids, is simply the difference between the composition of
the entire beet root and the beet pulp portion. It was assumed that 90% of the
solubles would be fermentable, which would result in 129.6 kg per tonne of
fermentable solids of beet roots.

Although these data are a reasonable approximate composition of sugar beets,
it must be realized that variations in composition will occur among sugar beet
varieties, with different cultural and harvesting practices, and with different efficiencies in beet processing operations.

Note also that everything in the dry-matter portion of sugar beet tops except the
ash has been categorized as combustible organic material. Actually, this material may contain significant quantities of fermentable solids, which would have
a higher dollar value than the fibrous portion.

Production Costs

Sugar beets are grown in widely dispersed regions throughout the United States;
thus we do not find the large contiguous areas of sugar beets for which sugarcane is noted. Any consideration of sugar beet biomass as an energy-chemicals
feedstock must include a production area sufficiently large to support a biomass
processing facility. With a beet-root yield of 45 tonnes/ha (fresh weight, average
U.S. yields), approximately 30,000 ha of land would need to be devoted to sugar
beets to supply a plant processing 500,000 tonnes of dry matter per year.

The USDA production cost budgets have been assembled for 19 sugar beet growing regions in the United States. These regions average about 27,500 ha with a
range of 5,250 ha in Arizona up to 62,750 ha in eastern Colorado. Naturally,
area within any given region varies from year to year.

The wide production area for sugar beets is characterized by different climatic
and cultural conditions; hence beet yields per hectare, as well as production costs,
vary considerably. Typical beet yields under current average technology range
from 31 tonnes/ha in the Red River Valley of Minnesota and North Dakota up
to 78 tonnes/ha in the coastal region of California. The range of production
costs is from $18.60/tonne in the California coastal region up to $31.50/tonne
in south-central Idaho. Median sugar beet production costs in the United States
are estimated at approximately $25.73/tonne.

Table 4.8 indicates a breakdown of estimated sugar beet production costs in three widely scattered geographic regions. These regions, although comprising less than 7% of the total U.S. sugar beet hectarage, represent some of the more efficient production areas in terms of cost per unit of output. Two of the three areas—the California coastal region and the Texas high plains region—represent irrigated production. In the humid area of southern Minnesota, sugar beets are grown under natural precipitation.

Table 4.8: Estimated 1976 Sugar Beet Production Costs in Selected Areas in the United States*

	Cost ($/tonne)		
	California Coastal Region	Texas High Plains	Southern Minnesota
Preharvest costs			
Seed	0.32	0.30	0.67
Fertilizer	1.21	2.71	4.66
Chemicals	0.47	0.42	1.47
Labor	2.79	3.68	1.89
Irrigation	1.07	1.23	—
Fuel and lubricants	0.17	0.41	0.43
Repairs	0.39	0.53	0.35
Interest on operating capital	0.31	0.45	0.41
Miscellaneous	—	—	0.14
Subtotal	6.73	9.73	10.02
Harvest costs			
Custom harvest	0.84	1.98	2.20
Labor	0.07	—	0.14
Hauling	1.51	1.32	2.20
Fuel and lubricants	0.02	—	0.10
Repairs	0.06	—	0.16
Interest on operating capital	—	—	—
Subtotal	2.50	3.30	4.80
Machinery ownership costs	1.56	2.52	2.27
Land charge	5.51	2.16	3.67
Management charge at 7% of gross receipts	2.31	2.70	1.93
Total costs	18.61	20.41	22.69
Yield per hectare, tonnes	84	49	36

*Based on data from unofficial U.S. Department of Agriculture FEDS crop budgets.

Source: NTIS TID-22781

Production costs and investment requirements in sugar beet production are affected by several factors: the quantity and quality of resources, such as land and management; institutional factors, such as the level of sugar beet allotments; and climatic factors.

For example, the size of a sugar beet farming operation is one important factor governing costs. Sugar beet production should be on a scale that will use efficiently the necessary machinery for production of the crop. A one-row harvester will harvest the beets on about 32 ha, and a two-row harvester, beets on 61 to 65 ha. A grower owning a two-row harvester with only 16 ha of beets would not be using his machinery efficiently unless he did custom work for neighboring growers. The same principle applies to planting, cultivating, and mechanical thinning equipment (Johnson et al 1971).

Sugar beet production is currently changing from the use of hand labor for thinning and weeding to the use of mechanical thinners and chemical weed control. The rate at which this new technology is adopted is one cause of variability in machinery, labor, and chemical costs among producing regions within the United States.

The wide disparity in land charges as a percentage of total costs is explained largely by regional variations. These variations are caused not only by the land's potential productivity for agricultural purposes but also by nonagricultural factors, such as potential for real estate development or recreation.

Alternative Uses of Sugar Beet By-Products: Beet Pulp — A by-product of sugar beet processing is dried beet pulp, a commonly used product for feeding dairy cattle. It is made by drying the residual beet chips from which the sugar has been extracted. Beet pulp is bulky and palatable and may be fed in either wet or dry form. Experiments have indicated that it is equal in ruminant nutrition to grain when used in moderate amounts (Communications Marketing, Inc., 1974).

Molasses beet pulp is another product consisting of beet pulp mixed with beet molasses and then dried. Molasses beet pulp is also very palatable for dairy cattle, beef cattle, and sheep.

On a solids basis the pulp produced from sugar beets will consist of (1) water and soluble components called marc, (2) water solubles left behind during the extraction or diffusion process, and (3) any solids added before drying. Molasses solids are added at nearly all American beet sugar plants.

The marc content can vary widely among sugar beet varieties. Although ranges from 2.7 to 6.1% have been reported, the normal range is 4.5 to 5%. Since the percentage values are based on the fresh weight of beets, any cultural or harvesting practice that affects moisture content will also affect the marc content.

The efficiency with which the diffusion operation is conducted will determine the quantity of soluble solids left in the pulp. The quantity will vary widely, although it is a relatively small portion of the total.

The quantity of added molasses has a marked effect on pulp production. The range will probably be from as little as 15% to as much as 35%.

The moisture content of the dried product will also bear on the quantity produced, as will any physical losses encountered in the pressing and drying processes. The normal recovery of pulp ranges from 5 to 8%, with 6.5% being a good average.

The average dry-matter content of dried beet pulp is listed at 91% (Communications Marketing, Inc., 1974). As of mid-August 1976, this product was selling at $165/tonne fob Chicago. The quoted price of molasses beet pulp at West Coast locations was about $110/tonne.

In recent years U.S. production of molasses and dried beet pulp has averaged about 1.36 million tonnes annually. Based on an average price of $116/tonne and an annual sugar beet production of 24.2 million tonnes, the average value of dried pulp has been approximately $6.50 per tonne of fresh beets.

Owing to the high value of beet pulp as an ingredient in cattle feed, it is not feasible to use it as a raw material for fuel produciton. Its current price as a feedstuff is five to ten times that of its potential value as a fuel raw material.

Sugar Beet Tops — If sugar beets are to be considered as a raw-material source for fuel production, beet tops must be collected; this, in turn, significantly reduces raw-material costs per tonne.

James, Weaver, and Reeder (1968) indicated that the yield of sugar beet tops bears little relationship to the yield of roots. In one experiment where the total dry weight of tops plus crowns was 12.3 tonnes/ha, root yields averaged 67.3 tonnes/ha. In another experiment the dry-weight yield of tops averaged 7.2 tonnes/ha, but the root yield averaged 82.9 tonnes/ha. Where high levels of nitrogen are available, much greater amounts of foliage are produced (Schmehl and James, 1971).

For the purposes of this review, it was estimated that 0.9 tonne of fresh beet roots (21.5% dry-matter content) would be associated with 0.82 tonne of fresh tops (15.9% dry-matter content); this is based on estimates by Hills, Albaugh, and Pearl (1955), who reported that tops produced per tonne of roots in California contain about 136 kg of dry matter. Data furnished by the National Academy of Sciences (1971) indicate a dry-weight content of tops of 15.9%.

Various ways of using beet tops include leaving them on the field as a source of organic material and fertilizer and as various forms of feed for cattle and sheep.

Increased yields of grain crops following sugar beets in a rotation have been noted. This is likely due to the plant nutrients, primarily nitrogen, that are returned to the soil in beet tops. The nitrogen content in dried beet tops is approximately 2.3%. This plant residue decomposes rapidly and thus releases nitrogen and other nutrients to the soil. A 67-tonne/ha beet yield is estimated to contain over 670 kg of nitrogen in the tops (Hills, Albaugh, and Pearl, 1955). In the fall of 1975, the average farm price paid for nitrogen fertilizers (approximately $0.40/kg) represented a nitrogen value for the beet tops of about $1.41 per tonne of fresh tops (about $8.82 per tonne of dry tops).

Although it is more common to simply leave the tops on the field, high-quality beet tops have a high feed value for cattle and sheep. Hills, Albaugh, and Pearl (1955) point out five main ways that beet tops can be used by livestock:

(1) Pasturing: Pasture feeding of beet tops by cattle and sheep is popular because of the low labor requirement. It is considered wasteful because weathering and trampling cuts the feed value considerably.

(2) Curing and hauling loose from the field: Beet tops will cure satisfactorily and can be stored under the same conditions as hay.

(3) Baling tops: Tops baled after curing in the field for 3 to 4 weeks contain about 20% moisture and can usually be stored successfully at this level.

(4) Silage: A tonne of beet-top silage (65 to 70% moisture content) will replace about 350 kg of alfalfa hay. At current

alfalfa hay prices, the value of beet-top silage is approximately
$22/tonne.

(5) Feeding green beet tops: There is limited information, however,
on this method of feeding. The high moisture plus the high
oxalic acid content of fresh leaves can cause some scouring
of cattle.

Fermentable Solids and Combustible Organic-Materials Costs: The steps followed
in deriving estimated raw-materials costs of fermentable solids and combustible
organic material from sugar beets are given in Figure 4.6.

Figure 4.6: Steps in Calculating Raw-Materials Costs from Sugar Beets

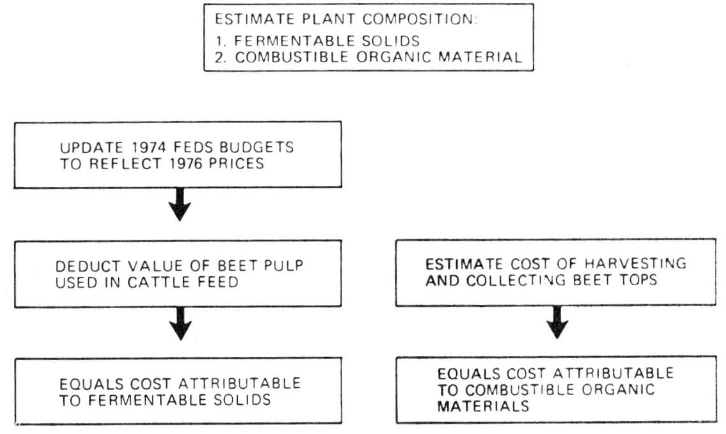

Source: NTIS TID-22781

The assumptions used in allocating total production costs between fermentable
solids and combustible organic material are indicated in the footnotes to Table
4.9. Note that an allowance for the value of beet pulp used as a cattle-feed
ingredient is deducted from total beet production costs. Also, an additional
charge is levied for harvesting, collecting, and transporting beet tops.

Even after credit is taken for beet pulp as an allowance against production costs,
the raw-materials cost of fermentable solids derived from sugar beets is generally
higher than that for fermentable solids derived from sugarcane. The estimated
raw-materials cost of fermentable solids contained in sugar beets is $0.101/kg
in the California coastal region, $0.115/kg in the Texas high plains, and $0.132/kg
in southern Minnesota. If we assume a desired raw-materials cost of $0.066/kg,
these estimates range from 50 to 100% above the desired level. In most other
sugar beet producing regions of the United States, these costs could be higher.

The raw-materials cost per kilogram of combustible organic residues ranges from
$0.022 in California to $0.042 in southern Minnesota. These values compare
with a desired cost of $0.0143/kg.

Table 4.9: Estimated 1976 Raw-Materials Costs of Fermentable Solids and Combustible Organics from Sugar Beets

Region	Production Costs ($/ha)		Yield (kg/ha)		Cost ($/kg)	
	Fermentable Solids*	Combustible Organics**	Fermentable Solids*	Combustible Organics**	Fermentable Solids*	Combustible Organics**
California coastal region	1,031	214	10,201	9,349	0.101	0.022
Texas high plains	734	185	6,412	5,874	0.115	0.031
Southern Minnesota	578	170	4,372	4,013	0.132	0.042

*Method of calculation: ~5% of total sugar beet production is pulp used in cattle feeds. Therefore 78 tonnes of sugar beets yields 3.9 tonnes of pulp, with a total value at $110/tonne of $429. Subtracting the value of pulp from total production costs per hectare determines fermentable sugars costs per hectare; e.g., California, total production cost of $1,460 – $429 = $1,031 costs attributable to fermentable sugars.

**Method of calculation: Beet yield per hectare x 0.94 = yield of fresh tops per hectare. Variable costs per tonne of harvesting beets x fresh tops yield per hectare + machinery ownership costs per hectare x 0.25 = total fiber costs per hectare; e.g. California, 78 tonnes x 0.94 = 73.3 tonnes of tops; (73.3 x $2.50) + ($123.28 x 0.25) = $214.07/ha

Source: NTIS TID-22781

On the basis of the preceding estimates, the California coastal region sugar beets appear to be competitive with most sugarcane-growing regions in terms of raw-material costs for producing fermentable solids. However, the quantity of fermentable solids obtained from beets, even in California, would be somewhat less than that from sugarcane in Texas, Florida, and Hawaii. The cost per pound of combustible organic material from the California coastal region sugar beets also would be comparable to that for sugarcane. Sugar beet costs for the Texas high plains and for southern Minnesota generally do not compare as favorably as those estimated for most sugarcane-producing areas.

It should be noted that the California coastal region accounts for only about 4% of current total sugar beet production in the United States. On the basis of these factors, sugar beets appear to have a lower priority than sugarcane as a potential energy source except in the highest yielding beet-production areas.

ENERGY BALANCE

The ultimate resources in classical economics are land, labor, and capital. Recently energy has joined the ranks of the ultimate resources. Because agriculture involves open systems in which irreversible reactions are carried out, the term energy balance is a misnomer. What is really meant by energy balance is the relationship between the energy inputs to the energy outputs of some more or less arbitrarily selected system. The methodology of energy input/output analysis is in a state of turmoil, in part because both the energy outputs and the energy inputs vary in both quality and scarcity. Many alternative accounting techniques measure both the input and the output for express profits and losses.

Starting from the premise that no fully satisfactory energy balances can be derived at this time owing to a combination of methodological problems and data availability-cost problems, this review uses data recently developed by USDA's Economic Research Service (ERS) under a cooperative agreement with the U.S. Federal Energy Administration (U.S. Department of Agriculture, 1974c). These data are based on information developed by ERS commodity specialists and are reviewed and updated by agricultural extension economists, engineers, and others in each production location.

Although this approach omits factors that may merit inclusion (the energy used to clear the land for farming and the energy required to manufacture farm machinery, for example), the ERS approach has the outstanding compensation that it involves an operationally meaningful procedure that can be repeated annually to create a useful time series. At present there is not sufficient information known about sweet sorghum agriculture to permit an analysis of its associated energy balance.

Sugarcane

The energy inputs for the production of sugarcane in Florida, Louisiana, Texas, and Hawaii are given in Table 4.10. In each of these geographical areas, the energy output (in terms of the energy contained in the biomass) far exceeds the energy inputs. The ratio of energy content generated to fuel value used ranges from 6.4 in Louisiana up to 9.3 in Texas, with a U.S. average of 8.4. These data are for millable cane only and do not include tops and leaves. These numbers need to be evaluated in light of the following factors:

- The Florida data are a composite of production on mucklands that require no fertilization and sandy and peat soils that require considerable fertilization.

- Florida production benefits from extensive flood control and water-table control by the U.S. Corps of Engineers.

- Hawaiian production is from a composite of irrigated and unirrigated land with quite different energy consumptions.

- Louisiana energy consumption data need to be averaged over a considerable period of time because of the impact of unusual weather considerations (e.g., hurricanes), which have a major impact on yields there.

- The Texas data represent a very small area only recently devoted to sugarcane production.

Despite the problems in these data, the energy contained in sugarcane grown where it grows well is so much more than the direct energy consumption that this crop appears to have some real potential as a source of fuel.

Invested energy (shown in Table 4.10) primarily is that used in manufacturing fertilizers. This represents a substantial portion (25% for all U.S. cane) of the energy input for the growth of sugarcane except for the Florida muckland area. In Hawaii over 36% of the total energy used is in the form of electricity used for irrigation purposes. Therefore the U.S. averages represent a smoothing of two extremes.

Table 4.10: Energy Used in Sugarcane Production by Form and by State*

| | Area Harvested (10^3 ha) | Yield (tonnes/ha) | Fuel Used (10^6 GJ**) | | | | | Total Energy Value Used | | |
State			Gasoline	Diesel Fuel***	Liquefied Petroleum Gas	Electricity	Invested Energy†	GJ/tonne (fresh weight)	(GJ/ha)	Total (10^6 GJ)
Florida	111	67.3	0.61	2.48	0.10	0.06	0.87	0.57	37	4.12
Louisiana	138	47.1	0.50	2.65	—	0.006	1.72	0.78	35	4.88
Texas	11.7	82.9	0.06	0.29	—	0.0005	0.15	0.54	43	0.51
Hawaii††	91	212.5	0.63	1.99	—	2.25	1.25	0.33	67	6.13
United States	350	77.3	1.80	7.43	0.10	2.32	3.98	0.60	45	15.63

*Based on preliminary data developed by the Economic Research Service, U.S. Department of Agriculture, under a jointly sponsored interagency agreement with the U.S. Federal Energy Administration, plus BCL estimates.

**1 gigajoule = 1,000 megajoules.

***Includes transportation from field to mill.

†Mostly fertilizers and pesticides.

††Calculated for an entire 24-month growing season. On an annual basis, the number of hectares harvested and total gigajoules would be one-half the designated values. Gigajoules per tonne and gigajoules per hectare remain unchanged.

Table 4.11: Energy Typically Used for Sugarcane Production, by Operation, in the United States*

| Operation | Fuel Source (MJ/ha) | | | | | | Percent of Total Energy Consumption |
	Gasoline	Diesel Fuel	Liquefied Petroleum Gas	Electricity	Invested Energy***	Total	
Preplant	—	1.30	—	—	—	1.30	7.3
Plant	—	0.55	—	—	—	0.55	3.1
Cultivate	—	0.83	—	—	—	0.83	4.7
Fertilizer application	—	0.11	—	—	—	0.11	0.6
Invested energy, fertilizer	—	—	—	—	5.01	5.01	28.4
Pesticide application	—	0.10	—	—	—	0.10	0.6
Irrigation	0.16	0.28	0.25	2.54	—	3.23	18.3
Harvest	—	2.35	—	—	—	2.35	13.4
Farm pickup	0.31	—	—	—	—	0.31	1.8
Farm auto	1.46	—	—	—	—	1.46	8.3
Electrical overhead	—	—	—	0.02	—	0.02	0.1
Transportation to mill	—	1.66***	—	—	—	1.66	9.4
Miscellaneous	0.17	0.54	—	—	—	0.71	4.0
Total	2.10	7.72	0.25	2.56	5.01	17.64	100.0

*Based on preliminary data developed by the Economic Research Service, U.S. Department of Agriculture, for the U.S. Federal Energy Administration, plus BCL estimates.

**Mostly fertilizers and pesticides.

***Assumptions: 77.3 tonnes/ha, 19.4 km average hauling radius, and 2.74 MJ/tonne-km.

Source: NTIS TID-22781

For the mainland states, over 57% of the total energy used in sugarcane production is for diesel fuel, whereas the usage of diesel fuel in Hawaii accounts for only about 33% of total fuel usage.

Table 4.11 shows the approximate energy use per hectare, by type of operation, for a typical U.S. sugarcane plantation. On this typical plantation (which is assumed to be irrigated), the major uses of energy are for fertilizer (28.4%), irrigation (18.3%), harvesting (13.4%), and transportation (9.4%). Energy consumed in cane planting, cultivation, and fertilizer and in chemical application accounts for only about 16% of total energy usage. Since planting is conducted only once every 3 to 4 years, energy consumption in this activity is reduced compared with that for an annual crop.

Sugar Beets

The energy consumption in the production of sugar beets in eight selected states is shown in Table 4.12. These states have been chosen because of their contribution to the total U.S. sugar output and to exemplify extremes in cultural practices. For example, the California data include considerable irrigated areas that achieve high yields and relatively low energy consumptions per tonne of beets produced.

In contrast, the Minnesota data represent production practices in which relatively less energy-intensive inputs (e.g., no irrigation) are used but lower yields are obtained and the fuel consumed per tonne of beets is higher. For the states selected, fuel consumption per tonne of beets produced ranges from 670 MJ in California up to 1,480 MJ in Nebraska with a U.S. average of 950 MJ. Part of the reason for the wide variation in energy consumption per tonne of beets is the difference in yields between producing regions.

Water availability (need for irrigation) is another factor, coupled with the availability of various types of fuels. Undoubtedly many other variations in climate and in cultural and harvesting practices affect energy consumption in sugar beet production. A closer examination of these differences was not possible in this study.

The estimated energy use per hectare, by fuel source, for various operations in sugar beet production is shown in Table 4.13. It is significant to note that fertilizer (31.3%), irrigation (27.9%), and transportation (18.8%) account for an estimated 78% of total energy consumed in sugar beet culture, harvest, and transportation. Thus the energy used is concentrated among fewer operations for sugar beets than for sugarcane.

The ratio of energy content generated to fuel used ranges from 2.4 in Nebraska to 5.3 in California, with a U.S. average of 3.8. This ratio is less than one-half that estimated for sugarcane (U.S. average, 8.4).

Although sugar beet production is less energy-intensive, its average yields (only about 50 to 55% as high as those for millable sugarcane) more than offset this advantage. Also, the dry-matter content of sugar beet roots is less than that of sugarcane (21.5% vs 27.2%); thus the costs of handling fresh plant material are increased.

Table 4.12: Energy Used in Sugar Beet Production, by Form, for Selected States*

State	Area Harvested (10^3 ha)	Yield (tonnes/ha)	Fuel Used (GJ) Gasoline	Diesel Fuel**	Liquefied Petroleum Gas	Natural Gas	Electricity	Invested Energy***	Total Energy Value Used (GJ/tonne)	(GJ/ha)	Total (10^6 GJ)
California	94.7	58.3	0.29	1.4	—	0.12	0.54	1.22	0.67	38	3.57
Washington	26.3	54.9	0.12	0.29	0.01	—	0.67	0.40	1.07	57	1.50
Idaho	38.0	45.1	0.18	0.43	0.03	—	0.42	0.74	1.06	48	1.80
Colorado	52.2	40.1	0.14	0.67	0.01	0.44	0.12	0.41	0.89	34	1.80
Minnesota	76.5	25.6	0.21	0.57	0.02	—	0.003	1.29	1.1	27	2.10
North Dakota	57.9	25.8	0.30	0.49	0.02	—	0.02	0.21	0.72	18	1.05
Nebraska	33.6	40.6	0.22	0.63	0.24	0.16	0.08	0.62	1.48	58	1.95
Michigan	33.2	38.1	0.10	0.30	0.01	—	0.001	0.59	0.81	30	0.99
United States (17 states)	506.7	41.0	2.13	6.18	0.43	1.4	2.29	6.59	0.95	38	19.07

*Based on preliminary data developed by the Economic Research Service, U.S. Department of Agriculture, under a jointly sponsored interagency agreement with the U.S. Federal Energy Administration, plus BCL estimates.
**Includes transportation from field to mill.
***Fertilizers and pesticides.

Table 4.13: Energy Used in Sugar Beet Production, by Operation, in the United States*

Operation	Fuel Source (MJ/ha) Gasoline	Diesel Fuel**	Liquefied Petroleum Gas	Natural Gas	Electricity	Invested Energy	Total	Percent of Total (MJ/ha)
Preplant	0.1803	1.1510	0.0506	—	—	—	1.38	8.6
Plant	0.0390	0.0896	0.0105	—	—	—	0.14	0.9
Cultivate	0.0095	0.0949	0.0032	—	—	—	0.11	0.7
Fertilizer application	0.0179	0.0148	0.0053	—	—	—	0.04	0.2
Invested energy fertilizer	—	—	—	—	—	5.02***	5.02	31.3
Pesticide application	0.0095	0.0116	0.0032	—	—	—	0.02	0.2
Irrigation	0.0685	0.3721	0.2656	1.2	2.6	—	4.50	27.9
Harvest	0.0032	0.3984	0.0011	—	—	—	0.40	2.5
Farm truck (on-farm use)	0.3310	—	—	—	—	—	0.33	2.1
Farm pickup (on-farm use)	0.2170	—	—	—	—	—	0.22	1.4
Farm automobile	0.7568	—	—	—	—	—	0.76	4.7
Electrical overhead	—	—	—	—	0.02	—	0.02	0.1
Transportation to processing plant	—	3.0071	—	—	—	—	3.01	18.8
Miscellaneous	0.0653	0.0580	—	—	—	—	0.12	0.8
Total	1.6980	5.1975	0.3395	1.2	2.62	5.02	16.07	100.0

*Based on preliminary data developed by the Economic Research Service, U.S. Department of Agriculture, under a jointly sponsored interagency agreement with the U.S. Federal Energy Administration, plus BCL estimates.
**Assumptions: 46.1 tonnes/ha, 16.1 km average hauling radius by truck, and 2.74 MJ/tonne for trucks, 242.5 km average hauling radius by rail at 0.48 MJ/tonne-km.
***Fertilizers and pesticides.

Source: NTIS TID-22781

Energy Input/Output

None of the preceding data on energy use and generation by sugar crops includes use of crop residues normally returned to the soil, namely, sugarcane tops and leaves and sugar beet tops. These residues obviously have some fuel value, but they would require additional use of fuel for their collection and transportation to a processing site.

As previously indicated, including sugarcane tops and leaves is assumed to increase cane yields, on the average, by 30%. The quantity of sugar beet tops accumulated per tonne of roots is known to be quite variable; however, the assumption was that the total fresh-weight yield of a sugar beet crop would be increased by 94% and the dry weight would be increased by approximately 70% if the tops were included (Hills, Albaugh, and Pearl, 1955).

Table 4.14 compares energy input/output for sugarcane and sugar beets with and without crop residues. The values, excluding residues, are based on yields of 31.3 tonnes of millable cane and 18.7 tonnes of sugar beets (roots) per hectare. In this instance the ratio of energy produced to energy used per acre is 8.6 for sugarcane and 4.1 for sugar beets.

Table 4.14: Comparison of Energy Balance of Sugarcane and Sugar Beets, Excluding and Including Use of Crop Residues

	Excluding Residues	Including Residues
. . . . Sugarcane.		
Energy content generated, MJ/ha	372,702	484,515
Fuel energy used, MJ/ha	43,505	47,003
Ratio of energy generated/energy used	8.6	10.3
. . . . .Sugar Beets		
Energy content generated, MJ/ha	161,216	258,360
Fuel energy used, MJ/ha	39,562	43,061
Ratio of energy generated/energy used	4.1	6.0

Source: NTIS TID-22781

It was assumed that, with residues included, the increase in fuel usage for the harvesting, transportation, and miscellaneous categories would be proportionate to the increased quantity of biomass to be handled. Under this assumption, only these categories of fuel charges were increased, by 30% for sugarcane, and by 70% for sugar beets. All other categories (such as planting, cultivating, and fertilizer) were assumed to remain unchanged in terms of fuel usage.

The increase in energy content generated was assumed equal to the increase in biomass resulting from residue collection. Based on these assumptions, the ratio of energy content generated to fuel value used for sugarcane becomes 10.3 when residues are included. The benefit was slightly greater for sugar beets, increasing from 4.1 to 6.0. It is important to again emphasize that these fuel input/output ratios can vary tremendously among producing regions. Under present cultural practices, sugarcane will generate significantly more energy per unit of area, with approximately the same energy input, than will sugar beets.

Although the USDA's energy-consumption data provide an excellent basis for a time series that will report the changing patterns of energy consumption with regard to source (diesel fuel vs gasoline vs natural gas vs coal) and the effects of energy conservation, these data must be used cautiously in forecasting the energy input/output of energy farms. The initial output of an energy farm is a large quantity of low-quality energy that requires further energy-intensive processing to achieve real use.

Battelle suggests that energy input/output be expressed as a ratio of products of comparable quality. Thus, if motor fuel is used to manufacture sugarcane, the comparison should be with either a motor fuel output or at least a synthesis gas output, not millable sugarcane. This concept is pursued further by Battelle Memorial Institute (1977a).

The USDA (1976c) and California (Cervinka, 1974) energy budgets include fuels but not the energy inputs to make the farm equipment. The Cornell and Kansas State researchers include farm equipment. The methodology is not advanced enough to make energy input/output results on equipment meaningful (Battelle Memorial Institute, 1977a).

GENERAL ROUTE FROM SUGAR CROP TO FUEL

The conversion of sugar crops into fuels and/or chemical feedstocks begins with the extraction of the sucrose-rich juice, leaving the fiber as a solid coproduct. So much energy is required to extract the juice that sugar mills have their own steam plants. Therefore the conversion section begins with a discussion of the capital and operating costs of juice extraction and the associated central steam and power station.

In the process cited below for making raw sugar, millable cane is produced by removing so-called tops and trash (both are highly fibrous and contain relatively little sugar). For maximum energy production, it is suggested that these components be retained and used for fuel production. Presently the millable cane is squeezed thoroughly by a capital-intensive and energy-intensive process to yield cane juice and bagasse. Sucrose is so much more valuable than bagasse that much energy and capital are expended for maximum recovery of sucrose. Capital investment could be reduced greatly and energy consumption within the sugar mill could be reduced substantially if the goal of sucrose maximization were replaced by a goal of optimizing sugar and fuel values.

In principle, the extraction of juice could be abandoned, with microbiological conversion conducted on chopped cane.

Two economic cases have been used in the calculation presented here (Battelle Memorial Institute, 1976b). The first is a low debt-to-equity case that assumes a debt of 30%, a borrowing rate of 8.5%, and a return of 15% on equity after taxes. This corresponds to the case of a large chemical company. Smaller, more highly leveraged firms are accommodated by the high-debt case, which assumes 60% debt, 8.75% interest cost and 14% yield after taxes. Many farm cooperatives in the fertilizer business would fit this case. In all cases straight-line depreciation over 25 years on steam plants and 11 years on other plants is assumed. Project life was taken as 25 years on steam plants and 20 years on other plants. A 24%

tax credit spread over 3 years has been assumed in all cases. The selection of a 24% investment tax credit and straight-line depreciation schedule leads to an annualized capital charge almost identical to the charge that arises from the selection of a 12% investment tax credit and sum-of-the-year's digits depreciation schedule. The 24% investment tax credit is about double the present credit but may be an incentive under consideration for a crisis situation.

Production of Juice and Fiber from Sugarcane

The production of fuels and chemical feedstocks from sugarcane is predicated on separate treatment of the sugarcane juice and the fiber so that the economic potential of sugarcane can be fully realized. Sugarcane is a highly seasonal crop except in Hawaii and Puerto Rico. Since land, water, and climate considerations have placed Florida's sandy soil region and Texas high on the priority list, a campaign method for the production of juice and fiber will be necessary. The maximum length of the campaign is likely to be about 180 days.

Obtaining enough juice and fiber to manufacture such chemicals as ethanol, methanol, and ammonia in quantities significant for U.S. energy goals would require a facility that could process approximately 7,000 tonnes/day on a bone-dry basis for 165 days per year. The composition of the sugarcane varies with location and variety. To obtain 7,000 tonnes/day on a dry basis would require 25,000 to 30,000 tonnes of sugarcane, tops, and leaves per day. Approximately 30,352 to 40,469 ha would be required to satisfy this facility.

The processing facility would consist of the elements indicated in Figure 4.7.

Figure 4.7: One Train of the Juice-Fiber Production Facility

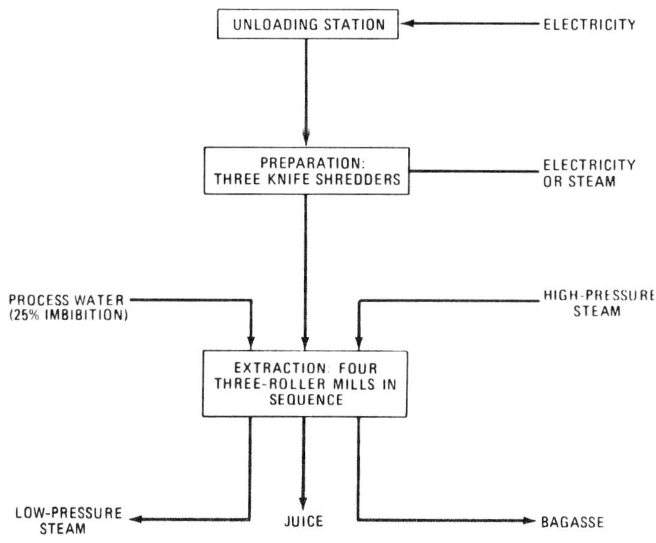

Source: NTIS TID-22781

There would be two such trains of 3,000 tonnes/day nominal bagasse capacity each because few economies of scale are obtained by making these individual elements larger. (Bagasse capacity has been used as the basis because the processing requirements are probably more nearly related to the fibrous content of the cane plus tops than to its sucrose content). The high-pressure steam would be generated in a separate central power station and steam facility, discussed later in this review. The low-pressure steam product is used to evaporate cane juice and/or operate the ethanol distillation facility. If the process water used in the extraction of the soluble solids from the sugarcane were wastewater from the ethanol plant, wastewater treatment costs might be reduced. Experimentation would be required to determine whether such a procedure is desirable if the reliability of the extraction process were at risk.

The aerial part of sugarcane would be harvested, loaded on trucks, and delivered through the unloading station at the processing facility. The sugar industry is generally not accustomed to handling tops and leaves in processing to obtain sugarcane juice. Experiments in Hawaii (Sloane and Rhodes, 1972) indicate that the succulent tops and leaves, which contain more fiber and chlorophyll and less soluble solids, cause problems. However, in Hawaiian areas where the cane leaves burn poorly, processors have learned how to modify operations to obtain reasonable efficiency. Therefore, although this use of sugarcane tops and leaves may cause some problems, successful modification of the equipment and operating practices is expected to surmount them.

The cane, tops, and leaves are prepared for extraction with three knife shredders (Meade, 1963), as is standard in processing millable cane. This approach uses less horsepower than busters and fiberizers. This facility is operated with the steam and electricity obtained from a power and steam station that operates on bagasse.

After the cane has been prepared, it is extracted by standard sugar-industry methods with the use of four three-roller mills in sequence. The quantity of imbibition water is well worked out for millable cane but will have to be determined for this mixture of cane, tops, and leaves. An important consideration in the selection of the amount of imbibition water is whether the juice is to be converted into food-grade sucrose through crystallization or to be converted into ethanol by fermentation. An extraction of 85% of the fermentable sugars with approximately 25% imbibition water is expected.

In accordance with standard sugar-industry practice, juice would be obtained for subsequent clarification and processing with about 16% solubles. Many uncertainties in the concentration and composition of this juice arise from the lack of commercial experience in processing tops and leaves along with normal cane. Also, the sugarcane fiber obtained would differ from conventional bagasse in its composition and properties in ways not currently known. This sugarcane fiber probably would have a moisture content of about 49%. If problems arise from the composition of the juice or from processing, the tops and leaves could be processed separately from the millable cane.

It is desirable to operate processes for converting sugarcane juice and sugarcane fiber into fuels and chemical feedstocks approximately 330 days per year to make efficient use of capital investment and labor. However, the sugarcane campaign can probably last only 165 to 180 days (unless lengthened with sweet sorghum). As described below, the potential feasibility of installing efficient storage capacity for sugarcane juice and/or sugarcane fiber to permit year-round operation has been considered. Another alternative would be to use sugarcane products during the harvesting and grinding season but to process wood products during the off-season.

Capital and Operating Costs: Estimates were made of the capital investment and operating costs for the sugarcane processing facility described. Estimates of the capital investment (see Table 4.15) were derived from information supplied in part by Hawaiian Agronomics Company. The battery limits plant (i.e., the main plant, excluding off-sites, storage, and service buildings) is based on the use of four of the three-roller mills in sequence, which may or may not give the estimated 85% extraction.

Table 4.15: Estimated Capital Costs for Cane Juice and Fiber Facility*

Scale: Feed (whole cane with tops and leaves) containing approximately 7,000 tonnes of solids on a bone-dry basis per stream day; 25,000 to 30,000 tonnes per stream day would be required on a wet basis.

Product: Bagasse, 6,000 tonnes, at roughly 49% moisture (3,060 tonnes of fiber with ash) and sugarcane juice containing 2,556 tonnes of fermentable sugars per stream day. In addition, 1,384 tonnes of nonfermentable solubles would be produced.

Basis: 165 stream days per year

Cost Element	Millions of 1976 Dollars	
Battery limits plant (BLP)	21.8	
Utilities (at 10% of BLP)	2.2	
Services (at 10% of BLP)	2.4	
Interest during construction	3.2	
Estimated total fixed capital	29.6	
Start-up expenses	0.3	
Working capital**	0.6	
Total estimated capital costs	30.5	
Annualized capital charge	4.84***	2.80†

*These are tentative estimates; they are intended to be only +50 or –30% accurate. Data based on information provided by Hawaiian Agronomics Company.

**Does not include inventory costs of products to be sold to other modules.

***Capital charges using low-debt economics.

†Capital charges using high-debt economics.

Source: NTIS TID-22781

The cost of the steam facility is in the central power station and is not included here. The working capital for this facility does not include storage of the products during the off-season; these stored products would be charged as raw-material inventories in the downstream facility for converting sugarcane juice and fiber into salable products.

Major operating and maintenance costs are listed in Table 4.16. High-pressure steam is assumed to be purchased from the central power and steam facility at $2.85/GJ. This transfer price covers fuel, depreciation, and labor and is competitive with fuel-oil steam plants. The spent low-pressure steam would be sold to the sugar crystallization or ethanol facilities for $1.90/GJ. Thus a net steam cost of approximately $1.7 million is incurred for this utility. If the sugarcane fiber used to raise steam were valued at incremental costs, the $2.85/GJ steam cost could be reduced, but the credit would have to be correspondingly reduced.

Table 4.16: Estimated Annual Operating and Maintenance Costs for Manufacturing Juice and Fiber*

Cost Element	Millions of 1976 Dollars
Steam (186,295 kg/m² at 263°C at $6.61/tonne)	5.0
Steam credit (24,605 kg/m² at 136°C at $4.41/tonne)	(3.3)
Net steam cost	1.7
Process water	**
Labor and supervision	0.2
Maintenance	2.1
Maintenance supplies	0.2
Payroll burden	0.3
Indirect costs	0.7
Property taxes and insurance	0.6
Total	5.8

*These are tentative estimates; they are intended to be only +50 or
   −30% accurate. Data based on information provided by Hawaiian
   Agronomics Company.
**A negligible cost element.

Source: NTIS TID-22781

Cost for utilities other than steam are negligible. Crushing, grinding, and extracting equipment is steam-operated, needing little electricity from the allied plant.

The labor and supervision part of operating costs assumes such employees will do maintenance work in off-season, according to industry practice. Supervision is calculated at 15% of operating labor. Work crews include at least 2 weighers, 2 unloaders, 4 workers on preparation and milling trains, 1 laboratory person during each shift, plus 6 men on shop backup on day shift. Wages and fringes are calculated at $8.10/hr, between sugar-industry and chemical-industry standards.

Maintenance is estimated at 7% of fixed capital, with 60% for materials, 40%, labor. Maintenance labor supervision is 20% labor. These costs, $2.1 million per year, reflect the importance of maintenance to good operation. Maintenance supplies, payroll, indirect costs, property tax, and insurance were estimated with standard cost engineering factors, with supplies at 20% of materials, payroll burden at 30% of payroll, and indirect costs, 40% of labor, supervision, maintenance, and maintenance supplies. Charges for property taxes and insurance are based on a standard 2% of total fixed capital investment.

The total of the cost elements for operation and maintenance is $5.8 million per year. Capital charges are $4.8 million for the low-debt economics and $2.8 million for the high-debt economics. As shown in Table 4.17, total processing costs are $8.6 to $10.6 million.

Table 4.17: Estimated Product Costs for Juice and Fiber Facility*

Cost Element	Low-Debt Economics	High-Debt Economics
	...... MM$ 1976	
Agricultural production		
Fermentable sugars at $102/tonne	43.2	43.2
Fiber at $29/tonne	14.5	14.5
Juice processing		
Operating costs	5.8	5.8
Annualized capital cost	4.8	2.8
Total annual costs	68.3	66.3
	.... $/tonne	
Initial allocation of costs		
Fermentable sugars**	128	128
Fiber	29	29
Reallocation of costs		
Fermentable sugars***	133	133
Fiber	25	20

*These are tentative estimates; they are intended to be only +50 or −30% accurate.

**Juice carries all processing costs.

***Juice carries all processing costs, and a transfer pricing policy makes fermentable sugars available at $133/tonne. The fiber transfer price is reduced correspondingly.

Source: NTIS TID-22781

Total costs can be estimated as follows: When the processing costs are allocated solely to the fermentable sugars, these sugars would have an interest transfer price of about $128/tonne. The fiber would have a cost of $29/tonne under present conditions. If a transfer price for fermentable sugars of $133/tonne were established, fiber costs could be brought close to $20/tonne in the high-debt case.

Central Station Power: The generation of electric power is considered feasible for sugar mills (Paturau, 1969); the mills need the electricity, the low-pressure steam for use in the boiling house, and medium-pressure steam for the steam turbines that drive the crushers. The combustion of the bagasse also serves to dispose of an otherwise nuisance material. Sugar-mill boilers usually are not designed for commercial electricity production, because the utility serving the area pays only for the fuel saved by obtaining the electricity from the mill. The reason is that, in temperate climates, the mill operates during only about half the year. The rest of the year the utility must supply the load, and thus the purchase of mill-generated power saves the utility little, if any, on its capital investment.

In Hawaii the situation is different because of nearly year-long grinding and lack of cheap, indigenous fossil fuel. In Hawaii, sugar mills supply 16% of the electricity distributed.

For this review the cost of producing electricity from 2,468 tonnes of dried bagasse per day for 330 days/year was determined (Tables 4.18 and 4.19). Operation for 330 days was chosen to match the ethanol plant described below which would use the low-pressure steam. In a sense the ethanol plant serves as a condenser for the system. Depending on the season, the cost was found to be 24 to 35 mills/kW-hr for a 28- to 36-MW power plant (a small but commercially viable size).

Table 4.18: Estimated Capital Cost for Central Power Station*

Scale: Feed (bagasse dried to 38% moisture), 2,468 tonnes per stream day, 1,530 tonnes per stream day on a bone-dry basis.
Basis: 330 stream days per year.
Products: Steam for crushers and ethanol plant.
Electricity: 28 to 36 MW.

Cost Element	Millions of 1976 Dollars	
Storage, conveyors	4.3	
Boilers	17.4	
Turbogenerators, including building	7.8	
Water-system piping	1.4	
Steam piping (main)	1.2	
Electrical	1.4	
Site (land and preparation)	1.2	
Makeup water	0.1	
Building services	0.1	
Process-steam system	0.1	
Stack	0.3	
Foundations	1.3	
Substation	1.2	
Transmission line	0.3	
Interest during construction	4.5	
Estimated total fixed capital	42.6	
Start-up costs, 1-month output	0.6	
Working capital	1.2	
Total estimated capital cost	44.4	
Annualized capital charge	8.2**	5.5***

Note: Footnotes are given at end of Table 4.19.

Table 4.19: Estimated Annual Operating and Maintenance Cost for Central Power Station*

Cost Element	Millions of 1976 Dollars
Water, raw for cooling	0.052
Water, boiler feed	0.002
Chemicals	0.055
Labor	0.686
Supervision (18.4% of labor)	0.126
Maintenance materials	0.630
Maintenance labor	0.350
Maintenance supervision	0.070
Payroll burden (30%)	0.370
Operating supplies	0.126
Administration and general overhead	0.795

(continued)

Table 4.19 (continued)

Cost Element	Millions of 1976 Dollars
Property taxes and insurance (2% of fixed cost)	0.700
Gross total	3.960
Credit for steam to crusher and ethanol†	(10.237)
Net cost	(6.277)

*Tentative estimates intended to be only +50 or -30% accurate, based on data from Hawaiian Agronomics Company.
**Capital charges using low-debt economics.
***Capital charges using high-debt economics.
†At \$6.61/tonne for 18.6 kg/cm^2; \$4.41/tonne for 2.46 or 4.57 kg/cm^2.

Source: NTIS TID-22781

The ethanol plant has a maximum need for 320,690 kg of steam per hour; the electric plant will produce 283,950 kg/hr. During the season when the fresh juice is sterilized and concentrated, some supplementary fuel may be needed. The steam is assumed to be cascaded from the electric plant to the crusher to the ethanol plant as shown in Figure 4.8.

Figure 4.8: Flow Diagram of the Central Power Station

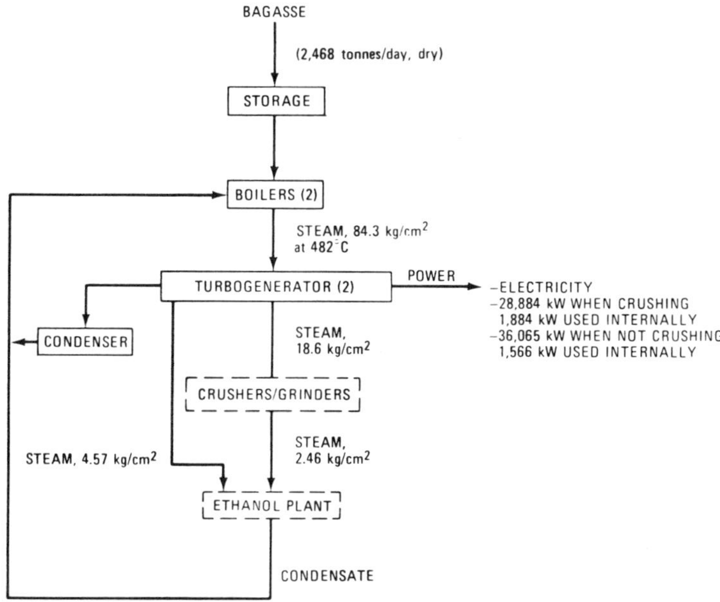

Source: NTIS TID-22781

The requirements of the crushers are small compared with those of the ethanol plant. Comments on this imbalance appear in the ethanol discussion.

The overall thermal efficiency is on the order of 80%. If only electricity is generated in an industrial boiler of the type contemplated, efficiency is about 25%.

Ethanol by Fermentation—Microbiological Conversion Process

When criteria other than costs are applied, ethanol ranks as a high-priority product for both fuel and chemical feedstock application. It is a clean fuel concentrating about 30% more energy than methanol. The conversion of simple sugars into ethanol represents only a small thermodynamic loss and concentrates almost 80% of the energy contained in the glucose in half the weight of the final product.

Alternative Sources of Sugars: Sugarcane, sweet sorghum, and sugar beets can be used to make fermentable sugars for ethanol manufacture via three raw materials: (1) extracted juices primarily from the cane stalk or beet root, (2) sugar-rich syrupy residue from the crystallization of sucrose known as molasses, and (3) total saccharides derived from hydrolysis of the cellulose and hemicelluloses of the fibrous portion of the sugar crop. Each of these sources has advantages and disadvantages.

The juice that is squeezed from the sugar crop as the first step in the preparation of crystalline sucrose represents a prime starting material for ethanol fermentation. The major disadvantage of using the juice as a raw material is that it is subject to spoilage unless it is concentrated to roughly molasses strength. Alternatives to the evaporation process were considered: heat sterilization, chemical sterilization, and ultrafiltration to remove microbes. However, the quantity of liquid stored would result in high storage costs, and contamination dangers would be substantial.

Molasses is a sugar-manufacture by-product that can be stored during the off-season. The industrial organization that makes both sucrose and molasses usually can obtain a higher return for its sucrose-rich juice by selling sucrose and still having molasses to use as a fermentation starting material.

Approximately 40% of the sugars contained in the sugar crops is represented by polymers in the fibrous residue. They are relatively difficult to hydrolyze. Thorough exploitation of the sugar value of the sugar crops logically would involve hydrolysis of the polymers and use of the combined mixture of all the sugars. The major point at issue is whether this logical plan can be executed economically.

In the following sections each of these potential raw materials is considered in turn. The scale of operation and time horizon for completion of development efforts are important considerations.

Sugar-Crop Juices — The compositions of the juices extracted from sugarcane, sweet sorghum, and sugar beets have subtle differences that affect crystallization properties for the manufacture of sucrose. For example, sweet sorghum has enough starch and aconitic acid to cause crystallization problems. However, these differences are not expected to have a major effect on fermentation. Therefore the three juices are considered interchangeable in the discussion of ethanol manufacture.

To the cost of the fermentable sugars must be added the costs of processing to make the juice. For sugarcane these costs are approximately $0.022/kg. The costs of processing sweet sorghum in existing sugarcane mills will probably be very nearly equal to those of sugarcane. The principal technical difference is the need to strip the leaves from the sweet sorghum stalks. It is unlikely that the costs will be lower. The extra costs reported (U.S. Department of Agriculture, 1976d) for processing sweet sorghum arise in preparing the juice for crystallization. Since the added processing is not necessary here, it seems appropriate to consider costs of sweet sorghum and sugarcane to be equal in estimating manufacturing costs. Similarly, in the estimation of beet juice fermentable-solids costs, it appears appropriate to take a value equal to that of sugarcane juice.

Molasses — Historically, molasses has been a raw material for ethanol manufacture both for rums and for industrial alcohol. The alternative use as animal feed gives molasses the greatest value.

The animal feeder probably will outbid the distiller up to the point where corn grain is as cheap as molasses on a delivered nutritional-equivalent basis. Even at that price the fact that molasses is a palatability aid for feeding urea to ruminants means that the cattle feedlot operator will be willing to pay still more if the price of soybean meal stays high.

A further problem in the use of molasses as a fermentation feed is that its price can fluctuate widely, depending on world sugar and U.S. grain prices. Since there is no futures market in molasses, hedging would require using sugar and/or corn futures without assurance of complete protection.

The future decisions of the ethanol venture need not be dogmatic with regard to juice vs molasses. Juice may be most economical during the grinding season, and molasses may be a very desirable supplement during the off-season.

Simple Sugars from Total Saccharides — Juice obtained from sugar crops contains only about half the simple sugars that could be derived from the entire biomass. The remaining sugars are in the form of polysaccharides, such as cellulose and hemicelluloses. The lignin and ash of sugar crops are the only major constituents that could not be converted to simple sugars. With the alternative use of bagasse primarily in low-value fuel applications, it appears appropriate to examine the conversion of the polymeric saccharides into simple sugars for fermentation into ethanol.

The potential production of glucose (and other simple sugars) from cellulosic natural products has been studied intermittently for more than 60 years (Meller, 1959; Harris, 1949). Early attempts in the United States and Germany used acids that depolymerized the polysaccharides. The fast rate was an advantage that was offset by overreaction of the simple sugars formed early in the depolymerization to yield sensitive furan intermediates that repolymerize into messy tars (Harris, 1975).

The field was rejuvenated when elegant research at U.S. Army Natick Laboratories led to the isolation of mixtures of cellulase enzymes that yield sugars without a tendency to form degradation products (Mandels et al, 1975; Spano et al, 1975). Following these basic discoveries, numerous research teams throughout the world have pursued the enzymatic approach with results that have great scientific interest (Wilke, 1975; Bailey et al, 1975).

The purity of product obtained from the *Trichoderma viride* process is gained at the cost of increasing reaction times from the few minutes required for acid hydrolysis to 1 or 2 days required for cellulase hydrolysis. Such slow reaction times are related to specific biochemical events that will be difficult to speed up. Thus one enzyme appears to convert crystalline cellulose into amorphous cellulose. Only then can a second enzyme reduce the degree of polymerization to two (cellobiose). A separate, highly specific enzyme is needed to convert cellobiose to glucose.

Many so-called glucose solutions are poor substrates for ethanol-producing microorganisms because the solutions contain a substantial quantity of cellobiose. Unless the enzyme preparation made from the fungus *Trichoderma viride* is strong in all three enzymes, a slow reaction will occur.

The extended reaction time of 1 to 2 days applies to cellulose that has been ball-milled to an extremely small particle size (approximately 25 to 50 μm). These small-size particles can be rapidly attacked by the appropriate cellulase, but the cost of ball-milling is considered prohibitive (Brandt in Wilke, 1975). Furthermore, bagasse is not pure cellulose; it consists of numerous polysaccharides, including cellulose, protected with lignin. Lignocellulose is even slower to convert to glucose enzymatically than is pure cellulose, and the yield is lower. Hemicellulose constituents of bagasse are not affected appreciably by the highly specific cellulases. Rather, xylanases, pectinases, and other enzymes need to be present in the mixture to depolymerize hemicellulose.

The production of a wider variety of enzymes in *Trichoderma viride* appears possible, at least to a limited extent, by growing the *Trichoderma viride* on bagasse instead of on purified cellulose (Andren, Mandels, and Medeiros, 1975). Nevertheless, the reactions become very involved, and the ability of ethanol-producing microorganisms to use the product diminishes. For example, research conducted by Natick researchers (Andren, Mandels, and Medeiros, 1975) showed that ball-milled sugarcane bagasse yielded approximately 20% glucose and 8% xylose in 48 hr of hydrolysis.

If the bagasse was used without ball-milling, only 5% glucose and 1% xylose were obtained. Sugarcane medulla yielded only 11% glucose and 4% xylose even when ball-milled. A cruder measure of simple sugar production indicated that more sugars are present, but bagasse still stands well below most types of wastepaper. The chemical constitution of bagasse, which is more complex than wood, increases the complexity of the process and reduces the yields and desirability of the products.

Enzymatic hydrolysis processes are at such an early stage of development that only speculative cost evaluations can be made. The Natick group has established a pilot plant to gather some of the pertinent information. A major breakthrough in enzymatic activity or in pretreatment processes to circumvent some of the slow enzymatic steps may be needed, however, before a commercial process can emerge. When the needed breakthroughs occur, so many factors may change that current estimates of costs would be obsolete.

Beginning from such a speculative base, Wilke and his coworker (Wilke and Yang, 1975) developed cost estimates (excluding raw-materials costs) of $0.145/kg. When a raw-material cost of $22/tonne is added, the costs of simple sugars rise to $0.23/kg. The base which was considered by Wilke and Yang does not include taxes and a number of other factors that would be considered in an industrial evaluation. When these are included and costs are updated to 1976, simple sug-sugars derived from a wastepaper source will probably have a manufacturing cost of at least $0.33/kg. These costs are two to three times those expected for cane-juice sugars. A bagasse starting material would have a higher unit cost because a lower yield for simple sugars at a lower rate would be expected.

This leads to the conclusion that, before the total-saccharides concept can be considered for sugar crops, pretreatment and/or enzyme technology must be substantially improved (Millet et al, 1975). Opportunities to apply current enzyme technology to highly selected wood-pulping operations and wastepaper products have different prospects because the percentages of lignin and hemicelluloses in such products are very low compared with those for bagasse.

This negative evaluation of the initially attractive enzymatic route from cellulose and hemicelluloses to simple sugars leads to a reconsideration of acid hydrolysis. Harris (1975) discusses the mechanism of acid hydrolysis of cellulose. Research at Dartmouth by Grethlein (1975) includes some experiments and speculative economics. Both authors discuss wood- or refuse-based (not bagasse-based) processes.

Grethlein (1975) reported a 50 to 60% conversion of cellulose to glucose by first raising the temperature of a slurry of paper to over 200°C and then adding strong acid. The reaction is conducted in a plug-flow device that makes it easy to quench the products within 1 min of initiation of the reaction. Thus knowledge of the mechanism is used to reduce the quantity of overreaction products. Unfortunately the reported reaction products are both impure and in dilute solution. Nevertheless, this approach appears at least as promising as the elegant enzymatic method because it is potentially much cheaper.

Therefore neither acid nor enzymatic hydrolysis of bagasse for making fermentable sugars for ethanol production seems likely to be attractive in the near future. The long-range opportunities (past 1985) are distinctly better, but near-term decisions regarding the sugar crops to sugars to ethanol chain should rely on cane juice for the simple sugars.

Flow Sheet: A sample ethanol process was conceptualized with the use of fermentation literature (Paturau, 1969; Underkofler and Hickey, 1954; Scheller and Mohr, 1976; and Kampen, 1975). The block flow diagram (Figure 4.9) illustrates schematically the conversion of approximately 11,067 tonnes of raw sugarcane juice per day at 15.79° Brix (1,278 tonnes of fermentable sugars per day) into 784,705 liters of 190-proof ethanol per day. The anhydrous ethanol needed for fuel use (especially in cold climates) will entail additional processing. In addition to producing ethanol, the facility manufactures, per day, about 139 tonnes of yeast (50% solids), approximately 1,088 tonnes of wet stillage (50% solids), about 566 tonnes of carbon dioxide, and approximately 2.2 tonnes of fusel oil.

Figure 4.9: Ethanol by Fermentation

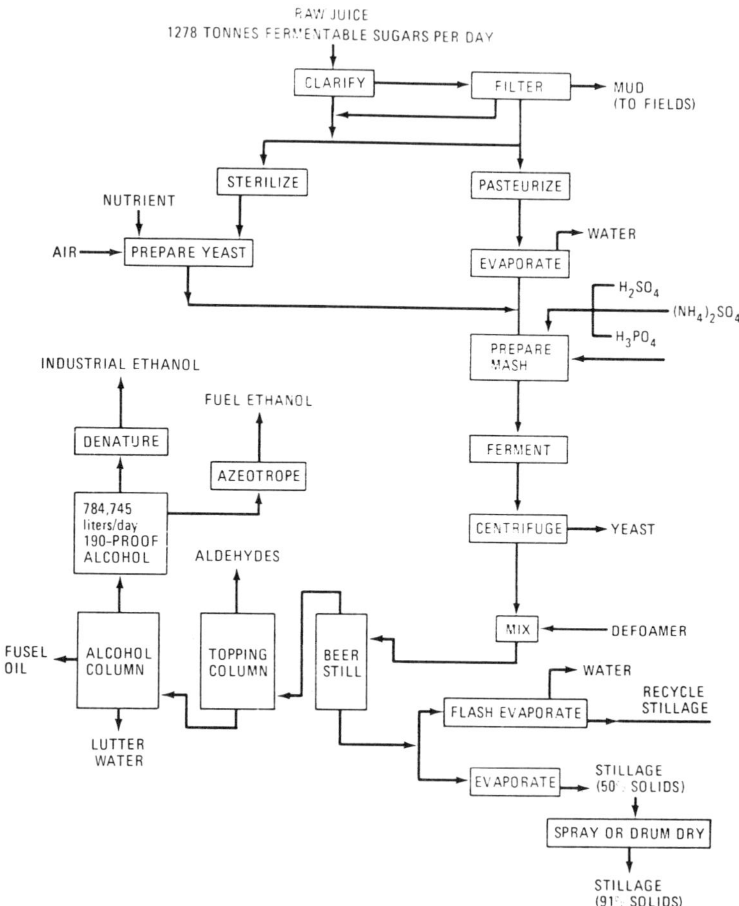

Source: NTIS TID-22781

Not all products are salable separately. For example, the yeast and wet stillage are combined and dried to 9% moisture and sold for animal-feed uses.

First, raw juice is clarified by a slight variation on standard sugar-industry practice. The variation consists in not neutralizing the somewhat acidic juice with lime (Meade, 1963), because the fermentation is run at approximately pH 4. Yeast is prepared under sterile conditions, but the mash is prepared and fermented with pasteurized, clarified cane juice. Because of the expense of distillation following fermentation, the concentration of the input to the mash is made as high as possible without impeding the rate of fermentation.

After approximately 40 hr of fermentation (Underkofler and Hickey, 1954), the product is centrifuged and subjected to initial ethanol concentration in a beer still. Since the fermentation does not use up all the nutrient values in the mash, some of the stillage is recycled into the fermenter. The exact percentage will be determined by experimentation. Ultimately, the spent yeast and stillage are converted into a product relatively high in ash and low in protein. The natural tendency of liquid to foam in the beer still will be combated with appropriate defoaming agents, and the ethanol will be concentrated and purified in separate topping and alcohol columns.

Estimated Capital Cost for Ethanol Plant: Based on this conceptual design, the estimated capital cost for the ethanol facility would be approximately $126.8 million. The usual contingencies, interest during construction, working capital, start-up expenses, and similar charges more than double the capital cost of the battery limits plant. This capital investment may be somewhat on the high side because of economies of scale in combining the utilities and services for the mill (the same roads, cafeteria, laboratory, etc.).

The fermentation and distillation sections of the plant are the major cost elements. The section involved in juice evaporation is the most amenable to elimination. Optimization of the imbibition process might achieve the right concentration at the outset.

Estimated Operating Costs: The annual operating costs for the ethanol facility are estimated at $22.2 million (Battelle Memorial Institute, 1976a).

The major chemicals cost for industrial ethanol is for denaturing so that the product can be sold only for nonbeverage uses. The major denaturing chemicals now used in industrial alcohol production are ethyl acetate, methanol, benzene, brucine salts, and Bitrex tertiary-butyl alcohol. On the average the cost of denaturation is approximately $0.005/liter. A fuel-ethanol product could be denatured with gasoline at the point of production for a much lower cost.

The largest single operating cost is the purchase of steam from the steam and power facility that operates on bagasse. Compared with steam, other utilities represent a small cost. In this calculation the steam is purchased at $4.41/tonne. Charges for steam in an integrated facility are somewhat arbitrary. The $4.41 per tonne is a projection of the cost of steam made from low-sulfur coal.

Approximately 480 man-hours per day are required for operating the 55 fermenters and relatively massive distillation unit. Supervision is taken at a standard 15% of operating labor. The cost of cooling water for the process depends greatly on water costs and on the specific design of the multifermenter facility. Fortunately, only the yeast-production unit needs sterilization. Cooling-water costs can be held to less than $0.5 million if heat economy is designed into the facility and if low-temperature cooling water is available.

Working Capital Estimates: The major elements of working capital are in the storage of the sugarcane-juice feedstock over the 165-day off-season, 30-day inventory of ethanol, and 30-day accounts-receivable period. Other working capital requirements are quite small. The major problem arises in the juice feedstock storage. This cost is a major penalty that a seasonal field crop must pay compared with municipal solid waste as a source of fuels.

Consolidated Manufacturing Costs: The net manufacturing cost of 190-proof ethanol is estimated at $78.5 to $88.2 million/year (depending on assumptions on debt level), or about $0.31 to $0.34/liter. The capital and operating charges are those derived in the previous sections. Sugarcane juice is estimated to cost $0.13 per kilogram of fermentable solids which is the equivalent of almost $56 million/year. The importance of agricultural improvements in production and efficiencies in steam production and use is apparent when the magnitude of juice costs is compared with that of the other cost elements. The next most costly input is steam. Fuel-blend use of the ethanol probably will necessitate production of anhydrous ethanol at an additional cost of $0.085/liter, according to Miller (1976).

Stillage Credit: In general, the manufacture of ethanol by fermentation includes coproduction of protein-rich vitamin-rich animal-feed ingredients. The successful isolation and marketing of these products are necessary conditions for achieving low net costs for ethanol. A by-product credit of about $13.4 million ($66/tonne) is estimated for the stillage derived from fermentation of sugarcane juice.

CONCLUSIONS

This comprehensive evaluation and review of sugar crops has resulted in the following conclusions.

Of the three sugar crops discussed, sugarcane is the most promising in the near term, sweet sorghum will gain promise in the future, and the sugar beet is so unpromising as to warrant dropping it from the fuels-from-biomass program. The primary attraction of sugar crops is that they can provide fermentable sugars at low cost, especially during the harvesting season. The fibrous residue (bagasse) also is extremely cheap during the processing season. It is unlikely that corn, trees, or other biomass sources will be as inexpensive when full collection and transportation costs are included.

The major drawbacks of sugar crops, especially sugarcane, are limited availability and seasonality on the mainland. Sweet sorghum has prospects for overcoming the limited availability problem if a long-term development program succeeds, but it still is highly seasonal. Sugar crops are capable of producing 0.1 to 0.5 quad of energy within a time horizon of several decades. Thus there is a cost-availability trade-off with sugar crops. When full correction is made for storage costs of fermentable sugars and/or bagasse for year-round conversion to chemical feedstocks to fuels, the cost advantage of sugar crops vs corn or trees is diminished.

New agricultural practices involving close-row spacing hold promise for increasing the sugarcane crop and production of fermentable sugars and bagasse by large margins. These crop increases would increase the availability and reduce the cost of this resource. Perhaps 20 to 40% of sugarcane biomass is now left in the field as trash or is burned. This additional biomass can be made available for processing into fuels and chemical feedstocks.

The U.S. sugar industry is developing a strong motivation to explore new markets because competition from foreign sugar producers and from domestic corn-sweetener manufacturers has led to unstable market conditions and to low table-sugar prices. Among the new markets that might alleviate the industry's oversupply problem is that of ethanol produced via fermentation.

The major unfinished business in the use of sugar crops as a resource for making fuels and chemical feedstocks is to increase the availability of sugar-crop biomass, to reduce the cost of fermentable sugars and bagasse, and to generate solid data (not speculations or extrapolations) on the feasibility of alternative processing and conversion routes.

REFERENCES

Alpert, S.B., et al, 1972, *Pyrolysis of Solid Waste: A Technical and Economic Assessment,* pp. 218-231, Stanford Research Institute.

American Petroleum Institute, 1976, *Alcohols—A Technical Assessment of Their Application as Fuels.*

American Public Works Association, 1963, *Municipal Solid Waste,* Proceedings of the National Conference on Solid Waste Research, December 1963, p. 37.

Anderson, J.E., 1973, U.S. Patent No. 3,729,298.

Andren, R.K., M.H. Mandels, and J.E. Medeiros, 1975, "Production of Sugars from Waste Cellulose by Enzymatic Hydrolysis. I. Primary Evaluation of Substrates," *Proceedings of the Eighth Cellulose Conference. I. Wood Chemicals—A Future Challenge,* T.E. Timell (Ed.), pp. 205-219, John Wiley & Sons, Inc., New York.

Anonymous, 1974, "LSU Promotes Protein-from-Cellulose Process," *Chem. Eng. News,* 52(7): 20.

Anonymous, 1976a, "Petro-Protein Stalled on Purity Issue," *Chem. Week.,* 118(17): 79.

Anonymous, 1976b, "Propylene Supply Tightens, Prices Rise," *Chem. Eng. News,* 54(38): 10.

Anonymous, 1976c, "Ammonia Outlook Brightens Considerably," *Chem. Eng. News,* 54(22): 8.

Baba, T.B., and J.R. Kennedy, 1976, "Ethylene and Its Coproducts: The New Economics," *Chem. Eng.,* 83(1): 116-128.

Bailey, M., T-M. Enori, and M. Linko (Eds.), 1975, *Symposium on Enzymatic Hydrolysis of Cellulose,* Aulanko, Finland, Mar. 12-14, 1975. Sponsored by the Finnish National Fund for Research and Development (SITRA).

Bailie, R.C., 1972, *Final Report on High Energy Gas from Refuse Using Fluidized Beds,* under Environmental Protection Agency Grant 5R01 EC 00399-03 EUH.

Baratz, B., et al, 1975, *Survey of Alcohol Fuel Technology,* Volume 1 (November 1975), prepared for the Office of Energy Research and Development Policy, National Science Foundation.

Barnes, A.C., 1974, *The Sugar Cane,* 2nd ed., Halsted Press, Division of John Wiley & Sons, Inc., New York.

Bartha, I., 1965, "The Production of Biogas and Biofertilizer from By-Products of Sugarcane Processing," *Sharkara,* 7: 70-76.

Battelle Memorial Institute, 1976a, *Systems Study of Fuels from Sugar Cane, Sweet Sorghum, and Sugar Beets, Agricultural Considerations,* ERDA Report BMI-1957 (Vol. 2).

Battelle Memorial Institute, 1976b, *Systems Study of Fuels from Sugar Cane, Sweet Sorghum, and Sugar Beets, Conversion to Fuel and Chemical Feedstocks,* ERDA Report BMI-1957 (Vol. 3).

Battelle Memorial Institute, 1977a, *Systems Study of Fuels from Sugar Cane, Sweet Sorghum, and Sugar Beets, Comprehensive Evaluation,* ERDA Report BMI-1957 (Vol. 1).

Battelle Memorial Institute, 1977b, *Systems Study of Fuels from Sugar Cane, Sweet Sorghum, and Sugar Beets, Corn Agriculture,* ERDA Report BMI-1957 (Vol. 4).

Battelle Memorial Institute, 1977c, *Systems Study of Fuels from Sugar Cane, Sweet Sorghum, and Sugar Beets, Comprehensive Evaluation of Corn,* ERDA Report BMI-1957 (Vol. 5).

Becher, P., 1976, "Costs Prohibit Cellulosic Use as Feedstock," *Chem. Eng. News,* 54(14): 12.

Blum, D.B., et al, 1975, "Liquid-Phase Methanation of High Concentration CO Synthesis Gas," *Methanation of Synthesis Gas,* R.F. Gould (Ed.), pp. 149-159, American Chemical Society, Washington, D.C.

Breaux, R.D., R.J. Mattherne, R.W. Millhollon, and R.D. Jackson, 1972, *Culture of Sugarcane for Sugar Production in the Mississippi Delta,* U.S. Department of Agriculture. Agricultural Research Service, Agricultural Handbook 417.

Bull, T.A., and K.T. Glasziou, 1975, "Sugarcane," *Crop Physiology,* L.T. Evans (Ed.), Cambridge University Press, New York.

Burke, D.P., 1975, "Methanol," *Chem. Week,* 117(13): 33ff.

Buswell, A.M., and C.S. Boruff, 1933, "Mechanical Equipment for Continuous Fermentation of Fibrous Materials," *Ind. Eng. Chem.,* 25(2): 147-148.

Callihan, C.D., and C.E. Dunlap, 1971, *Construction of a Chemical-Microbial Pilot Plant for Production of Single-Cell Protein from Cellulosic Wastes,* U.S. Environmental Protection Agency.

Campbell, J.R., 1968, Louisiana Report, Western Regional Project WM-51, prepared for annual meeting of the Technical Committee, Bozeman, Mont.

Cervinka, V., et al, 1974, *Energy Requirements for Agriculture in California,* California Department of Food and Agriculture, University of California, Davis, Calif.

Coe, G.E., and H. Menser, 1976, "Selecting Sugar Beets for Zone Resistance and the Unexpected Production of a High Purity Line," paper presented at the 19th General Meeting of the American Society of Sugar Beet Technologists, Phoenix, Ariz., February 1976.

Coleman, O.H., 1970, "Syrup and Sugar from Sweet Sorghum," *Sorghum Production and Utilization,* J.S. Wall and W.M. Ross (Eds.), Air Publishing Co., Westport, Conn.

Communications Marketing, Inc., Edina, Minn, 1974, *Feed Industry Red Book,* 1974 edition.

Cover, A.E. et al, 1973, "Coal Gasification: Kellogg's Coal Gasification Process," *Chem, Eng.,* 69(3): 31-36.

Cowley, W.R., and B.A. Smith, 1972, "Sweet Sorghum as a Potential Sugar Crop in South Texas," *Proceedings of the 14th Congress,* International Society of Sugar Cane Technology.

Curran, G., J. Clancey, B. Posek, and M. Pell, "Production of Clean Fuel Gas from Bituminous Coal," Report PB-232 695/7GA, Research Division, Consolidation Coal Company.

Cushing, R.L., and D. Heinz, 1975, Hawaiian Sugar Planters Association, Honolulu, Hawaii, personal communication with Battelle.

Dunlop, A.P., and F.N. Peters, 1953, *The Furans,* Reinhold Publishing Company, New York.

Faith, W.L., et al, 1965, *Industrial Chemicals,* John Wiley & Sons, Inc., New York.

Feldmann, H.F., et al, 1976, "Syngas Process Converts Waste to SNG," *Hydrocarbon Process.* 55(11): 201-204.

Fick, G.W., R.S. Loomis, and W.A. Williams, 1975, "Sugar Beet," *Crop Physiology,* L.T. Evans (Ed.), Cambridge University Press, New York.

Finney C.S., and D.E. Garrett, 1973, "The Flash Pyrolysis of Solid Wastes," paper presented at the American Institute of Chemical Engineering Meeting, Philadelphia, Pa., Nov. 11–15, 1973.

Fisher, R.A., et al, 1975, *Feasibility of Separating Carbon Monoxide from Producer Gas,* Contract No. NSF C-902, prepared for Office of Energy Research and Development Policy, National Science Foundation, Washington, D.C.

Fisher, T.F., M.L. Kasbohm, and J.R. Rivero, 1976, "Waste Treatment Advances: The Purox System," *Chem. Eng. Prog.,* 72(10): 75-78.

Frank, M.E., M.B. Sherwin, D.B. Blum, and R.L. Mednick, 1976, "Liquid Phase Methanation/ Shift–PDU Results and Pilot Plant Status," *Eighth Synthetic Pipeline Gas Symposium,* Chicago, Ill., Oct. 18–20, 1976, American Gas Association, Arlington, Va.

Freeman, K.C., D.M. Broadhead, and N. Zummo, 1973, *Culture of Sweet Sorghum for Syrup Production,* U.S. Department of Agriculture, Agricultural Research Service, Agricultural Handbook No. 441, Meridian, Miss.

Freeman, K.C., D.M. Broadhead, and N. Zummo, 1975, *Cooperative Sweet Sorghum Variety Tests for Syrup During 1971 in Five Southeastern States,* Report ARS-S-58, U.S. Department of Agriculture, Agricultural Research Service, Meridian, Miss.

Ghosh, S., and D.L. Klass, 1976, "SNG from Refuse and Sewage Sludge by the Biogas Process," paper presented at the Symposium on Clean Fuels from Biomass, Sewage, Urban Refuse, Agricultural Waste, Orlando, Fla., Jan. 27–30, 1976, Institute of Gas Technology, Chicago, Ill.

Ghosh, and D.L. Klass, 1977, "Anaerobic Digestion of Macrocystis Pyrifera Under Mesophilic Conditions," paper presented at the Second Symposium on Clean Fuels from Biomass and Wastes, Orlando, Fla., Jan. 25–28, 1977, Institute of Gas Technology, Chicago, Ill.

Grethlein, H.E., 1975, "The Acid Hydrolysis of Refuse," *Cellulose as a Chemical and Energy Resource,* C.R. Wilke (Ed.), pp. 303-318, John Wiley & Sons, Inc., New York.

Harris, E.E., 1949, "Wood Saccharification," *Advances in Carbohydrate Chemistry,* pages 153-188, Academic Press, Inc., New York.

Harris, J.F., 1975, "Acid Hydrolysis and Dehydration Reactions for Utilizing Plant Carbohydrates," *Proceedings of the Eighth Cellulose Conference. I. Wood Chemicals—A Future Challenge,* T.E. Timell (Ed.) pp. 131-144, John Wiley & Sons, Inc., New York.

Herbert, L.P., and E.R. Dire, 1972, *Long-Time Testing of Sugarcane Varieties in Southern Florida,* Report ARS-34-131, U.S. Department of Agriculture, Agricultural Research Service.

Hills, F.J., R. Albaugh, and R. Pearl, 1955, *Value and Use of Sugar Beet Tops,* Agricultural Extension Service, University of California, Davis, Calif.

Hills, F.J., and S.S. Johnson, 1973, *The Sugar Beet Industry in California,* Circular No. 562, California Agricultural Experiment Station Extension Service, University of California, Davis, Calif.

Hills, F.J., and A. Ulrich, 1971, "Nitrogen Nutrition," *Advances in Sugar Beet Production; Principles and Practices,* R.T. Johnson, J.T. Alexander, G.E. Rush, and G.R. Hawkes (Eds.), Iowa State University Press, Ames, Iowa.

Humbert, R.P., 1968, *The Growing of Sugarcane,* 2nd ed., Elsevier Publishing Co., New York.

Jackson, R.D., 1972, "Insects," *Culture of Sugarcane for Sugar Production in the Mississippi Delta,* U.S. Department of Agriculture, Agricultural Research Service, Agricultural Handbook 417, Superintendent of Documents, U.S. Government Printing Office, Washington, D.C.

James, D.W., W.H. Weaver, and R.L. Reeder, 1968, *Soil Test Index of Plant Available Potassium and the Effects of Cropping and Fertilization in Central Washington Irrigated Soils,* Washington State University, Agricultural Experiment Station, Bulletin No. 697.

Johnson, R.T., J.T. Alexander, G.E. Rush, and G.R. Hawkes (Eds.), 1971. *Advances in Sugar Beet Production: Principles and Practices,* Iowa State University Press, Ames, Iowa.

Kampen, W.H., 1975, "Technology of the Rum Industry," *Sugar Azucar,* 36-43.

Kendrick, J.G., 1975, *The Development of a High Protein Isolate from Selected Distillers By-Products,* prepared by the University of Nebraska for the National Science Foundation, U.S. Department of Commerce; Phase II—Incremental Investment and Operating Costs for Appropriate Process Facilities Modification—Part A (William Scheller).

Kispert, R.G., et al, 1975, *Fuel Gas Production from Solid Waste,* Final Report, Report NSF/RANN/SE/C-827/PR/74/5, National Science Foundation.

Krenz, R.D., 1975, "The USDA Firm Enterprise Data System Capabilities and Applications," *South. J. Agric. Econ.,* 7: 33-38.

Lill, J.G., 1964, *Sugar Beet Culture in the North Central States,* U.S. Department of Agriculture Farmers Bulletin No. 2060, Superintendent of Documents, U.S. Government Printing Office, Washington, D.C.

Livingston, R.C., and D. McNeil, 1974, "Beyond Petroleum. Volume I, Biomass Energy Chains, Final Report," Study prepared by Engineering 235 for the Institute for Energy Studies, Stanford University.

Loomis, R.S., and P.A. Gerakis, 1975, "Productivity of Agricultural Ecosystems," *Photosynthesis and Productivity in Different Environments,* J.P. Cooper (Ed.), Cambridge University Press, New York.

Loomis, R.S., A. Ulrich, and N. Terry, 1971, "Environmental Factors," *Advances in Sugar Beet Production: Principles and Practices,* R.T. Johnson, J.T. Alexander, G.E. Rush, and G.R. Hawkes (Eds), Iowa State University Press, Ames, Iowa.

Lowenheim, F.A., and M.K. Moran, 1975, *Faith, Keyes, and Clark's Industrial Chemicals,* 4th ed., John Wiley & Sons, Inc., New York.

Mandels, M., et al, 1975, *Enzymatic Hydrolysis of Cellulosic Wastes to Glucose,* U.S. Army Natick Development Center.

Maugh, R., 1976, University of California at Los Angeles, personal communication.

McCarty, P.L., et al, 1976, "Heat Treatment of Organics for Increasing Anaerobic Biodegradability," Progress Report on Grant ERDA-F (04-3)-326-PA-44, prepared for the September 16-17, 1976, Coordination Meeting, Greeley, Colo.

McCutchan, W.M., and R.J. Hickey, 1954, "The Butanol-Acetone Fermentations," *Industrial Fermentations,* L.A. Underkofler and R.J. Hickey (Eds.), Chemical Publishing Company, Inc., New York.

Meade, G.P., 1963, *Cane Sugar Handbook,* 9th ed., John Wiley & Sons, Inc., New York.

Meisel, S.L., J.P. McCullough, C.H. Lechthaler, and P.B. Weisz, 1976, "Gasoline from Methanol in One Step," *Chemtech.*, 6(2): 86-98.

Meller, F.H., 1959, *Conversion of Organic Solid Wastes into Yeast,* U.S. Health, Education, and Welfare Public Health Service, Rockville, Md.

Miller, D.L., 1976, "Fermentation Ethyl Alcohol," *Enzymatic Conversion of Cellulosic Materials: Technology and Applications,* E.L. Gaden, Jr., et al (Eds), pp. 307-312, John Wiley & Sons, Inc., New York.

Millet, M.A., et al, 1975, "Pretreatments to Enhance Chemical, Enzymatic and Microbiological Attack of Cellulosic Materials," *Cellulose as a Chemical and Energy Resource,* C.R. Wilke (Ed.), pp. 193-219, John Wiley & Sons, Inc., New York.

Millhollon, R.W., 1972, "Weed Control," *Culture of Sugarcane for Sugar Production in the Mississippi Delta,* U.S. Department of Agriculture, Agricultural Research Service, Agricultural Handbook No. 417, Superintendent of Documents, U.S. Government Printing Office, Washington, D.C.

National Academy of Sciences, Washington, D.C., 1971, *Atlas of Nutritional Data on United States and Canadian Feeds.*

National Academy of Sciences, Washington, D.C., 1972, *Atlas of Nutritional Data on United States and Canadian Feeds.*

Nichol, G.E., L.M. Burtch, and D.J. Traveller, 1971, "Seedbed Preparation, Planting, and Thinning," *Advances in Sugar Beet Production: Principles and Practices,* R.T. Johnson, J.T. Alexander, G.E. Rush, and G.R. Hawkes (Eds.), Iowa State University Press, Ames, Iowa.

Norman, Lloyd W., 1975, Utah-Idaho Sugar Company, Moses Lake, Wash., personal communication.

Paturau, J.M., 1969, *By-Products of the Cane Sugar Industry,* p. 51, Elsevier Publishing Co., Amsterdam.

Pfeffer, J.T., and J.C. Liebman, 1975, *Bioconversion of Organic Refuse to Methane,* Annual Progress Report, Report NSG/RANN/SE/GI-39191/PR/75/2, National Science Foundation.

Scheller, W., et al, 1975, *The Development of a High Protein Isolate from Selected Distillers By-Products,* Final Report at the University of Nebraska on National Science Foundation Grant No. AER 74-10456 A01, pp. 68-105.

Scheller, W., and B.J. Mohr, 1976, "Net Energy Analysis of Ethanol Production," paper presented at the 169th American Chemical Society Meeting, Philadelphia, Pa., April 1976.

Schmehl, W.R., and D. W. James, 1971, "Phosphorus and Potassium Nutrition," *Advances in Sugar Beet Production, Principles and Practices,* R.T. Johnson, J.T. Alexander, G.E. Rush, and G.R. Hawkes (Eds.), Iowa State University Press, Ames, Iowa.

Seeley, D.B., 1976, "Cellulose Saccharification for Fermentation Industry Applications," *Enzymatic Conversion of Cellulosic Materials: Technology and Applications,* E.L. Gaden, Jr. (Ed.), John Wiley & Sons, Inc., New York.

Sherwin, M.B., and D. Blum, 1975, "Methanol Synthesis in a Three-Phase Reactor," paper presented at the American Chemical Society Meeting in Chicago, Ill., Aug. 24, 1975.

Sloane, G.E., and L.J. Rhodes, 1972, "A Comparison of the Processing of Burned and Unburned Sugarcane," *Hawaii. Plant. Rec.,* 58(14): 173-182.

Spano, L., et al, 1975, *Enzymatic Hydrolysis of Cellulosic Wastes to Glucose,* U.S. Army Natick Development Center.

Tamers, M.A., 1976, "Total Synthesis Benzene and Its Derivatives as Major Gasoline Extenders," *Science,* 193: 231-232.

Underkofler, L.A., and R.J. Hickey, 1954, *Industrial Fermentations,* Chemical Publishing Company, Inc., New York.

U.S. Bureau of Mines, 1974, *Consolidation Coal Company CO_2 Acceptor Process Pipeline Gas from Lignite Gasification 250-Million SCFD Plant—An Economic Analysis,* Report No. 75-5 by the Process Evaluation Group—Minerals Research and Environmental Development, Morgantown, W. Va.

U.S. Department of Agriculture, 1972 and 1973, *Soil Survey of the Islands of Kauai, Oahu, Maui, Molokai, and Lanai, State of Hawaii,* Soil Conservation Service, Washington, D.C.

U.S. Department of Agriculture, 1974a, *1974 Agricultural Statistics,* Superintendent of Documents, U.S. Government Printing Office, Washington, D.C.

U.S. Department of Agriculture, 1974b, *Sugar Program, Mainland Sugarcane Program,* Annual Report, Agricultural Stabilization and Conservation Service, Alexandria, La.

U.S. Department of Agriculture, 1974c, *Federal Energy Data Systems,* Commodity Economics Division, Economic Research Service, in cooperation with Oklahoma State University, Stillwater, Okla.

U.S. Department of Agriculture, 1974d, "Agricultural Prices," press release, Crop Reporting Board, Statistical Reporting Service, Washington, D.C.

U.S. Department of Agriculture, 1975a, "Agricultural Prices," press release, Crop Reporting Board, Statistical Reporting Service, Washington, D.C.

U.S. Department of Agriculture, 1975b, "Farm Real Estate Market Developments, CD.," Economic Research Service, Washington, D.C.

U.S. Department of Agriculture, 1975c, "Feed Market News," Vol. 59, No. 51, Grain Division, Agricultural Marketing Service, Independence, Mo.

U.S. Department of Agriculture, 1976a, "Agricultural Prices," press release, Crop Reporting Board, Statistical Reporting Service, Washington, D. C.

U.S. Department of Agriculture, 1976b, *1975 Annual Summary, Crop Production,* CrPr-1(76), Crop Reporting Board, Statistical Reporting Service, Washington, D.C.

U.S. Department of Agriculture, 1976c, *Costs of Producing Food Grains, Feed Grains, Oilseeds, and Cotton,* 1974, Agricultural Economic Report No. 338, Economic Research Service, Washington, D.C.

U.S. Department of Agriculture, 1976d, *Agric. Res.,* 25(2).

U.S. Department of Commerce, Weather Bureau, 1965, *Climatic Summary of the U.S.,* Supplement for 1951 Through 1960, Texas, Climatography of the U.S. No. 86-36, Superintendent of Documents, U.S. Government Printing Office, Washington, D.C.

Universidad de Puerto Rico, Estacio Experimental Agricola, Rio Piedras, Puerto Rico, 1975, *El Cultivo de la Cana de Azucar en Puerto Rico,* Bulletin No. 237.

Walker, C., 1972, *Costs and Returns from Sugarcane in South Florida,* Florida Cooperative Extension Service, University of Florida, Gainesville, Fla., Circular No. 374.

Wilke, C.R., 1975, *Cellulose as a Chemical and Energy Resource,* John Wiley & Sons, Inc., New York.

Wilke, C.R., and R.D. Yang, 1975, "Process-Development Studies of the Enzymatic Hydrolysis of Newsprint," *Proceedings of the Eighth Cellulose Conference. I. Wood Chemicals—A Future Challenge,* T. E. Timell (Ed.), pp. 175-188, John Wiley & Sons, Inc., New York.

Wise, D.L., et al, 1977, "Biomethanation of Coal Gasifier Product Gases," paper presented at 69th Annual Meeting of American Institute of Chemical Engineers, Chicago, Ill., Nov. 28-Dec. 2, 1976.

AVAILABILITY OF BIOMASS GRAINS AND GRASSES

The information in this chapter is based on *Systems Study of Fuels from Grains and Grasses*, prepared by W. Benson, A. Allen, R. Athey, A. McElroy, M. Davis and M. Bennett of Midwest Research Institute for the U.S. Department of Energy (DOE Report ALO/3729-1), February 1978.

INTRODUCTION

The production of fuels from biomass is one of the many possibilities of extracting energy from the sun. When one considers energy production in the context of its three major functions, extraction, conversion, and distribution, the concept of bioconversion is attractive. The distribution function is partially accomplished by the very nature of solar radiation. Further, grains and grasses are grown in most states in quantities sufficient to warrant their consideration as a source of biomass material. Thus, the distribution function of energy production is largely accomplished by this existing dispersion. The photosynthetic process, in part, accomplishes the extraction function by collecting the sun's energy and storing it in plant matter available for use upon harvest. The location of a suitable conversion facility in close proximity to the biomass source, and convenient to the point of demand for the energy produced, completes the production cycle.

The prospect of placing millions of acres of marginal land into grass production as energy plantations is attractive as a painless solution to the energy problem. The use of residues from food and fiber production has a similar appeal.

The scope of this project includes the members of the grass family (Gramineae) including the many species of sod crops which provide cured forage or pasturage for farm animals and the great food crops of wheat, rice, corn, soybeans, barley, oats, etc. The members of the legume family (Leguminosae), the clovers, lespedezas, alfalfas, and many others, are included within the scope of grains and grasses. Also included for consideration of their residue, or possible use as whole plant biomass, are those current agricultural food and fiber crops which occur regionally in sufficient quantity to be important sources of biomass.

CROP ENVIRONMENT

The focus of an analysis of the potential for biomass production for energy is the majority of the land devoted to agriculture in the United States. The acreage includes land in pasture, range, and federal range. Excluded from active consideration are the acreages devoted to sugar-producing crops (sugar canes and beets), cotton and tobacco, fruits and vegetables, and a few miscellaneous crops. The acres in forests are excluded because forested acreages have by definition been excluded from consideration as candidates for clearing and substitution of grass and grain crops. The acreages devoted to various agricultural crops are reported in numerous publications by state and federal agencies. State reports list crop production by county on a yearly basis. The sources of most of the data used in this study are the 1967 Conservation Needs Inventory (CNI) (USDA, 1969a), which details land capability classes (LCC) throughout the country on a county basis, land use (LUC) classifications by major uses such as row crops, close grown crops, pasture, range, etc., and needs for conservation, the 1969 Census of Agriculture (U.S. Bureau of Census, 1972), and *Agricultural Statistics* (USDA, 1969 to 1976), an annual publication of the United States Department of Agriculture (USDA). LRR refers to land resource regions and LRA indicates land resource areas.

Overall, the acreage which falls within the scope of the program is about 1.2 billion acres (0.49 billion hectares) categorized approximately as follows: 365 million acres (148 million hectares) in range; 106 million acres (43 million hectares) in permanent pasture; 450 million acres (182 million hectares) defined as cropland in tillage rotation; 300 million acres (121 million hectares) in federal range; and 25 to 30 million acres (10 to 12 million hectares) classified as other land not in farms. The total agricultural acreage includes 25 to 30 million acres (10 to 12 million hectares) in farmsteads, rural roads, and like uses which are not considered to be part of the productive resource. Also, agricultural land includes 130 to 150 million acres (53 to 61 million hectares) defined as farm woodland and forests; these acres have been excluded from the analysis.

The net resource, which can be considered to be potentially available for grass and grain biomass and food production is accordingly the 450 million acres (182 million hectares) of cropland, 106 million acres (43 million hectares) of permanent pasture, 365 million acres (148 million hectares) of range, 300 million acres (121 million hectares) of federal range, and some 30 million acres (12 million hectares) of land which apparently is devoted to rights-of-way, investment properties, and like uses, a total of about 1.25 billion acres (505 million hectares). This figure is reduced by 20 to 30 million acres (8 to 12 million hectares) by the exclusion of cotton, tobacco, sugar cane and beets, vegetables and fruits, and miscellaneous crops, so that the net potential resource is slightly more than 1.2 billion acres (485 million hectares).

Cropland acreage is without question potentially the most productive land which might be considered for biomass production. Of this acreage, 1975 agricultural statistics (USDA, 1976) show that 347 million acres (140 million hectares) (see Table 5.1) were harvested for the major and minor crops—corn, wheat, soybeans, grain and forage sorghums, other small grains (chiefly oats, barley, rye, and rice), hay and cotton, tobacco, fruits and vegetables, and miscellaneous crops. In the early 1970s, fewer acres were harvested for these crops, on the order of 300 million acres (121 million hectares). These figures point to an available acreage within the cropland category of 100 to 150 million acres (40 to 60 million hectares).

However, some of the nominally idle acreage is in fallow rotation, notably in dryland wheat farming, and this cannot be considered to be available. Additional acreages are temporarily under conservation practices or may be used as temporary pasture; such acreages are in principle available for biomass production, but they likely should be viewed as an essential part of the cycle for food and fiber production and thus will not be available without causing a disruption in the utilization of land for food and fiber production. One can nevertheless estimate that from 50 to 100 million acres (20 to 40 million hectares) of this land could be diverted to biomass production, with the lower figure being appropriate at present.

Table 5.1: 1975 Crop and Hay Acreages

	10^6 Acres	10^6 Hectares	% of Total
Wheat	75	30	22
Corn	78	32	23
Hay	62	25	18
Soybeans	55	22	16
Other small grains	38	15	11
Sorghums	18	7	5
Cotton and tobacco	11	4	3
Vegetables and fruit	6	2	2
Miscellaneous	4	2	1
Total	347	139	101

Source: DOE ALO/3729-1

According to the Conservation Needs Inventory of 1967, cropland distribution among land capability classes is as shown in Table 5.2. About 8%, 36 million acres (15 million hectares), is Class I, and 85%, 380 million acres (154 million hectares), is Classes II to IV. Somewhat surprisingly, some 22 million acres (9 million hectares) of cropland is Classes V to VIII land, which by definition is essentially unsuited for tilled crops.

Table 5.2: Land Use Distribution by Capability Class

	................LCC................			
	I	II to IV	V to VII	VIII
Cropland	36	380	22	0.1
Pasture	4	77	25	0.1
Range	1	97	263	4.5
Other land	0.6	12	7	–
Total	42	566	317	5

Source: DOE ALO/3729-1

In the table, LCCs show in a general way the soil and landscape most suitable for agriculture. Class I lands have few limitations that restrict their use; Classes II to IV lands have increasingly severe limitations that reduce the choice of plants or require special conservation practices or both; Classes V to VIII lands have severe limitations that restrict their use largely to pasture, grazing, woodland, or wildlife; Class VIII lands have limitations which may preclude commercial plant production. Classes II to VIII lands may be further classified into subclasses of erosion problems (e); wetness, drainage, or overflow (w); root zone limitations (s); or climate (c).

The 106 million acres (43 million hectares) of permanent pasture are an essential part of livestock production capabilities, and in this sense are not available for biomass production. The permanent pasture is for the most part difficult to farm with cultivated crops. However, most of it is in higher rainfall regions, and one accordingly is strongly tempted to consider its use for biomass, within limits dictated by the problems inherent in pasture land usage. Four million acres (1.6 million hectares) of permanent pasture is Class I land, which should pose no problems. The majority, 77 million acres (31 million hectares), is Classes II to IV land, which could in principle be brought into cultivated crop production, with increasing attention being given to conservation practices as one proceeds from Classes II to IV. The remaining pasture is Classes V to VIII land, which should be suitable for biomass harvesting only if it is used for perennial crops or permanent grasses.

Only about 1 million acres (0.4 million hectares) of range is Class I land, 97 million acres (39 million hectares) are Classes II to IV, and the remainder, 267 million acres (108 million hectares), is Classes V to VIII land. Approximately 100 million acres (40 million hectares) can thus be considered tillable and the remainder suited for perennial species.

Of the other land not in farms, about 600,000 acres (242,000 hectares) are Class I, 12 million acres (5 million hectares) are Classes II to IV, and 15 million acres (6 million hectares) are Classes V to VIII.

If one assumes that about 60 million acres (24 million hectares) of Classes II to IV cropland is idle insofar as crop production is concerned, and that all pasture, range, and other land not in farms are candidates for biomass production, the acreages shown in Table 5.3 are the result: 6 million acres (2 million hectares) of Class I; 246 million acres (100 million hectares) of Classes II to IV; and 295 million acres (119 million hectares) of Classes V to VIII land. The total is approximately 550 million acres (223 million hectares), of which about 85% is devoted to the livestock industry as pasture and range.

Table 5.3: Underutilized Land

	Class I (10^6 acres)	Classes II to IV (10^6 acres)	Classes II to IV (10^6 hectares)	Classes V to VIII (10^6 acres)	Classes V to VIII (10^6 hectares)
Cropland	–	~60	~24	–	–
Pasture	4	77	31	25	10
Range	1	97	39	263	106
Other	1	12	5	7	3
Total	6	246	99	295	119

Source: DOE ALO/3729-1

The distribution of the approximately 500 million acres (202 million hectares) of pasture, range, and other land not in farms is shown in Figure 5.1 by LRR, a designation used by the USDA Soil Conservation Service (Austin, 1972). About 25% of these acres is located in the eastern part of the country where annual rainfall exceeds 25 inches (64 cm), about 50% is located in areas with 16 to 25 inches (41 to 64 cm) of rain, and the remaining 25% receives less than 15 inches (38 cm). If rainfall were the only limiting factor, the production potential for biomass of the underutilized land would be on the order of 3 to 5 tons/acre (6.7 to 11 MT/ha) (higher in certain areas) for 140 million acres (57 million hectares), 1 to 3 tons/acre (2.2 to 6.7 MT/ha) on 275 million acres (111 million hectares), and substantially less than 1 ton/acre (0.907 MT/ha) on the remainder.

With regard to the problems associated with better utilizing the marginal land, approximate percentages of marginal land in different land classes are as follows where e = an erosion problem, w = a water or wetness problem, s = a soil problem (e.g., stony or shallow soils), and c = a climatic problem (e.g., temperature or moisture).

> Classes II to IV (e), 28%;
> Classes II to IV (w), 8%;
> Classes II to IV (s), 5%;
> Classes II to IV (c), 5%;
> Classes V to VIII (e), 28%
> Classes V to VIII (w), 2%;
> Classes V to VIII (s), 22%; and
> Classes V to VIII (c), 3%.

The Classes II to IV (e) and (w) lands thus comprise about 36% of all underutilized land, and these would likely be principal candidates for utilization for biomass production. The Classes V to VIII (e) and (w) land comprise about 30% of underutilized land and would be more difficult to bring into production. The lands with soil (s) and climate (c) limitations would not likely be considered as candidates for biomass production.

In summary, some 500 to 600 million acres (202 to 243 million hectares) of the U.S. agricultural resource is currently not in crop or hay production, much of it deservedly so. These acres are, for the most part, seriously limited when viewed in terms of cultivated crop production. The shifting reservoir of cropland not in crops, 100 to 150 million acres (40 to 60 million hectares), should most readily adapt to biomass-for-energy production; these acres are inventoried as part of the crop/hay production resource and there is the danger that indiscriminate shifting into biomass production will impair crop production capabilities.

The 25 to 30 million acres (10 to 12 million hectares) classified as other land not in farms is a potential resource whose ultimate value is lessened by its diffusivity, but which presumably would be brought into production without impacting crop and hay production. Permanent pasture, 100+ million acres (40+ million hectares), of which 80 million is Classes I to IV land, is a significant potential resource located mostly in higher rainfall areas. Finally, about 100 million acres (40 million hectares) of rangeland is potentially tillable Classes I to IV land, but these acres are located primarily in arid to semiarid parts of the country. Diversion of the potentially available underutilized land into biomass production will be in conflict with other beneficial uses, both for crop production and for livestock production, and this basic issue must be carefully analyzed.

Figure 5.1:   Underutilized Acres by Land Resource Region (Pasture, Range, and Other Land Not in Farms, Million Acres)

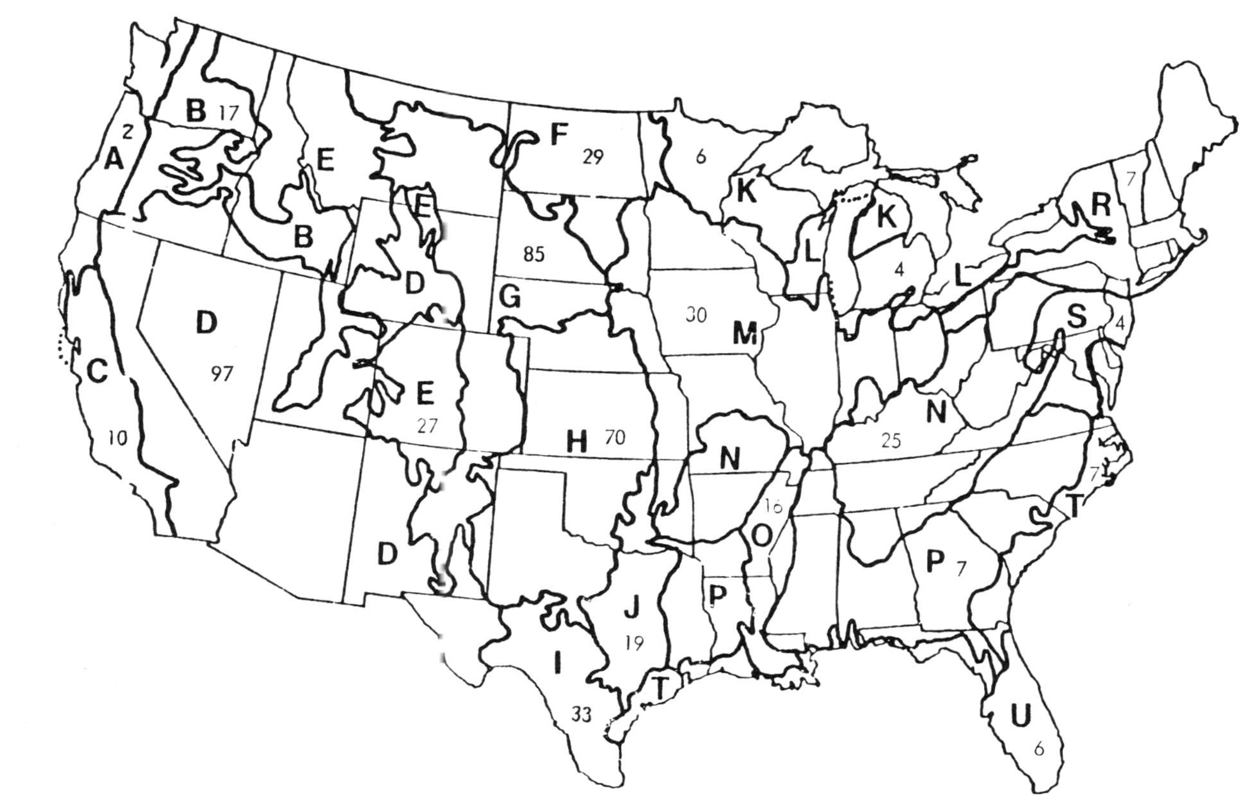

Source: DOE ALO/3729-1

The principal technical problem consists of finding and adopting combinations of biomass crops and accompanying cultural practices necessary to insure significant and lasting productivity, since a majority of the underutilized land has one or more problems.

Factors Pertinent to Marginal Land Use

The land which has been defined as marginal or underutilized is so categorized for very good reasons when viewed from the standpoint of tilled crop production. The basic consideration is, of course, economics: the land is not used for tilled crop production because it is not needed to meet current demands for grain, fiber, hay, and other crops; and/or because of the cost penalties associated with its use for those purposes. A primary economic factor is the reduced productivity compared to more productive land. The marginal land is apt to have less productive soils on steeper slopes and to be less effective in utilizing incident rainfall. It may be too wet, and thus be less productive.

A second factor is the added costs associated with its use, e.g., higher fertilizer requirements, costs for building terraces, or incremental costs for conservation-oriented farming and less effective use of machinery. The U.S. farmer has accordingly used such lands primarily for livestock-oriented operations which do not require that the land be tilled and which utilize the market product as the means to harvest biomass. Since livestock production is an important and economically viable part of our agricultural production capability, the land is being underutilized only to the extent that one or more of the following is true:

(1) A fraction of the land is not actually needed for livestock or as cropland reserves;
(2) An alternate use might, for at least a fraction of the land, be higher priority in the context of the national objectives and priorities; and
(3) The economic return to the farmer might be increased by alternate uses, of which biomass for energy is the alternate use of interest in this study.

The latter factor, the economic return to the farmer, will be the ultimate governing factor which determines whether marginal land will be diverted to biomass production, although diversions will undoubtedly have to be assessed in terms of possibly conflicting needs for energy and food.

One option, which has been excluded in this study, is the diversion of irrigation water to biomass production. If this were feasible (i.e., if sufficient irrigation water were available), there is little doubt that extensive acreages could be made available for biomass production, principally in semiarid regions (the Southwest and High Plains) where the adequacy of rainfall has inhibited crop production.

Probably the one constraint to use of marginal land of greatest importance falls under the category of conservation. It is generally recognized that much of the land in cropland use is in need of improved conservation practices, with erosion control being of foremost importance. Drainage and basic fertility or productivity are other problems. The marginal lands will generally pose a greater problem in this regard than does the cropland currently in production. Three basic approaches can be employed for erosion control: (a) structural measures such as terraces, waterways, and sediment retention basins; (b) cultural practices such as contour farming, strip and contour strip farming, and mulch tilling; and (c) selection of crops and crop rotations for soil conservation. In the latter category falls production of perennial or permanent crops rather than row crops or small grains. These three approaches apply also to the problem of improving the productivity of marginal land.

The structural approach is the most costly option. Costs for terraces and other structures vary with the slope of the land, soil erodibility, rainfall characteristics, and intended crop use and range from $50 to more than $200/acre ($125 to $500/ha). Maintenance of such systems is an added and continuing expense, and costs for farming terraced lands are higher than for nonterraced land. The added costs are compensated by improved yields, upgraded water retention, and long-term enhanced productivity. Nevertheless, acceptance of such conservation practices is far from universal, and the increasing trend to large machinery/less labor is in some conflict with implementation and preservation of good conservation practices.

Contouring, strip cropping and contour strip cropping are low-cost methods. These may be the source of added production costs and be unattractive in the sense that they are incompatible with single crop/high land utilization farming. Mulch tilling and similar practices can be effective in preventing erosion; they require special equipment and are not equally adapted to all crops. Mulch tillage is basically incompatible with harvesting of aerial biomass for energy purposes, since it requires retention of residues on soil surfaces.

The Conservation Needs Inventory of 1967 specifies that permanent cover be used for large acreages not deemed at that time to be adequately protected. With permanent cover, both the root systems and above ground cover of vegetative matter serve as an effective deterrent to soil erosion by water or wind. This method of utilizing marginal land is compatible with harvest of aerial biomass for energy. With the exception of extremely difficult situations, most of the aerial biomass could be harvested, with protection against erosion being adequately effected by roots and a small fraction of unharvested above-ground biomass.

The utilization of marginal land for biomass will demand greater attention to sound land management and soil conservation practice than is customary for land now used for crop production. The marginal land will be more easily mismanaged than is the better land now in crop production, and mismanagement will be accompanied in the very short term by significant losses in productive capability. It is essential then that care be exercised in selection of biomass crops for specific regions and types of land, and that cultural practices are at the same time selected with care and adhered to.

The first choice, in the conservation sense, would probably be permanent or perennial crops, with the root systems left undisturbed and a fraction of the above-ground biomass left unharvested to help retain moisture and protect against erosion. The choice of biomass crops is limited, with this option, to permanent grasses and perennial grasses and legumes. The opportunities for utilization of perennial species are further limited by the fact that certain perennials winterkill in less temperate regions. Nevertheless, numerous species of grasses and legumes are candidates for use on marginal lands throughout the country.

As one considers options for utilization of other than permanent or perennial plants on marginal land, the need for structural or cultural practices increases. Small grain, annual or biennial grass and legume, and row crop production on marginal land will present difficulties with regard to soil conservation and maintenance of long-term productivity, and obviation of these difficulties will require care in selection of the overall cultural process, which can in principle consist of combinations of several options. The amount of produced biomass which can be harvested for biomass will be a function of the specific method selected. With well-terraced fields, for example, it should be possible generally to remove a greater proportion of the aerial biomass than with contouring or strip cropping since the residue will be less needed for erosion protection in the former case than in the latter.

Permissible Levels of Residue/Biomass Harvesting

It is not mechanically feasible to harvest all of the aerial biomass of any of the grasses and grains. Perhaps 90% is the amount which can be conveniently harvested with present equipment. With hays and forages, this amount is generally harvested. With grain crops, current practice usually calls for cutting no more than is necessary of the above-ground biomass and leaving all but the grain in the field. However, with grain crops, a follow-on harvest of residues can be effected with hay/forage harvesting equipment.

A key question which in many instances takes precedence over the mechanical feasibility of total or 90% harvesting of biomass is the need to leave residues on the soil. The issues are as follows:

(1) The organic residues contain nutrient elements which, if removed, must be replaced with synthetic fertilizer;

(2) The organic residues are in varying degree essential to maintenance of the basic productivity characteristics of the soil, namely, mechanical properties, moisture retention and water absorptivity and permeability, aeration properties, and what is called tilth. Productive soils must contain a certain amount of organic matter which must be replenished as natural forces result in decomposition of residual organic matter; and

(3) The organic matter plays a significant role in moisture retention (e.g., in retaining snowfall) and in preventing soil and water erosion.

The organic farming disciple will insist that none of the residue be harvested and perhaps that auxiliary organic matter be applied. The vast majority of farmers are accustomed, however, to harvesting salable and consumable grains and other biomass, the latter to an extent determined by overall economics including the economic and convenience factors associated with erosion prevention and fertility maintenance. There is no uniformly applicable rule which determines the amount of biomass which can be harvested, and conversely the amount which should be left on the field. The discussion below addresses various factors in abbreviated form.

(1) Any evaluation of permissible levels of biomass removal must encompass both the above-ground and below-ground biomass,

the proportions of which vary substantially. Both contribute to the recycled organic and nutrient budget, and both contribute also to erosion control.

(2) The permissible level is a function of the ability of the soil itself to maintain an acceptable level of productivity with varying levels of recycle of residues. This ability varies both with the soil and its location in the country. For example, soils in the North Central region are less sensitive to organic removal than are soils in the East and Southeast. Temperature, rainfall, slope, and soil depth are some of the factors which determine how sensitive a soil is to organic removal.

(3) The type of crop and the methods used for cultivation and harvest are important. Row crops will generally demand return of greater amounts of residue, particularly if water or wind erosion is a problem. Perennial or permanent crops with well-developed root systems can generally be nearly completely harvested.

(4) Three types of hazards exist, namely, resistance to erosion, ability to absorb and retain moisture and maintain other soil properties at optimal levels, and infertility. Which of these is most important depends on climate, topography, soil types, and crop types.

(5) The task of analyzing residue removal must be essentially site and crop specific, although certain generalizations can be made for regions and crop types. These broad generalizations are as follows:

    (a) Soils in the East and Southeast can be easily mismanaged and require either return of a high percentage of residues to the soil or a production system geared to perennial or permanent plants.

    (b) In the North Central and High Plains regions, residue return (generally) is only marginally required for fertility/productivity maintenance, and the need to return residues is principally governed by requirements for wind and water erosion control and water retention. In the High Plains, for example, residues help to retain snowfall and prevent wind erosion.

    (c) In the arid/semiarid regions, penalties from a productivity or fertility standpoint are usually negligible, and there would appear to be no valid reason not to remove most of the aerial biomass.

    (d) Nutrients supplied or recycled with the organic matter can be supplied synthetically. The limiting factor is thus the amount of organic or carbonaceous matter which must be maintained, by residue return, to maintain the soil productivity factors associated with soil organic content, to pre-

vent erosion, and to maintain water holding
and absorptive capacity in the soil. Replacement of nutrient elements is, of course, an
economic penalty which must be borne by
the biomass producer.

This study has not addressed the question of the permissible level of aerial biomass removal in detail. The question will, as a matter of fact, be satisfactorily
resolved only on a site-by-site basis, and even then possibly not without controversy. Furthermore, the question is so closely allied to production economics
on a sustained basis that it probably is not possible without more detailed study
of costs and without benefit of research and field experience to reach definitive
answers. The study team has accordingly opted to use a uniform format or formula for various crop types, which has been applied country-wide. The factors
for percentages of biomass (both whole plant and crop residues) harvested for
energy are shown in Table 5.4.

Table 5.4: Percentage of Aerial Residues Harvested for Biomass

	% Harvested
Corn and grain	75
Sorghum grain	85
Wheat	75
Soybeans	50
Rice	90
Oats, rye and barley	75
Peanut hay	100
Sunflower	85
Safflower	75
Hops	80
Forages and hays	100*
New crops	100*
Grasses	**

  *100% of normally harvested biomass
**Normal hay harvesting practice

Source: DOE ALO/3729-1

Allocation of Underutilized or Marginal Land to Biomass Crop Production

Since this study is national in scope, the methods used to allocate marginal land
to biomass production were nationally, rather than locally, oriented. Marginal
land use was restricted to the growth of kenaf in higher rainfall areas (Iowa,
Missouri, Arkansas, and Louisiana east), cattail in peatbogs and wetlands (principally Minnesota, Michigan, Wisconsin, Florida, and Georgia), guayule in the
semiarid Southwest, guayule/mariola hybrid in less temperate semiarid regions,
and giant reed in the Mid-South (Texas and Oklahoma). This approach left un-
covered the Central to Northern High Plains and the Intermountain regions, i.e.,
no new crops were proposed for use in these regions of the country.

Two types of formulas for marginal land allocation have been used. The first
provided arbitrarily for a certain percentage of all pasture, range, and other land

not in farms to be diverted to biomass production with new crops. Yields have been estimated for 10, 25, and 50% utilization of the marginal land.

A second formula was developed which is more sensitive to differences in production capabilities and other properties of the marginal land. This formula is shown in Table 5.5.

Table 5.5: Percentages of Marginal Land Allocated to New Crops

LCC	Kenaf	Giant Reed	Cattail	Guayule	Guayule/Mariola
I					
Pasture	90	-	-	10	10
Range	80	-	-	10	10
Federal range	50	-	-	10	10
Other land	80	-	-	10	10
II and III (e, s, c)					
Pasture	70	-	-	10	10
Range	50	-	-	10	10
Federal range	10	-	-	10	10
Other land	60	-	-	10	10
IV (e, s, c)					
Pasture	50	-	-	-	-
Range	10	-	-	-	-
Federal range	10	-	-	-	-
Other land	40	-	-	-	-
II to V (w)					
Pasture	-	-	10	-	-
Range	-	-	10	-	-
Federal range	-	-	10	-	-
Other land	-	-	10	-	-
VI and VII (e, s, c)					
Pasture	-	10	-	10	10
Range	-	10	-	10	10
Federal range	-	10	-	10	10
Other land	-	10	-	10	10

Source: DOE ALO/3729-1

Allocations of new crops were further modified by assigning each crop to only those parts of the country in which they were adapted for growth. An element of conservatism in the yield data was introduced prior to final calculation of yields, by downgrading yields on less productive land: pasture, range, federal range, and other land not in farms. Since cattail is proposed for growth in peat or wetlands, yields were assumed to be the same in all categories of land. Kenaf yields, as percentages of yields on prime agricultural land, were taken to be 75% on pasture land, and 50% on range, federal range, and other land. Yields of giant reed were downgraded similarly for pasture, range, federal range, and other land. Guayule yields were taken to be the same for all types of land.

The formula in Table 5.5 yielded a total of 73,800,000 acres (29,866,860 hectares) diverted from present pasture, range, and idle use to production of kenaf, giant reed, cattails, guayule and guayule/mariola.

CROP RESOURCE

Presented in this section are criteria for selection of plants, a discussion of some of the factors which pertain to biomass production from grasses and grains, a summary discussion of grasses and grains of current economic importance in the United States, and discussion of potential new crops which might be considered in biomass agriculture.

Criteria for Crop Selection

This study focused on biomass production from grasses and grains of current economic significance in the continental United States, with exclusion as noted previously of sugar cane and sugar beets, cotton, tobacco, and certain crops which are either of very minor economic significance or are classified as fruits and vegetables. In addition, it was desired to include in the analyses representative plants which presently are of no economic significance, but which conceivably could be used to augment biomass production from current grasses and grains, or be substituted for them on significant acreages. With regard to new crops, a limited number of candidates, from among the multitude of available candidates, were selected for evaluation in the systems analysis. These few new crops were selected for one or more of the following reasons:

(1) The existence of a sufficient body of information on adaptability, cultural requirements, and yields to permit reasonable assessments of their potential.

(2) The extent to which they are adapted to representative agricultural production conditions in various parts of the United States, with emphasis on marginal land use. In this connection, plants were sought which might be suitable for production in wetlands; in the South and Southeast where agriculture is only moderately intensive but where rainfall is not a limiting factor; in the Southern High Plains where rainfall is somewhat limiting, insolation is high, and large noncropland acreages exist; and in the Southwest where rainfall is a definite limiting factor but land and insolation are not limiting.

(3) Supplemental water (irrigation) use was not considered; both current and new crops were, therefore, considered in terms of their potential with rainfall-derived moisture. (Analyses of biomass production from current crops, however, included existing irrigated acreages and the yields which are reported for irrigated crops.)

(4) The ability to produce reasonably well with moderate fertilization was considered to be an important criterion. Thus, high yields observed for some plants with fertilization rates of 200 to 800 lb N/acre were discounted. Such a limiting criterion should admittedly be further examined in both economic and energy/resource terms in future studies.

(5) For marginal land use, particularly on erodible soils, both the capability of the plant to utilize marginal resources and its compatibility with soil conservation needs were important criteria. Thus, perennial, sod-forming species

were preferred for marginal situations to annuals which
demand land-disturbing cultural practices.
(6)   Production costs are an important criterion. Economic considerations were included in the analysis only perfunctorily,
however (e.g., by discounting plants with high fertilization,
requirements). More detailed economic analyses are planned
in Phase II of this study.

In addition to the so-called new crops, current agricultural practice encompasses
a large spectrum of plants and species or hybrids with known characteristics
which are candidates for inclusion in biomass production scenarios. These include perennial sorghums (or other perennials) which are candidates for growth
on erodible soils; hybrids or species of corn, wheat, and other grain crops which
are not presently grown because stalk growth is needlessly excessive in proportion
to grain yields (accompanied by difficulty in harvesting or grain losses due to
stalk instability); species of sorghum excluded from current forage production
for reasons of toxicity or low nutritional values; plants which might logically be
included in biomass or biomass-grain multiple cropping scenarios with the objective of more fully utilizing year-round production resources for biomass or
biomass and grain; species which might be planted in different climatic regions
in order to more fully utilize the growing season for biomass production; and
species which will tolerate planting densities optimized for biomass yield or for
increased biomass residue yield at little sacrifice in grain yield.

From the foregoing discussion, it is readily apparent that several options or approaches can be taken to enhance biomass production. These range from only
modest intrusions into present agricultural practice and current food-forage-fiber
production capabilities (i.e., utilize current crop residues for biomass and divert
relatively small acreages of idle land and pasture/range to biomass crops), to very
substantial intrusions which demand very significant deviations from present practice and might, in addition, seriously impede our capability to produce food
crops. The present study deals primarily with the modest intrusion into present
practice, but lays a foundation upon which analyses of other options can be
based.

Within the context of the definition of the program as presented above, the following approach was taken to survey information on plants and select those considered in the present study.

(1)   Plants and crops of current economic significance in the
United States have been summarily surveyed and their characteristics synopsized.
(2)   The productivity of these plants and crops under current agricultural practices has been documented from existing statistics on an LRA by LRA basis. This documentation has not
been detailed by hybrid or other species variation, rather the
data have been accumulated on the basis of the composite
varieties grown in the region of interest. The breakdown,
thus, goes no further than wheat, corn (and other grains),
alfalfa-alfalfa mixtures, other hay (which includes managed
as well as unmanaged stands of grasses), small grain hay,
corn-sorghum-other silage, corn forage, sorghum forage,
other forage crops, and grasses not harvested for hay (these

may be managed or unmanaged stands). Grass yields were
estimated as a fraction of other hay yields. Yields of grain
crop residues were estimated as a fraction of grain yields.
(Alich, 1976). Yields of hay and forage crops were obtained either directly or indirectly (e.g., dry weight forage
crop yields were calculated from green crop weights) from
Census of Agricultural Statistics.

(3) Symposia and conference proceedings and special reports or
reference works on special topics were surveyed to obtain
lists of plants which have been described and evaluated relative to expanding the scope of agriculture in the United
States through consideration of new crops for special purposes.

(4) Visits or telephone contacts were made with organizations
such as the USDA North Central Research Laboratory in
Peoria, Illinois; the USDA Northern Great Plains Range
Management Research Facility in Mandan, North Dakota;
the USDA Southern Range Management Facility in Wood-
ward, Oklahoma; the USDA Plant Introduction Station in
Savannah, Georgia; and with plant breeding/agronomic specialists to obtain information on potential new crops and
data as on yields of grasses and other crops of interest to
the respective organizations.

(5) Consultants in agronomy at Iowa State University and the
University of Minneosta were retained to contribute the
benefit of their knowledge.

(6) Finally, candidate lists of crops, either new or presently
known and documented for use in the United States,
were developed through the above sources, and supporting literature was obtained in order to more fully docu-
ment the selected plants. The initial list was expanded
upon as the literature was retrieved and additional plants
were indicated.

(7) The new plants were evaluated in terms of the criteria specified, and a limited number selected for inclusion as new
crops in the systems study.

(8) Several plants were identified, largely from the plants now
known to American agriculture, and have been discussed
briefly in relation to their utility for marginal land use,
in double or multiple cropping systems. These were not,
however, included in the various scenarios which were
analyzed and are reported in subsequent sections of this
report.

This overall approach and the resultant data and information reported herein do
not constitute an exhaustive survey of the tremendously large number of species
and varieties which can be considered as potential candidates for grass and grain
biomass. Rather, they constitute what should be viewed as a cross-sectional representation of current U.S. agriculture augmented by opportunities that are reasonable and either available now or close to being available. For present purposes,
the study group has further taken a conservative approach to the question of
yields. For example, kenaf yields reported as possible under ideal conditions
were discussed by reference to actual reported yields in various parts of the

country and through discussions with research staff currently working with kenaf. Wheat, corn, hay, sorghum forage, etc., yields employed in the study are those being experienced in actual production rather than higher yields obtained in isolated cases or research studies.

The above rather conservative approach is appropriate for the moment, since it gives a current, reasonable picture as opposed to a perhaps unachievable futuristic picture. However, allowances do need to be made for improvements in the future, and it is quite reasonable to expect that they will occur.

Many plant species possess genetic attributes which, through intensive selection processes and hybridization may be developed into desirable biomass production species. Further, biomass production as a goal offers the opportunity to tailor species to less productive land and adapt cultural practices suited to such land. However, for purposes of this report, the primary interest is in what is available and discussions are limited to species which have been addressed over fairly broad geographical areas and in sufficient detail to determine their useful potential.

Overview of Grass and Grain Crops

Grass and grain production in the United States is a function of four important variables, namely, topography, soil, water (rainfall and irrigation), and temperature. These factors interact to determine which plant species are adapted to a particular region, and are an important consideration with regard to disease problems. Rainfall or water supply and soil characteristics, together with temperature and growing season length, help to determine yields, and the supply or acreages of suitable soils on land of reasonable topography available in a region determine the magnitude or intensity of the agricultural operation.

The United States is so diverse in these characteristics that it is not easy to generalize, except to state that agriculture has adapted remarkably well to the resources at hand. The North Central states are blessed with generally adequate rainfall, large acreages of fertile well-drained soil on fairly level topography, and a climate which is favorable to a corn-soybean dominated crop system. This region is also well suited to other crops, wheat, for instance, and the current predominance of corn and soybeans is a consequence of crop economics, as well as crop suitability.

The High Plains Area from the Dakotas through Texas generally has a good resource of soil and topography, but is less fortunate than the North Central Region with regard to moisture. The High Plains, therefore, has a higher proportion of pasture and rangeland, emphasizes the more drought-resistant crops such as wheat and sorghum, and augments rainfall with irrigation where possible to grow moisture-demanding crops such as corn and soybeans. Since livestock is an important commodity, forage crops, e.g., sorghum, alfalfa, and native or other cultivated hay/grass crops, are significant.

The more western states vary considerably in rainfall and, thus, are heavily dependent on irrigation, except in certain areas. Where irrigation is practiced heavily, the emphasis tends to be on high value crops such as vegetables, fruit, and rice, but crops in these areas still generally are fairly representative of the cross section of crops grown in the United States.

The Southwest, with a long growing season, lacks rainfall and has extensive acreages with very little productivity in the absence of irrigation.

The East-Northeast regions of the country have ample rainfall coupled generally with a dearth of land with suitable soils and topography with the result that agriculture varies widely in intensity and magnitude in these states. Certain areas are, nevertheless, highly productive.

The South-Southeast also has ample annual rainfall, but is generally less fortunate than the Midwest in the availability of suitable soils on reasonable topography, has greater problems with regard to maintaining soil fertility, and faces some of the problems inherent in agriculture in warm, humid areas. This region grows a considerable variety of crops, including crops such as tobacco, cotton, and peanuts, which are not important further north, and is generally diversified in its agriculture.

The extent and diversity of U.S. agriculture is generally depicted in Table 5.6, which lists acreages of harvested grass and grain crops in several states in different regions of the country for 1975, as well as for all the states. The statistics were derived from *Agricultural Statistics* 1976, a compilation published annually by the U.S. Department of Agriculture.

In addition to acreages of grain and forage crops, yields of grain and forage or hay (dry matter basis) are recorded, and estimated yields of aerial grain crop residues are given. The yield data are for composites of irrigated and nonirrigated land. Worthy of note is the fact that the yields reported in the tables are averages representing 1 year, and the considerable variation in farm management skills used in crop production. While the better managed farms produce at better than the average, the reported results are nevertheless a good representation of what can be expected currently from U.S. agriculture if crop residues or forage crops were to be utilized for energy purposes on a fairly large scale.

Table 5.7 presents data on yields reported in numerous publications and papers for a variety of fairly common grasses in various parts of the United States. These yields are generally obtained from research studies; they reflect the conditions which existed during a specific growing season, and they are, of course, sensitive to the extent of fertilization and other management practices. They, thus, do not necessarily represent the yields which might be achieved on the average under field conditions.

Table 5.8 presents data on less common grass and grain crops. These data tend to be sparse, and one generally needs to exercise considerable caution in their use and interpretation.

A few generalizations are appropriate at this point. It is clear that corn, wheat, and soybeans are the major grain crops, and that in terms of magnitudes of acreages, total grain yields, and total biomass yields, these three constitute the greatest potential biomass for energy resource. The sorghums, small grains, and cultivated hays (alfalfa in particular) are grown on smaller, but still fairly large, acreages, and constitute in the aggregate a substantial portion of the potential biomass for energy resource.

Table 5.6: Representative Statistics on Grass and Grain Crop Production for 1975

	IL	WI	PA	GA	AR	TX	CA	OR	WA	NE	ND	KS	U.S.
Corn grain													
10^3 acres	10,710	2,390	1,080	1,380	38	1,100	38	11	12	5,920	132	1,640	77,902
10^3 ha	4,354	967	437	761	15	445	15	4	13	2,396	53	–	–
Grain, ton/acre	3.25	2.32	2.30	1.54	1.40	2.88	3.05	2.38	2.86	2.38	1.43	2.35	2.41
MT/ha	7.28	5.20	5.15	3.45	3.14	6.45	6.83	5.33	6.41	5.33	3.20	5.27	5.41
Residue, ton/acre	3.58	2.55	2.53	1.69	1.54	3.17	3.36	2.62	3.15	2.62	1.57	2.59	2.65
MT/ha	8.02	5.71	5.67	3.79	3.45	7.10	7.53	5.87	7.06	5.87	3.52	5.79	5.94
Total residue													
Ton, 10^3	38,342	6,095	2,732	3,177	59	3,487	128	29	101	15,510	207	4,248	206,440
MT, 10^3	34,783	5,529	2,478	2,382	54	3,163	116	26	92	14,070	188	3,853	187,282
Wheat													
10^3 acres	1,500	250	345	135	520	5,700	986	1,110	2,740	3,070	123	11,100	51,544
10^3 ha	607	101	140	55	210	2,307	399	449	1,109	1,242	50	4,897	20,860
Grain, ton/acre	1.17	0.93	0.99	0.81	0.90	0.69	1.86	1.41	1.47	0.96	0.77	0.87	0.96
MT/ha	2.62	2.08	2.22	1.81	2.02	1.55	4.17	3.16	3.29	2.15	1.72	1.95	2.15
Residue, ton/acre	2.13	1.69	1.80	1.47	1.64	1.26	3.39	2.57	2.68	1.75	1.40	1.58	1.75
MT/ha	4.77	3.79	4.03	3.29	3.67	2.82	7.59	5.76	6.00	3.92	3.14	3.54	3.92
Total residues													
Ton, 10^3	3,195	423	621	198	853	7,182	3,343	2,853	7,343	5,373	172	19,118	90,200
MT, 10^3	2,899	393	563	180	774	6,516	3,033	2,588	6,662	4,874	156	17,344	81,830
Soybeans													
10^3 acres	8,220	191	1,380	1,260	4,700	370	–	–	–	1,230	149	1,080	53,606
10^3 ha	3,327	77	558	510	1,902	150	–	–	–	498	60	437	21,694
Grain, ton/acre	1.07	0.77	0.84	0.75	0.72	0.74	–	–	–	0.81	0.59	0.62	0.85
MT/ha	2.40	1.72	1.88	1.68	1.61	1.66	–	–	–	1.81	1.32	1.39	1.90
Residue, ton/acre	0.92	0.66	0.72	0.65	0.62	0.64	–	–	–	0.70	0.51	0.53	0.73
MT/ha	2.06	1.48	1.61	1.46	1.39	1.43	–	–	–	1.57	1.14	1.19	1.64
Total residue													
Ton, 10^3	7,562	126	994	819	2,914	237	–	–	–	861	76	572	39,130
MT, 10^3	6,860	114	902	743	2,644	215	–	–	–	781	69	519	35,500

(continued)

Table 5.6: (continued)

	IL	WI	PA	GA	AR	TX	CA	OR	WA	NE	ND	KS	U.S.
Sorghum grain													
10^3 acres	60	–	–	47	200	7,200	207	–	–	1,880	–	3,430	15,484
10^3 ha	24	–	–	19	81	2,914	84	–	–	761	–	1,388	6,266
Grain, ton/acre	1.90	–	–	1.01	1.37	1.46	2.02	–	–	1.54	–	1.18	1.37
MT/ha	4.26	–	–	2.26	3.07	3.27	4.52	–	–	3.45	–	2.64	3.07
Residue, ton/acre	1.20	–	–	0.64	0.86	0.92	1.27	–	–	0.97	–	0.74	0.86
MT/ha	2.69	–	–	1.43	1.93	2.06	2.84	–	–	2.17	–	1.66	1.93
Total residue													
Ton, 10^3	72	–	–	30	172	6,624	263	–	–	1,824	–	2,538	13,316
MT, 10^3	65	–	–	27	156	6,009	239	–	–	1,655	–	2,302	12,080
Oats													
10^3 acres	490	1,350	375	90	60	650	113	80	55	570	1,370	150	13,650
10^3 ha	198	546	152	36	24	263	46	32	22	231	554	61	5,524
Grain, ton/acre	0.86	0.88	0.82	0.72	0.8	0.48	0.85	0.80	0.86	0.78	0.66	0.64	0.77
MT/ha	1.93	1.97	1.84	1.61	1.79	1.08	1.91	1.79	1.93	1.75	1.48	1.43	1.73
Residue, ton/acre	2.33	2.38	2.22	1.95	2.17	1.30	2.30	2.17	2.33	2.11	1.79	1.73	2.09
MT/ha	5.22	5.34	4.98	4.37	4.86	2.91	5.16	4.86	5.22	4.73	4.01	3.88	4.69
Total residue													
Ton, 10^3	1,142	3,213	833	176	130	845	260	174	128	1,203	2,452	260	28,529
MT, 10^3	1,036	2,915	756	160	118	767	236	158	116	1,091	2,224	236	25,882
Rye													
10^3 acres	17	15	16	105	–	40	–	11	10	55	119	15	814
10^3 ha	7	6	6	42	–	16	–	4	4	22	48	6	329
Grain, ton/acre	0.48	0.42	0.45	0.42	–	1.12	–	0.31	0.28	1.54	3.33	0.42	22.79
MT/ha	1.08	0.94	1.01	0.94	–	2.51	–	0.69	0.63	3.45	7.46	0.94	51.09
Residue, ton/acre	0.86	0.76	0.81	0.76	–	2.02	–	0.56	0.50	2.77	5.99	0.76	41.02
MT/ha	1.93	1.70	1.82	1.70	–	4.53	–	1.26	1.12	6.21	13.43	1.70	91.95
Total residue													
Ton, 10^3	15	11	13	80	–	81	–	6	5	152	713	11	33,390
MT, 10^3	14	10	12	73	–	73	–	5	5	138	547	10	30,291

(continued)

Table 5.6: (continued)

	IL	WI	PA	GA	AR	TX	CA	OR	WA	NE	ND	KS	U.S.
Barley													
10^3 acres	14	35	155	3	–	70	1,060	177	400	35	1,990	55	8,711
10^3 ha	6	14	63	3	–	28	429	72	162	14	805	22	3,525
Grain, ton/acre	1.01	1.03	1.20	0.91	–	0.82	1.37	1.20	1.27	0.86	0.91	0.84	0.89
MT/ha	2.26	2.31	2.69	2.04	–	1.84	3.07	2.69	2.85	1.93	2.04	1.88	2.00
Residue, ton/acre	2.30	2.35	2.74	2.07	–	1.87	3.12	2.74	2.90	1.96	2.07	1.92	2.03
MT/ha	5.16	5.27	6.14	4.64	–	4.19	6.99	6.14	6.05	4.39	4.64	4.30	4.55
Total residue													
Ton, 10^3	32	82	425	17	–	131	3,307	485	1,160	69	4,119	106	17,683
MT, 10^3	29	74	386	15	–	119	3,000	440	1,052	63	3,737	96	16,042
Sorghum silage													
10^3 acres	14	–	–	22	14	56	11	–	–	65	–	260	729
10^3 ha	6	–	–	9	6	23	4	–	–	26	–	105	295
Biomass, ton/acre	4.0	–	–	4.3	3.5	4.0	6.0	–	–	3.2	–	3.0	3.3
MT/ha	8.97	–	–	9.64	7.85	8.97	13.45	–	–	7.17	–	6.73	7.4
Total biomass													
Ton, 10^3	56	–	–	35	49	224	66	–	–	208	–	780	2,406
MT, 10^3	51	–	–	36	44	203	60	–	–	189	–	708	2,183
Corn silage													
10^3 acres	255	1,000	440	30	9	70	162	32	47	490	294	260	9,713
10^3 ha	103	405	178	32	4	28	66	13	19	198	119	105	3,931
Biomass, ton/acre	5.50	3.50	4.67	4.67	3.00	5.17	6.73	7.00	6.50	4.00	1.80	3.67	3.90
MT/ha	12.33	7.85	10.47	10.47	6.73	11.59	14.19	15.69	14.57	8.97	4.04	8.23	8.74
Total biomass													
Ton, 10^3	1,403	3,500	2,055	374	27	362	1,025	224	306	1,960	529	954	37,881
MT, 10^3	1,273	3,175	1,864	339	24	328	930	203	278	1,778	480	865	34,366
Hays: Alfalfa													
10^3 acres	740	3,020	820	–	–	200	1,120	420	500	1,730	1,650	980	27,057
10^3 ha	299	1,222	332	–	–	81	453	170	202	700	668	397	10,950
Biomass, ton/acre	2.94	2.54	2.4	–	–	4.18	5.25	2.85	3.12	3.03	1.69	2.54	2.55
MT/ha	6.59	5.69	5.38	–	–	9.37	11.77	6.39	6.99	6.79	3.79	5.69	5.72
Total biomass													
Ton, 10^3	2,176	7,671	1,968	–	–	836	5,880	1,197	1,560	5,242	2,789	2,489	68,995
MT, 10^3	1,974	6,959	1,785	–	–	758	5,334	1,086	1,415	4,756	2,530	2,258	62,592

(continued)

Table 5.6: (continued)

	IL	WI	PA	GA	AR	TX	CA	OR	WA	NE	ND	KS	U.S.
All other hay													
10^3 acres	530	950	1,150	470	730	2,050	530	620	377	2,320	1,890	1,300	34,806
10^3 ha	214	384	465	190	295	830	214	251	153	939	765	526	14,086
Biomass, ton/acre	1.87	1.87	1.69	2.23	1.56	1.87	1.74	1.51	1.51	0.76	0.89	1.34	1.41
MT/ha	4.19	4.19	3.79	5.00	3.50	4.19	3.90	3.38	3.38	1.70	2.00	3.00	3.16
Total biomass													
Ton, 10^3	991	1,777	1,944	1,048	1,139	3,834	922	936	569	1,763	1,682	1,742	49,076
MT, 10^3	899	1,612	1,764	951	1,033	3,478	836	849	516	1,599	1,526	1,580	44,522

	GA	AR	TX	CA	LA	MI	MO	AL	FL	NC	OK	VA	U.S.
Rice													
10^3 acres	-	882	548	525	658	171	18	-	-	-	-	-	2,802
10^3 ha	-	357	222	212	266	69	7	-	-	-	-	-	1,134
Grain, ton/acre	-	2.27	0.27	0.26	0.33	0.09	0.01	-	-	-	-	-	2.28
MT/ha	-	5.09	0.61	0.58	0.74	0.20	0.02	-	-	-	-	-	5.11
Residue, ton/acre	-	2.59	0.31	0.30	0.38	0.10	0.01	-	-	-	-	-	2.61
MT/ha	-	5.81	0.69	0.67	0.85	0.22	0.02	-	-	-	-	-	5.85
Total residue													
Ton, 10^3	-	2,284	170	158	250	17	0	-	-	-	-	-	7,313
MT, 10^3	-	2,072	154	143	227	15	-	-	-	-	-	-	6,635
Peanuts													
10^3 acres	524	-	304	-	-	-	-	206	55	165	115	102	1,504
10^3 ha	212	-	123	-	-	-	-	83	22	67	47	41	609
Grain, ton/acre	1.65	-	0.76	-	-	-	-	1.30	1.62	1.13	1.01	1.40	1.28
MT/ha	3.70	-	1.70	-	-	-	-	2.91	3.63	2.53	2.26	3.14	2.87
Residue, ton/acre	1.95	-	0.90	-	-	-	-	1.53	1.91	1.33	1.19	1.65	1.52
MT/ha	4.37	-	2.02	-	-	-	-	3.43	4.28	2.98	2.67	3.70	3.40
Total residue													
Ton, 10^3	1,022	-	274	-	-	-	-	315	105	219	137	168	2,286
MT, 10^3	927	-	249	-	-	-	-	286	95	199	124	152	2,074

Source: DOE ALO/3729-1

Table 5.7: Reported Yields of Forage Crops

| Crop | Site | Year | . . . . Yield . . . . . | | Source |
			t/acre	MT/ha	
Orchard Grass/Ladino Clover	VA	1953-1954	1.3-2.2	2.9-4.9	Taylor, 1959
Orchard Grass	PA	1969	2.2	5	Berg, 1971
Alfalfa	WA	1947-1949	3-6	7-13	Jackobs, 1956
Kentucky Blue Grass	WV	1971	2.2	5	Jung, 1974
Tall Fescue	WV	1971	2.7-5.3	6-12	Jung, 1974
Tall Fescue	PA	1969	2.2	5	Berg, 1971
Orchard Grass	WV	1971	2.7-5.8	6-13	Berg, 1971
Smooth Bromegrass	WV	1971	2.7-6.2	6-14	Berg, 1971
Crown Vetch	WV	1969-1970	2.7-4.5	6-10	Brann, 1974
Switchgrass	PA	1969	2.7-3.1	6-7	Berg, 1971
Switchgrass	LA	1970s	3-5	7-11	Schaller, 1976
	MD				Matches, 1976
	CA				Anderson, 1976
Pearl Millet	GA	1968-1970	3.6-5.4	8-12	Burton, 1974
Forage Sorghum	LA	1960s	7.8	17	Burns, 1964
Sudan	LA	1970s	3	7	Burns, 1964
Sorghum SXS	LA	1960s	3-8.5	7-19	Wedin, 1970
					Adigun, 1969
Reed Canary Grass	LA	1960s	1.6-6.3	3.6-14	Wedin, 1966
Sorghum Almum	LA	1970s	11	25	White, 1974
	IN	1970s	8-9	18-20	White, 1974
	GA	1970s	4-7	9-16	White, 1974
	TX	1970s	4-7	9-16	White, 1974
Sorghum bicolor	LA		7-12	16-27	White, 1974
	IN		7-10	16-22	White, 1974
	GA		6-8	13-18	White, 1974
	TX		4-6	9-13	White, 1974
Bermuda Grass	GA	1947-1949	3-7	7-16	Burton, 1952
Alfalfa/Crested Wheatgrass	WA	1946-1949	3-5	7-6	Woods, 1953
Alfalfa/Timothy	WA	1946-1949	2-5	4.5-11	Woods, 1953
Bromegrass	MT*	1946-1947	3.7	7-16	Stitt, 1949
Ladino Clover	WI	1947	2.2-2.5	4.5-5.6	Tesar, 1950
Crested Wheatgrass	ND	1949-1959	1-3.8	2.2-8.5	Schaaf

*Irrigated

Source: DOE ALO/3729-1

The remaining cultivated or harvested crops are of relatively minor significance when viewed in the context of a currently available resource. This does not, of course, mean that any one of the minor crops could not be a significant resource on a regional or site-specific basis.

With regard to yields of biomass on a per acre or hectare basis, corn stands out for crop residues, grain, and total biomass. The forage sorghums yield close to corn on a total biomass basis. Alfalfa and alfalfa mixtures yield well (2.5 to 3 tons/acre; 5.5 to 6.5 MT/ha). The other hay class is generally inferior to alfalfa. Wheat is of interest because of large acreages and intermediate residue yields. The sorghum grains have clearly been bred for low residue to grain ratios, a statement that is generally true for grain crops. The residue yields per acre from soybeans are relatively low, and the potential of soybeans for biomass is further diminished by the need to return soybean residues to the soil for erosion prevention and nutrient recycle.

Table 5.8: Selected Yields of Less Common Plants

Crop	Site-Year	Yield t/acre	Yield MT/ha	Source
Kenaf	MD 1967	4.4–7.2	10–16	White 1970
	NC 1967	3–4	7–9	White 1970
	SC 1967	1.7–3.2	4–7	White 1970
	GA 1964	3.2–4.6	7–10	White 1970
	1967	5.1–8.2	11–18	White 1970
	KS 1967	4.3–5.1	10–11	White 1970
	TX 1967	5.1–6.8	11–15	White 1970
	TX 1966	4.1–6.0	9–13	White 1970
	LA 1964	7.5–8.8	17–20	White 1970
Giant Reed (Arundo Donax)	Texas (Rio Grande)	5–8	11–18	Perdue 1958
	Italy	15	34	Perdue 1958
	India	3–6	7–13	Perdue 1958
	Argentina	4–8	9–18	Perdue 1958
Guayule	TX-Beaumont, irrigated	2.8*	6*	Retzer 1947
	CA-King City, irrigated	2.9*	6.5*	Retzer 1947
	CA-Bakersfield, irrigated	5–6*	11–13*	Retzer 1947
	TX-Beaumont	1.6*	3.6*	Retzer 1947
	CA-Chico	2–2.6*	4.5–5.8*	Retzer 1947
Roselle	GA 1968	6–7.5	13–17	White 1970
Jojoba beans	Southwest	0.25–0.75	0.5–1.5	Yermanos 1974
Hemp	North Central	2–2.5	4.5–5.6	Robinson 1943
Jute fiber	Pakistan	0.6	1.3	Nelson 1960
Flax	MN 1945-46	2.1	4.7	Robinson 1949
Kudzu	AL	3.3**	7.4**	Richardson 1945
	AL, fertilizer	6**	13**	
Sunflower	MN	5	11.2	Drage 1952
	MN, fertilizer	8.8	20	Drage 1952
Aeschynomene	Mexico	5	11	White 1971
Sesbania exaltata	Brazil	7	16	White 1971
Urena lobata	AL	5	11	Ergle 1944

*(2 yr)
**green wt

Source: DOE ALO/3729-1

It is evident that a variety of grasses are available for possible expanded use, particularly within a framework of utilizing the more marginal lands for greater productivity than with unmanaged stands of native grasses. Candidates here include Sorghum almum, perennial Sudans, switchgrass, Bermuda grass, and reed canary grass.

Of the so-called new crops, giant reed (*Arundo donax*), kenaf, guayule, guayule/mariola, and cattails have been selected as meeting criteria of quality of supporting data, adaptability to specific areas or situations, and biomass yield potential.

Characteristics of Plant Species Relative to Biomass Production for Energy

General Criteria for Selecting Plants: There are a number of characteristics desirable in selection of plants for conversion to utilizable energy. Prior to selection of criteria, however, a decision must be made as to which form of energy is sought. For instance, plants which produce a large proportion of complex polysaccharides such as cellulose and lignin, are not as suitable for production of alcohols through fermentation as species which produce and maintain a high proportion of simple sugars. Conversely, if the vegetative material is to be burned to produce electrical energy, then molecular complexity is of no consequence; rather, the species, which produces the greatest caloric value, would be sought, provided other useful criteria are met.

Naturally, there will be no one species which can be grown successfully over the wide environmental gradients found in the continental United States. Therefore, a number of species must be sought which will adequately provide the desired characteristics of high useful productivity with nominal soil depletion rates.

There are a number of characteristics which are desirable in considering species suitable for biomass production for conversion into readily available energy. Generally, these characteristics may be categorized into four areas of concern: genetic characteristics; environmental concersn; adaptability to mechanized production, harvest, and storage; and efficiency of conversion to desired energy form.

Genetic Characteristics — The genetic characteristics of plants featured for biomass production are highly complex. First, the species selected should have a degree of genetic plasticity. The genetic code, which regulates expressed characteristics, appears to be constant throughout the range of conditions that the organism can tolerate, in some species. Other species, conversely, display a variety of characteristics, phenotypes, over the same environmental gradient. When the expression of a given genotype can be altered through environmental influences in a number of polymorphs, it is then considered to have a high degree of genetic plasticity for the given characteristic (Bradshaw, 1965). Genetic plasticity not only provides for adaptability of the species over a wide range of environmental gradients, but also allows more latitude in plant breeding and selection research efforts. In general then, a species, which exhibits a number of polymorphisms for various environmental gradients, will be highly desirable.

Another highly important genetic factor is the method by which plant species propagate. A perennial species may propagate either by seed, or in many instances, vegetatively by stolons, rhizomes, layering, etc. In fact, a number of species are confined to vegetative reproductive processes; of note are polyploid species, which produce infertile seed. Conversely, annuals are limited to reproduction by seed since mortality occurs for parent plants annually.

A plant species, which is perennial and has the ability to reproduce without seed, is highly desirable both environmentally and economically. Species, which reproduce vegetatively, are generally sod formers, and have a complex root system which reduces soil losses in comparison to species with less extensive root systems. Economically, species which reproduce vegetatively are superior. First, there is no cost for revegetating on an annual basis. Secondly, harvest may be achieved at any stage of phenological development desired. With a species which reproduces by seed, harvest must be deferred until after seed development, or stand density will be reduced.

Another favorable genetic characteristic, when considering interbreeding programs, is the propensity of certain species to produce allotetraploid zygotes (Levine, 1969). For instance, the wheat, *Triticum aestivum*, has a diploid number of 42. This is achieved by crossing *T. monococcum* (which has a haploid chromosome of 7) with *T. turgidum* (haploid number of 14). In this manner, viable hybrids may be formed which probably will have several characteristics dissimilar from either parent stock. Other characteristics, which are genetically controlled and have bearing on selection of species to be featured in biomass programs, are:

> Morphology: The particular form of a plant is highly important. Whether it is grown upright with an unbranching stem or prostrate and filiform is extremely important in harvesting procedures.
>
> Production of toxins: Species, which produce toxic or other deleterious material, should be avoided if this material is to be disseminated into the environment.
>
> Energy conversion efficiency: Plants may vary widely in their ability to assimilate energy into metabolites and, thus, exhibit wide variances in growth rates. Species, which have high photosynthetic capabilities, should be chosen over those which are less efficient.
>
> Disease resistance: Plant species, which are subject to a number of diseases which reduce growth rate and biomass accumulation, should be avoided. Also, species or varieties, which are highly susceptible to a particular disease resulting in mortality or severe decrease in growth and in which the geographical range of the species and contagia are similar, should be avoided.
>
> Other: There are a great number of other genetic characteristics which must be considered. However, the foregoing illustrate the importance of this factor in selection of species to be considered in biomass production-energy productivity.

For any given species or variety of plant, there is a range of tolerance to environmental parameters under which they are able to survive. Similarly, there is also an internal range of variability within plant species, genetic plasticity, which enables certain individuals to succeed, or be more productive than other individuals within various segments of the range of environmental difference. Thus, the genetic variability within one variety of corn, for instance, may permit greater biomass productivity than may be achieved by other corn varieties with a given set of environmental parameters. Yet this same corn variety may be less productive than other varieties when the environmental conditions are changed.

The mechanisms, which regulate the rate at which biomass is accumulated by any species, may be categorized into two broad areas—environmental variables and physiological resources; Etherington (1975) lists these as shown on the following page.

Environmental Variables

Energy
  Quantity, quality and duration of 0.4 to 0.7 μm
    photosynthetic energy input
  Quantity and duration of $>$0.7 μm input influenc-
    ing heating and water loss effects
  Other photomorphogenic and photoperiodic effects
CO_2 supply—concentration and leaf ventilation
Air and leaf temperature
Soil water
  Potential
  Quantity available
Mineral nutrition
Seasonal cycle
Pathological condition

Genotypic-Phenotypic Variables (Physiological Resources)

Leaf diffusion resistance
  Stomatal and cuticular
  Internal
Carbon pathway in anabolism and catabolism (e.g.,
  C_3 versus C_4 fixation pathway; presence or ab-
  sence of photorespiration)
Composition of photosystems
Chloroplast shape, structure, and distribution in cells
Leaf structure
  Anatomical
  Optical
Plant leaf area
Leaf display
  Angle
  Phyllotaxy, spacing, and overlapping (self-shading)
Concentration of photosynthesate and translocation rate
Endogenous rhythms (e.g., in stomatal aperture)
Developmental stage
Leaf stage
Adaptation and pretreatment effects
Factors relating to environmental conditions, e.g.,
  plant water potential, nutrient status, temperature,
  etc.

To discuss each of these factors in depth is beyond the scope of this report, and
would be somewhat rhetorical. Factors, such as CO_2 supply, net radiation of various wavelengths, and a number of other variables, are generally relatively constant for any given locality on an annual basis. Other factors, such as available
soil moisture, mineral nutrition, etc., are highly important considerations in selecting genotypes for given regions of the United States as potential energy sources.

An important consideration is the energy content of various plants and plant
parts on an energy per unit weight basis. As one would expect, the energy contents of different plant components differ markedly. Morrison (1951) reports
that peanut oil (fat) contained 8,800 cal/g dry weight, while wheat gluten (protein) contained 5,800 cal/g, but that starch, corn kernels, timothy hay, and
wheat straw contained 4,100, 3,975, 3,995, and 4,070 cal/g, respectively. The
caloric values of a wide variety of herbaceous and woody plants from all over
the world are remarkably alike. Leith (1968) determined caloric values for

leaves, stems, and grains of corn and found them to be 4,045, 4,155, and 4,197 cal/g, respectively. Thus, dry crop residues of any form that will be found in the corn or small grain fields can be considered to contain about 4,100 cal/g dry weight or 7,400 Btu/lb.

Environmental Factors — One of the greatest potential detrimental effects of biomass production is the effect on soil properties. Not only is there the probability that erosion rates will be increased, due to larger areas under cultivation, but there will also be a reduction in soil nutrient levels. The effect will naturally be site-specific and species-specific. Agronomists, botanists, and members of other discipline areas are well aware that the nutrient uptake of different species varies, and, therefore, the same crops are not usually planted on the same acre year after year.

Regardless of which biomass species is planted, there will be mineral and nutrient uptake by the plant. These minerals will not be returned to the soil in the usual form by detrital decomposition, if biomass is harvested. Therefore, to maintain soil fertility, it will be necessary to replace macronutrients, and in many instances, micronutrients to maintain productivity.

Plant species, which make less demand on soil nutrient factors per unit of converted energy, are preferable. Legumes and a number of species from various plant families have the ability, through symbiotic association with nitrifying bacteria, to change inorganic nitrogen to organic nitrogen. This then is a preferred attribute of those species which maintain soil nitrogen levels.

Other environmental concerns will vary with the ability of selected species to thrive under site-specific variables such as available water, possible air pollutants, high soil salinity, growing season, temperature, and insolation.

Adaptability to Cultural and Harvesting Practices — The life form of the species is important in adaptability to harvest. It is much easier to harvest plant material which grows in a straight, upright form. Decumbent forms present problems in harvest as do diffuse branching forms. Exceedingly short plants also present harvest problems. Therefore, the majority of the weight of the aerial portion of the plant should be sufficiently elevated so that conventional equipment may be utilized in harvest procedures.

The species selected should be adaptable to storage for substantial periods of time. Any plant material which deteriorates rapidly will not be acceptable in most instances. However, the exact nature of the conversion process must be determined prior to determining criteria for storage life.

Energy Conversion Efficiency — Another major factor in species selection is the efficiency of conversion to ultimate energy forms. A species with high lignin content will not be efficient in a fermentation process, for instance; therefore, total pounds per acre should not be the sole criterion for selecting one species over another. The criterion should be calories of converted energy per acre.

In the United States, there are some 1,398 numbered species of grasses belonging to 169 genera according to Hitchcock (1951). A number of these species have been introduced from other sections of the world, particularly those featured in agriculture. Presently, there is no firm basis for excluding species strictly on the basis of the efficiency of a species in converting photosynthetic energy and other resources (water, nutrients, etc.) into useful energy. The ultimate energy yield per acre is, of course, an important criterion, for this factor is related to energy production costs, to the overall energy balance in production and conversion (the net energy yield) and to the quantity of biomass energy achievable from given acreages in biomass-for-energy production systems.

It could be concluded that viable economics and significant contributions to energy needs will result only if dry weight yields are 11 MT/ha (5 tons/acre) or greater. Such a conclusion is presently not warranted, for it automatically excludes grain crop residues, which are potentially available on large acreages at less than 11 MT/ha (5 tons/acre); the economics of residue use are attractive in that a majority of the cost burden will be borne by the grain crop. Also excluded would be very large acreages in the United States in which rainfall is not normally adequate for production at 10 to 11 MT/ha. In any event, the cost of biomass is a fairly complex function of yield, intensity of management and cultural practices, harvesting, storage, and transportation.

Photosynthesis: The ultimate test of any system of biochemistry is the rate at which carbon dioxide is fixed by field canopies of leaves. Obviously, this field rate is a function, not only of the biochemical pathway, but also of such factors as the amount of leaf tissue present, its orientation in space, the conditions of the environment, and the radiant flux density. Measurements of photosynthesis in the field have been made on wheat and barley in Minnesota (Moss, 1976), wheat in Australia (Puckridge and Ratkowsky, 1971), corn in New York (Moss et al, 1961) and Minnesota (Moss, unpublished), and cotton in Mississippi (Baker, 1965). These experiments have given results which are amazingly similar in many ways; they are summarized in Table 5.9, which shows the considerable variability for maximum canopy photosynthesis.

Table 5.9: Midday Rate of Photosynthesis by Various Field Crops

Species Type	Location	Photosynthesis $(g\ CO_2/m^2\text{-hr})$
Soybeans (C_3)	Iowa	6.5-8.5
Corn (C_4)	New York	10
Corn (C_4)	Minnesota	11
Cotton (C_3)	Mississippi	6
Wheat (C_3)	Australia	5
Wheat (C_3)	Minnesota	10
Barley (C_3)	Minnesota	10

Source: DOE ALO/3729-1

For several years, weekly or continuous measurements of photosynthesis of field grown wheat and barley were made in Minnesota. The seasonal pattern with time is bell-shaped, and the maximum can only be measured for a few days (Moss, 1976). Thus, most field measurements of photosynthesis will show something less than the seasonal maximum for a crop.

It has become clear that individual leaves of C_4 species have higher rates of photosynthesis in high solar radiation flux densities, but that crop canopy rates do not differ significantly. This is due to the fact that leaves in grass-type canopies are oriented in all directions and only a small percentage of the leaf area is exposed perpendicular to the sunlight. At low light flux density, as found on most of the leaf area, there is little difference between C_3 and C_4 crop canopies when the leaf area index (LAI) is great.

It should be pointed out the highest rates of canopy photosynthesis for wheat and barley in Minnesota were measured in mid-June when the weather was cool. The maximum for maize was found in mid-August when it was warm, but when the midday radiant flux density was lower than in June. All maxima coincided with maxima in LAI of the particular crop.

Photosynthesis of a crop canopy is nearly a linear function of the radiant flux density in all crops under a wide variety of conditions, but the slope of the photosynthesis versus flux density relationship (grams CO_2 taken up per calorie light absorbed) depends on the leaf area index, the temperature, the health of the tissue, the fertility of the soil, the availability of water, and other factors. Growth analysis shows that seasonal photosynthesis is described best in many crops and natural stands by the multiple of LAI times time, the leaf area duration (LAD) (Cooper, 1975).

Taken together the results of the many experiments on photosynthesis and the literally hundreds of growth analyses delineate clearly the principle of high productivity by green plants. It can be stated simply: Conditions must be favorable for rapid growth and the growth must be covered with a leaf canopy of a species adapted to the temperature of the particular environment. This suggests that C_3 species will be superior in cool weather and C_4 species will do better at high temperatures. Also, growth rates will be slow when leaf area is slight. These simple facts have many implications for managing photosynthesis to obtain maximum capture of solar energy.

Plant stands seldom function anywhere near the theoretical limiting efficiency of photosynthesis. For short periods of time, many different carefully managed crop species produce about 50 g dry matter per square meter per day (Loomis and Gerakis, 1975). This happens during the maximum growth phase for maize, sugar cane or sorghum, for instance, when vegetative growth is most rapid and biomass productivity is not being sacrificed to form a particular product.

During the last decade, there has been a worldwide cooperative effort among plant scientists to study the biomass yield potential of plants. These results have recently been summarized (Cooper, 1975). The results taken together give us the best information that has ever been available on potential of plants as solar collectors. In that summary, Loomis and Gerakis (1975) give the growth rate results shown in Table 5.10.

C_4 species gave the greatest daily growth rates, as might be expected. From their survey, they concluded that there is a strong latitude dependence for optimum adaptation of C_3 and C_4 species, which correlates with peak radiation levels and the occurrence of chilling temperatures. C_4 species apparently excel in maximum growth rates at low latitudes, but are inferior to C_3 species above $40°$ to $50°F$.

It is interesting to note the maximum solar collection efficiencies in Table 5.10. Although the daily maximum growth rates of the C_4 species are appreciably greater than those of the C_3 species, the maximum solar radiation collection efficiencies are about equal, at slightly more than 4%. This means, of course, that the high daily growth rates of the C_3 species occurred on shorter (and probably cooler) days when the total daily solar flux was lower.

Table 5.10: Maximum Crop Growth Rates and Efficiency of Utilization of Solar Energy

	Maximum Growth Rate (g/m^2-day)	Maximum Solar Collection (%)
C_4 Species		
Bulrush millet	54	4.2
Corn	52	4.6
Sudan grass	51	–
C_3 Species		
Rice	36	3.2
Sugar beet	31	4.5
Soybean	27	4.4
Potato	37	–

Source: DOE ALO/3729-1

It is clear from the worldwide experience in managed plant stands that the greatest biomass production on lands suitable for intensive agriculture will likely result from light being absorbed by a leaf canopy composed of species adapted to the particular local environment. One can postulate with a fair degree of certainty that a C_3 canopy in the spring and fall (and winter where temperatures permit winter growth) and a C_4 canopy in summer would give the highest yearly biomass production in the United States.

The concept that combination crop canopies could yield greater than a monoculture was tested in Minnesota by measuring biomass production of winter rye (C_3) combined with corn or sorghum (both C_4) (Fox, 1975). The rye was planted in the fall. It germinated and established a root system and some ground cover before winter. In the spring, the rye started growth early and quickly established a dense leaf canopy. On May 20, the green, immature rye was harvested for biomass yield and followed with either densely planted corn or sorghum. Late maturing varieties that do not mature before frost in Minnesota were used. After a record early killing frost in Minnesota in 1974, the immature corn and sorghum were harvested and biomass yields measured. The two C_4 species were similar in yield.

The biomass yield (aerial parts only) of the rye was 3 tons per acre (6.7 MT/ha) and the corn and sorghum planted either alone or following the rye yielded 8.5 tons per acre (19 MT/ha). Thus, the C_{3-4} combination yielded 35% more than did the C_4 crops alone. Thus, the principle seems clear and was proven to be true for the single year in which C_{3-4} species combinations were tested to maximize biomass productivity.

Grain-Residue Relationships: A substantial fraction of the biomass produced in U.S. agriculture is crop residues, which generally have little real market value as food or fiber, and are for the most part recycled to the soil where nutrient values are at least partially recycled, and where they contribute importantly to maintenance of soil physical properties and their organic content, and additionally, play a role in minimizing wind and water erosion. Some fraction of the residues is potentially available as an energy resource. In this study, average values (ratios) have been employed to estimate residue yields as a function of grain yields, recognizing that the ratios of grain to residue may be quite variable. The following discussion presents a limited amount of data on crop residues which illustrate the variability which may exist and which additionally present information which can be interpreted as an indication that opportunities exist for enhancing residue yields in dual-purpose crops.

Crop residues, such as cornstalks or wheat straw, have traditionally been a problem rather than an asset to agriculture. Modern wheats are short plants which have a high ratio of grain to straw. Much of the impressive progress that has been made in yield of corn has come about by converting more of the biomass into grain rather than through increasing the amount of biomass produced. In the small grains, the reduction in straw has resulted in varieties which are much less susceptible to lodging and it is unlikely that the yield of straw can be greatly enhanced without causing renewed lodging problems. In corn, however, there appears to be much genetic diversity for stover to grain ratios which can possibly be exploited to develop dual-purpose crops.

Stover to Grain Ratio in Corn — Kipps (1970) gives the value 1 for the ratio of stover to grain for corn as an average for all varieties and conditions in the U.S. This agrees well with the results from experimental work at the University of Minnesota for many varieties in years of adequate rainfall. Stover to grain ratios are, however, subject to considerable variation with change in growing conditions or cultural practices, as illustrated in the following paragraphs.

In 1976, rainfall was not adequate in many places in Minnesota. In 1976, Richard Deloughery and R. Kent Crookston at the University of Minnesota grew two varieties of corn of each of five different maturity classes (75, 90, 105, 120, and 135 relative maturity), at populations of 5-, 10-, 20-, 40-, and 80-thousand plants per acre (12-, 25-, 50-, 100-, and 200-thousand per hectare) at Waseca, Rosemount, and St. Paul. Tables 5.11 through 5.13 give their results by maturity class, plant population, and location, respectively.

Table 5.11: The Effect of Maturity Class on Biomass, Grain and Crop Residue Yield, and Crop Residue to Grain Ratio*

Relative Maturity	Grain Yield (MT/ha)	(t/acre)	Residue Yield (MT/ha)	(t/acre)	Ratio Crop Residue/Grain
75	4.0	1.8	7.2	3.2	1.8
90	4.7	2.1	8.1	3.6	1.7
105	4.7	2.1	8.1	3.6	1.7
120	4.7	2.1	8.5	3.8	1.8
135	4.0	1.8	9.6	4.3	2.4

*Yields at 15.5% H_2O.

Source: DOE ALO/3729-1

Table 5.12: The Effect of Plant Population on Biomass, Grain, and Crop Residue Yield, and Crop Residue to Grain Ratio*

Population (per hectare)	(per acre)	Grain Yield (MT/ha)	(t/acre)	Residue Yield (MT/ha)	(t/acre)	Ratio Crop Residue/Grain
12,000	5,000	4.3	1.9	5.6	2.5	1.3
25,000	10,000	4.9	2.2	7.4	3.3	1.5
50,000	20,000	5.1	2.3	8.5	3.8	1.6
100,000	40,000	4.3	1.9	9.4	4.2	2.2
200,000	80,000	3.4	1.5	10.75	4.8	3.2

*Yields at 15.5% H_2O.

Source: DOE ALO/3729-1

Table 5.13: The Effect of Location on Biomass, Grain, and Crop Residue Yield, and Crop Residue to Grain Ratio*

Site	Grain Yield (MT/ha)	(t/acre)	Residue Yield (MT/ha)	(t/acre)	Ratio Crop Residue/Grain
St. Paul	2.7	1.2	6.9	3.1	2.7
Rosemount	3.6	1.6	7.6	3.4	2.2
Waseca	7.2	3.2	10.1	4.5	1.4

*Yields at 15.5% H_2O.

Source: DOE ALO/3729-1

Waseca had reasonably adequate rainfall during 1976 while Rosemount and St. Paul were very dry. At Waseca, the crop residue to grain ratio averaged 1.4 for all varieties and populations, while the ratios at the dry sites were 2.2 and 2.7. Thus, it is obvious that drought can have a much greater depressing effect on grain yield than on residue yield. In this example, the grain yield in Waseca was 3.2 tons per acre (7.2 MT/ha) and in St. Paul was only 37% as great, 1.2 tons per acre (2.7 MT/ha). In contrast, the residue yields were 4.5 tons (10.1 MT) in Waseca and were reduced by the drought in St. Paul to 3.1 tons per acre (6.9 MT/ha), fully 70% of the Waseca yield.

From these data, it is clear that grain and stover yields are affected differently by different environments. Other work suggests that large differences exist among genotypes in their stover to grain ratios and suggest that varieties could be selected for enhanced stover yields.

Perry and Olson (1975) grew two corn hybrids in Nebraska at four rates of nitrogen fertilizer in 1972 and 1973. Their results are shown in Table 5.14. At fertilizer rates similar to those used by Minnesota farmers, they found the residue to grain ratio to average 1.1.

Hanway and Russell (1969) grew 11 single-cross corn hybrids of varying maturation at different populations in two different years. They found the crop residue to grain ratio to vary among the hybrids (averaged over populations and years) between 1.1 and 1.6.

Table 5.14: Crop Residue to Grain Ratios for Corn in Nebraska

| Nitrogen Fertilizer | Crop Residue to Grain Ratio | |
(kg/ha)	1972	1973
0	1.5	1.2
90	1.3	1.2
180	1.1	1.2
270	1.0	1.1

Source: DOE ALO/3729-1

In data on stover to grain ratios for the different maturity groups and different plant populations in near adequate rainfall in Minnesota in 1976, it is interesting to note that the biomass yields and grain yields appeared to pass through an optimum as the population increased. The residue yield was greatest at the highest population. The relative maturity had a surprisingly small affect on biomass yield. All of these varieties mature normally in Minnesota, however. Greater biomass yields might have been obtained had a variety been included which did not mature. Corn does not do well in cool weather, however, and the cool evenings in September probably cause all varieties to slow production. From the results given, it is clear that there is much that could be done to enhance residue yields in corn, both by way of breeding new varieties and by management of existing varieties.

Soybeans — Soybeans also show genetic diversity for residue to grain ratio. Kipp (1965) gives the average value of 1.5 for soybeans. Auckland (1974) grew many lines of soybeans from different crosses in various planting patterns in the field. He found an average crop residue to grain ratio for all lines and populations of 1.25. Lawn and Brun (1974a; 1974b) report values of total plant and seed yields for Chippewa 64 and Clay varieties of soybeans from which the crop residue to grain ratios can be derived as 1.15 and 0.75, respectively. In another study, they found the ratios to be 1.24 and 0.80 for the same two varieties. Hardman (1970) found the crop residue to grain ratio for field grown soybeans at St. Paul to be 1.7. Hanway and Weber (1971) found ratios of residue to grain yield among eight varieties of soybeans grown in Iowa to vary between 1.8 and 2.4, while the residue yield varied between 1.9 and 2.8 tons per acre (4.3 and 6.3 MT/ha). It is clear that residue yields in soybeans vary markedly among genotypes and that, in fact, the residue yield could be fairly large. Soybeans are grown in the Corn Belt in acreages nearly as great as corn. If an energy recovery system were available that utilized cornstalks, then soybean residues might provide an additional source of energy.

Straw to Grain Ratio in Small Grains — Much of the wheat and barley grown in the United States is in areas of low rainfall. The per acre biomass yields are low relative to crops in more humid parts of the country. Since water is the major limiting factor to yields, it is unlikely that straw yields can be changed greatly from those now found. These residue yields are about 1 to 2 tons per acre (2.2 to 4.5 MT/ha). Thus, the collection of straw for an energy feedstock from much of the country where small grains are grown presents special problems and the potential appears less attractive than utilization of cornstalks.

Grasses and Grains Presently Grown in the United States

Current crops include the following: corn, produced both for grain and forage; sorghum, for grain and forage; soybeans; wheat; other small grains—oats, rye, and barley; rice; sunflower; safflower; hops; peanuts; alfalfa; other legumes produced for hay or forage; native grasses for hay, pasture, or range; and improved/cultivated grasses grown for hay or pasture.

Data on national acreages in various grains and grasses and on yields of grain, residues, and whole plant biomass are summarized in Table 5.15. These constitute the bulk of the resource which has been evaluated in this program. A limited discussion of major classes follows.

Table 5.15: National Summary of Grains and Grasses

Crop	Acres (millions)	National Average/Acre Yields (tons)			Total U.S. Production (million tons)		
		Grain	Residue	Whole Plant	Grain	Residue	Whole Plant
Corn	55–64	2.5	2.8	5.3	110–150	120–170	230–320
Corn forage	8–9			3.4–3.8			33
Wheat	47	0.9	1.6–1.8	2.5–2.7	42	80	122
Rye	4						8
Grain	1–1.5	0.7	1.25	1.95	0.7–1	1.3–1.9	
Rice	1.8	2.4	2.7		4.3	4.9	9.2
Oats	19–25	0.8	2.2	3.0	10–14	29–37	57–75
Grain	13–17						
Barley	~11	0.9–1.0	2.4	3.4	9–10	24	37
Grain	~10						
Sorghum grain	13–17	1.7	1.1	2.8	22–29	14–19	36–48
Sorghum forage	~4			3.5			14
Peanuts	1.5	1	1.2	2.2	1.5	1.8	3.3
Soybeans	43–47	0.8	0.7	1.5	34–38	30–33	64–71
Hops	0.06					0.08	
Sunflower	0.6		2.5			1.5	
Safflower	0.2		0.9			0.2	
Alfalfa	27–28			2.8			75–78
Clover/Timothy	13			1.8			23
Wild hay	0.9			0.9			8
All hay	63			2.0			125
Pasture grass	106			1.1			117
Range grass	365			0.6			220
Federal range	~300			0.5–0.6			150

Source: DOE ALO/3729-1

Current Grain Crops — The grain crops of significance in the United States include wheat, corn, sorghum, soybeans, oats, rye, rice, barley, hops, sunflower, and safflower. Each of these are represented by different species, and several have been extensively hybridized. At this point in the study, no effort has been made to describe the numerous species which exist within the grain crops.

The study has instead been based primarily on grain crop production with species utilized within the current decade and on the yields experienced and reported on an area-by-area basis. The study thus is based for present crops on ground-truth on-farm yields. The major uncertainty posed by this approach involves residue and whole plant biomass yields, since residue weight and yield information is not ordinarily obtained when grain crops furnish the principal or sole economic return. For this information, primary reliance has been placed on residue factors (mass of residue/mass of grain) developed by Stanford Research Institute (Alich, 1976). While these factors are based on a substantial amount of data, they are average or representative values developed for each crop and applied uniformly throughout the country. This approach is not sensitive to species-to-species variations, nor to the variations which occur due to effects such as drought, disease, and varying fertilization and planting densities.

Corn is the major U.S. grain crop and is grown as well for silage on 65 to 75 million acres (26 to 30 million hectares) in the early 1970s. The major producing area is the North Central Region, but corn is grown widely throughout the U.S. where moisture is adequate or irrigation water is available. The average yield of corn grain is about 90 bushels/acre in the U.S. with higher yields (110-bushel average in Illinois in 1972; 150 to 200 bushels/acre maximum) in the Corn Belt. The yield of oven dry residue (stover, etc.) for an average yield of grain is about 2.8 tons/acre (6.3 MT/ha), with average whole plant biomass yields being of the order of 5.3 tons/acre (11.9 MT/ha). Whole plant and residue biomass yields can be substantially higher than these values in heavily fertilized stands with optimum moisture and growing conditions. A fraction of the planted corn (13 to 15%) is harvested for livestock feed for which national average yields on an oven dry basis are in the 3.4 to 3.8 tons/acre (7.6 to 8.5 MT/ha) range.

Corn is grown as a row crop and is best suited for production on better lands, preferably where the danger of soil erosion is minimal. Soil losses are high, however, on much of the corn production acreage and the need for conservation practices is a limiting factor to more widespread production.

The yields of total biomass from corn are among the highest of the grass and grain species. The yields of residues in fact rival whole plant biomass yields obtained with hay crops. Factors of considerable additional significance are the large acreages presently devoted to corn production, which translates to a current biomass resource of considerable magnitude, its widespread familiarity to the American farmer and the accompanying established production technology, and the potential versatility of corn as a biomass crop. With regard to the latter, a considerable background of experience has been accumulated about production of ethanol from corn grain, and the cellulosic content of corn stover is a potential resource for production of petrochemical type products.

Wheat ranks close to corn in importance as a grain crop. The High Plains states are the major producers of wheat, but like corn, adapted species are grown throughout much of the U.S. Kansas, North Dakota, Montana, Nebraska, Oklahoma, and Washington are the major wheat producing states. Wheat will tolerate lower rainfall than corn, and substantial acreages are grown in areas in which land is in production and lying fallow in alternate years.

The U.S. yield of wheat in the early 1970s was about 1.5 billion bushels from about 47 million harvested acres (19 million hectares). These yields translate to 45 million tons (41 million metric tons) of grain, and 80 to 90 million tons (73 to 82 million metric tons) of oven dry aerial residue. Grain yields of 30

bushels/acre, a representative national average, are accompanied by residue yields
of 1.6 to 1.8 tons/acre (3.6 to 4.0 MT/ha).

Wheat production in the major wheat producing areas involves large acreages per
farm and is machinery intensive. Wheat land is susceptible to rainfall erosion,
and particularly in dry years to wind erosion. The latter in particular places a
practical limit on the amount of residue which might be harvested for energy
purposes, as wheat stubble and straw is managed so as to minimize wind as well
as water erosion. With current wheat combining practice, as little as possible of
the straw is cut, and residue harvesting would entail a significant deviation from
current practice. However, the straw can be readily harvested, after grain harvest, with conventional haying machinery.

Rye is planted on about 4 million acres (1.6 million hectares) and grain harvested
from 1 to 1.5 million acres (0.40 to 0.61 million hectares). Grain yields are
about 25 bushels/acre and residue yields about 1.25 oven dry tons/acre (2.8 MT/
ha). Whole plant biomass yield (entire crop harvested) is about 8 million tons
per year (7.3 million metric tons per year) for the 4 million acres (1.6 million
hectares) planted to rye. Rye production statistics are reported for 31 states,
from Oklahoma/Tennessee north.

Rice production is reported for six states, Missouri, Mississippi, Arkansas, Louisiana, Texas, and California, on about 1.8 million acres (0.73 million hectares).
Production is limited by water requirements, either natural or irrigation. Approximately 4.25 million tons (3.85 million metric tons) of grain are produced
each year. The accompanying oven dry aerial residue (excluding hulls) is about
5 million tons per year (4.5 million metric tons), or about 2.7 tons/acre/year
(6.0 MT/ha).

Production statistics for oats are reported for 42 states, with major exclusion
being the southeastern states. Planted acreages varied in 1970 to 1972 from 19
to 25 million acres (8 to 10 million hectares). Approximately 70% of the planted
acreage was harvested for grain, with yields averaging about 50 bushels/acre. The
remainder is utilized principally for livestock feed and bedding. In 1972, 13.6
million acres (5.5 million hectares) harvested for grain yielded 695 million bushels. The aerial residue associated with the grain is estimated to be about 30
million oven dry tons (27 million metric tons) (residue:grain factor, 2.71). The
biomass yield of grain plus residue was thus about 3 tons/acre (6.7 MT/ha).

Barley production statistics are reported for 38 states, with the southern tier of
states being excluded. Planted acreages ranged from 10.5 to 11.1 million acres
(4.25 to 4.49 million hectares) in 1970 to 1972, with approximately 90% being
harvested for grain at average yields of 43 to 46 bushels/acre. In 1972, 10.6
million acres (4.3 million hectares) were planted, and 9.7 million acres (3.9
million hectares) harvested to yield 423 million bushels. The total (grain plus
oven dry residue) and residue biomass yields estimated from these data are 33
and 23 million tons (30 and 21 million metric tons), respectively. Average residue yields are accordingly about 2.4 tons/acre (5.4 MT/ha), and total biomass
yields 3.4 tons/acre (7.6 MT/ha).

Sorghums were planted in 1970 through 1972 on 17 to 21 million acres (6.9 to
9.0 million hectares), with 78 to 80% being harvested for grain. The remainder
was harvested for silage or forage or grazed. Production figures are reported for

24 states. The Northeast, Northwest, and Northern Great Lakes states are not included in statistics on producing states, but otherwise the distribution is widespread. Four states, Kansas, Nebraska, Texas, and Oklahoma, are, however, by far the largest producers; these states planted about 14 million out of 17 million acres (6 to 7 million hectares) in 1970. In 1972, 13.5 million acres (5.5 million hectares harvested for sorghum grain yielded 822 million bushels (23 million tons) (21 million metric tons), and an estimated 14.5 million tons (13 million metric tons) of oven dry residue for a total biomass yield of 37.5 million tons (34 million metric tons). The average yield of residue is accordingly 1.1 tons/acre (2.5 MT/ha), and of total biomass, 2.8 tons/acre (6.3 MT/ha).

Data on yields from the about 4 million acres (1.6 million hectares) of sorghums devoted to silage, forage, and grazing exist only for silage. If one uses yield data for silage, the following data are obtained: the average green ton yield of sorghum silage was 11.8 tons/acre (26 MT/ha), or about 3.5 oven dry tons (7.8 MT/ha). The biomass yield from 4 million acres is thus estimated at 14 million tons (12.7 million metric tons).

Approximately 1.5 million acres (0.6 million hectares) are planted to peanuts in 10 states located in the Southeast, Southwest, and the Virginia-North Carolina area. Yields of peanuts in the early 1970s were about 1 ton/acre (2.2 MT/ha). Peanut hay residues, oven dry basis, are about 1.18 times nut yields, or about 1.2 tons/acre (2.7 MT/ha). The estimated residue yield from 1.5 million acres (0.61 million hectares) is thus 1.8 million tons/year (1.6 million metric tons per year).

Soybeans were planted in 1970 through 1972 on 43 to 47 million acres (17 to 19 million hectares), in essentially all states from Kansas eastward except in New England. About 98% of the planted acres were harvested for beans, at average yields of 27 to 28 bushels/acre. Average yields of oven dry residue are estimated to be 0.7 tons/acre (1.6 MT/ha), and of total biomass 1.5 to 1.6 tons/acre (3.4 to 3.6 MT/ha). Annual yields of residue and total biomass are accordingly 30 to 35 million and 65 to 70 million tons (27 to 32 million and 59 to 63 million metric tons), respectively.

Hops are grown on a small acreage (about 6,000 acres) (2,428 ha) in West Coast states. Yields of residue are of the order of 8,000 tons/year (7,256 MT).

The sunflower is grown on about 600,000 acres (243,000 ha), primarily in Minnesota and the Dakotas. The residues potentially available from sunflower production average about 2.5 tons/acre (5.6 MT/ha), oven dry. Current total production is thus about 1.5 million tons (1.36 million metric tons).

Safflower is grown chiefly in California, on about 200,000 acres (81,000 ha). Yields of residue per acre are low, about 0.9 tons (2 MT/ha), and total tonnage is thus less than 200,000 tons/year (181,000 MT/year).

Hay and Forage Crops in Cropland — Hay and forage production, excluding corn and sorghum forages, in the U.S. is distributed throughout the 48 states on about 60 million harvested acres (24 million hectares). About 15% of the harvested acres are native grasses. Alfalfa and alfalfa mixtures, and clover/timothy are grown on about 65% of the hay acreage. Lespedeza, soybeans, cowpeas, peanuts, and small grain hay make up the remainder. The dominant hay producing states

are Wisconsin, Iowa, Missouri, New York, North and South Dakota, Kansas, Nebraska, California, Texas, Minnesota, and Pennsylvania. Alfalfa yields nationally
are close to 3 tons/acre (6.7 MT/ha); clover/timothy, 2 tons/acre (4.5 MT/ha);
wild hay about 1 ton/acre (2.2 MT/ha); and the remainder generally of the order of 1 ton/acre (2.2 MT/ha). Small grain hay (oats in particular) yields approach those of alfalfa. The average national yield on an oven dry basis is just
slightly less than 2 tons/acre (4.5 MT/ha), with alfalfa being a major contributor
to this average.

If one includes sorghum and corn acreages harvested for livestock feed (excluding
livestock grazing on corn fields after grain harvest), 70 to 75 million acres (28
to 30 million hectares) are consigned to hay and forage production with yields
on an oven dry basis averaging slightly above 2 tons/acre (4.5 MT/ha). The hay
and forage crops thus are a significant fraction of the biomass production resource represented by harvested cropland.

Grasses in Pasture, Range, and Idle Land — Native or improved grasses occupy
quite large acreages of permanent pasture and range and idle land. In 1969,
449 million acres (182 million hectares) were reported to be in permanent pasture and range, and an additional 88 million acres (36 million hectares) classified
as cropland were used for pasture. Relatively standard figures for permanent
pasture and range are about 106 and 365 million acres (43 and 148 million hectares), respectively. Yields of harvestable aerial biomass are not readily available.
Statistics are available from research stations such as the USDA Northern Great
Plains Research Station at Mandan, North Dakota. These indicate variations due
to grazing patterns or frequency of cutting, and differences between species, as
well as variations due to yearly climatic differences.

For this study it was arbitrarily assumed that permanent pasture will produce
at 75% of the yield of other hay as reported in the 1969 Census of Agriculture
(COA), and that rangeland would produce at 50% of other hay yields for the
specific area. Since the majority of land classified as pasture is in higher rainfall
areas, yields per acre from pasture averaged about 1.1 oven dry tons/acre (2.5
MT/ha) with yields approaching 2 tons/acre (4.5 MT/ha) in a very few LRAs in
the Southeast. Rangeland yields averaged about 0.6 tons/acre (1.3 MT/ha), re-
flecting the fact that rangeland occurs chiefly in lower rainfall areas. A notable
exception is the rangeland in the Southeast (chiefly Florida) where yields were
estimated at 1.1 to 1.3 tons/acre (2.5 to 2.9 MT/ha).

Federal rangeland (land in the public domain) occupies about 300 million acres
(121 million hectares), with most of it occurring from the High Plains westward.
Yields from these acreages were estimated in the same manner as for nonfederal
range, with 0.5 to 0.6 tons/acre (1.1 to 1.3 MT/ha) being about the national
average.

Most of pasture and rangeland is subjected to only minimal management, i.e.,
it seldom is fertilized and native grasses are relied upon heavily. Brush and weed
control is an important management practice, particularly for sagebrush and
mesquite. For the most part, however, the land is producing at a rate commensurate with the natural resources; soil, soil fertility, climate, rainfall, and naturally adapted dominant grasses. Much of the land would be capable of increased
yields with other crops in higher rainfall area or with irrigation. On the other
hand, much of the land has characteristics (erodibility, poor soil, and topography) which limit its potential to be more productively utilized. In spite of these

inherent drawbacks, the sheer magnitude of the acreage in pasture and range means that it currently produces biomass in quantities which rival those by cultivated grass and grain crops. The results of this study indicate indeed that 760 million acres (308 million hectares) of pasture and range produce about 40% of the total theoretical grass and grain biomass yield (including federal range).

Documented Plant Candidates for Expanded Utilization

Possibly the soundest base upon which to increase or enhance biomass production consists of information on hand about plant species closely related to those currently used in U.S. agriculture, plus information about plants which grow in this country but may not now be significantly utilized in any productive sense. Such plants are of particular interest with regard to marginal land utilization, or as components of double or multiple cropping systems which might include a grain or forage crop as one component. Some of the possibilities are briefly discussed in this section.

Grasses and Legumes for Cooler Regions — Grasses and legumes for biomass production add an additional use to their present, multiple-use characterization. In contemporary agriculture and closely related activities, it is known that these important plant species are of value for pasture and forage, soil and water conservation, wildlife cover, recreational areas, roadbank stabilization, and their aesthetic qualities.

Production of grasses and legumes on marginal sites (marginal for row crop production) is dependent on soil productivity (including fertilizer applied), plant growth factors (solar radiation, temperature, carbon dioxide, and water), genetic potential, plant pest control (weeds, insects, and diseases), and the cropping system. Land use considerations are very important, and this constraint in itself cautions against an intensive monoculture of annual forages (grasses or legumes), regardless of their productive capacity for biomass or other uses.

Acreages in the U.S. are extensive for rangeland, pastureland, cropland used only for pasture, and grazed forests that are privately owned. In addition, federally owned lands, managed by the Forest Service and the Bureau of Land Management, are extensive.

The USDA Conservation Needs Inventory of 1967 (USDA, 1967) cited extensive needs for lands in all LCCs; i.e., from I to VIII, and Blakely and Williams (1974) pinpointed these for the grazing land resource. They further pointed out that hay or pasture cropland represents 11% of the total forage resource and much of this occurs on soil sites of high productivity. That land in the higher capability group may be pushed into row crop production at the expense of depleting the forage resource is a prime concern. This concern has been discussed by Long (1974) and Wedin, et al, (1975).

Considered herein in terms of yield potentials are sites presently used in perennial or annual grass-legume production in the North Central Region of the U.S. Reported for this region in 1968 was 48% of all the hay produced in the U.S. (Wedin and Vetter, 1970). The area includes over 100 million acres (40 million hectares) in pasture.

Climatic Conditions — The states in the North Central Region vary markedly in average annual temperature, average annual precipitation, frost-free days, and relative humidity. Increased precipitation, as compared to areas more western, was a main factor in causing the tall grasses to dominate the prairie vegetation. The prairie vegetation was associated with soils which were highly fertile, resulting from organic matter buildup and decreased leaching.

Range of Grass and Legume Species — Grasses and legumes considered for their biomass yield, and as used as forage in many instances, have been drawn from a wide range in germ plasm. For the grasses, they are represented by 600 genera, of which 150 occur in the U.S. These 150 genera include 1,500 species. For legumes, there are approximately 400 genera and over 12,000 species. There are about 80 grasses and legumes to which significant attention has been given. In the North Central Region, this list is narrowed considerably to orchardgrass, wheatgrass (international, crested, western, and tall), smooth bromegrass, reed canarygrass, creeping foxtail, tall fescue, timothy, sudangrasses and switchgrass.

When grown for forage purposes, grasses and legumes are harvested by grazing or machine (hay or silage). While yield per se has often been used as a criterion of worth, the forage quality (nutritive value and implied intake) has been a desired factor, often being chosen in lieu of added yield (biomass production). From the data available, then, one must speculate, theorize, or extrapolate to what yields are biologically possible in biomass production. What is economically feasible is also of prime concern, but can perhaps be answered once the yield potential is known.

While only a few of the grasses used presently would be candidates for economic biomass production, selection and improvement within the grasses could evolve better cultivars to meet the requirements of a crop grown for its biomass only on marginal soils. For example, perennial grasses are largely fibrous rooted and may propagate vegetatively to form soil-conserving sods through vigorous rhizomatous, stoloniferous, and tillering characteristics. Within their areas of adaptation, they are long-lived plants, and they tolerate disturbance.

Search for grasses for biomass should not exclude too quickly the various types of grasses. Nonetheless, from an ecological standpoint, Harlan (1959) has cautioned that there are distinct differences between native or introduced grasses and legumes. He pointed out that natives are climax species and the introduced subclimax. As such, the subclimax or introduced ones, thrive under disturbance, respond to higher fertility levels, have efficient means of seed production, withstand grazing and mowing, compete well under use, establish readily in clean seedbeds, and tend to disappear when not used.

Offsetting this thinking is the fact that warm season crop plants, which include perennial grasses such as switchgrass (*Panicum virgatum* L.) are C_4 plants; i.e., they fix carbon dioxide more efficiently in terms of light utilization in photosynthesis. Reed canarygrass (*Phalaris arundinacea*), a cool season perennial grass is a C_3 species and has a reduced efficiency in light utilization because photorespiration occurs, which dissipates up to 50% of the fixation of these C_3 plants. Nelson has pointed out that the C_3 species has the advantage of a longer growing season.

The two aforementioned factors, i.e., ecological characterization and photosynthetic efficiency characterization, have immediate and long-term advantages, respectively. Stated another way, among the perennial grasses ready to be used soon in biomass production, the search should be for those introduced species with associated technologies that are workable, while the longer look should be at the potentially more efficient species. For this comparison data relative to biomass production of reed canarygrass and switchgrass has been examined. In addition, a C_4 annual sorghum species is considered.

Reed canarygrass: This grass is well adapted to poorly drained soils, tolerating flooding for more than a month. It has also been shown to be productive on uplands when adequately fertilized with nitrogen. It does well under dry conditions. It is very winter-hardy, but frost sensitive. Its perennial nature permits early spring growth. (Eight inches of growth is common by May 10 in central Iowa.) It may reach a height of 6 ft or more when fully headed. Stems of reed canarygrass are stout and resist lodging. It grows well on most soils and will tolerate a pH range of 4.9 to 8.2. Yield data, representative of the potential production of reed canarygrass, are presented in Table 5.16.

Table 5.16: Yield Potential of Reed Canarygrass

Location	Cuts	Dry Matter MT/ha	tons/acre	N kg/ha	lb/acre	Reference
Connecticut	4	10.7	4.79	213	190	Decker et al (1967)
New York	4	13.7	6.11	213	190	Decker et al (1967)
Pennsylvania	4	9.5	4.23	213	190	Decker et al (1967)
Maryland	4	9.2	4.10	213	190	Decker et al (1967)
Iowa	3	14.2	6.32	540	480	Wedin (1974)
Iowa	3	3.8	1.69	67	60	Wedin (1974)
Iowa	3	4.3	4.13	270	240	Wedin (1974)

Source: DOE ALO/3729-1

Reed canarygrass responds markedly to nitrogen fertilization. Yields near Ames, Iowa, were in direct proportion to rate of application and split applications were beneficial. Other tests on reed canarygrass in southern Iowa were compared to data from an Ames study, and a marked residual nitrogen response was observed, which increased the next year's yield. When the yields of a late fall harvest (November) plus the early June and late July harvests of the following year were totaled, it was evident that the first increment of N [120 lb/acre (135 kg/ha)] was the most efficiently used. Response of species differed with reed canarygrass using nitrogen more efficiently, particularly at higher levels of nitrogen application (Wedin, 1974).

Sorghum: The common sorghums used for forage are sudangrass, the sorgos and grass sorghums, and grain sorghums. These grasses are coarse and erect; height will sometimes reach 10 ft. The forage sorghums will be taller yet. They are well adapted for marginal sites throughout the country, but of course must be reseeded each year. Of the sorghums grown, approximately 25% are for forage.

Because the sorghums contained a glucoside (dhurrin) which under conditions in the rumen converts to hydrocyanic acid, considerable selection and numerous management practices have been aimed at reducing this potential. Yield per se may have thus been selected against in some cases.

Some yields of sorghums which have been obtained are presented in Table 5.17. Notable is the highest yield from forage sorghum (7.8) and a sorghum x sudangrass cross (SXS) at 8.5. Forage sorghum was reduced in yield when two cuts were taken but sudangrass yielded more with two cuts (4.8) than one cut (3.0).

Table 5.17: Yield Potential of Sorghum in Iowa

| Species | Cuts | . . Dry Matter . . . | | N Rate | Reference |
		MT/ha	tons/acre	(lb/acre)	
Forage sorghum	1	17.5	7.8	90	Burns and Wedin (1964)
Sudan	1	6.7	3.0	90	Burns and Wedin (1964)
Forage sorghum	2	13.0	5.8	90	Burns and Wedin (1964)
Sudan	2	10.6	4.8	90	Burns and Wedin (1964)
SXS	1	14.1	6.3	120	Wedin (1970)
SXS	3	6.3	2.8	150	Wedin (1970)
SXS	1	19.0	8.5	150	Adigun (1969)

Source: DOE ALO/3729-1

There is a strong possibility that other species within the diverse sorghum germ plasm would be excellent biomass producers. White, et al, (1974) described the potentials of sorghums as a source of pulp. They studied nine accessions representing three sorghum species.

These were grown at six locations in the U.S., and greatest yields were in Iowa, Indiana, and Georgia. Yields exceeded 10 tons of dry matter per acre (22 MT/ha) [12.2 in Iowa (27 MT/ha)]. One accession of *Sorghum almum* yielded 11.0 tons (25 MT/ha) of dry matter per acre. These were 12-inch row plantings. The researchers pointed out that performance varied considerably within and among accessions and among locations. Also, pointed out is the wide variation which results from effects of environmental conditions on annuals.

A further consideration must be made in relation to possibilities, that a vigorous sorghum may develop its perennial characteristics to the extent that it would become a weed; i.e., a plant out of place. Appropriate caution would be in order if it were to be grown for biomass.

Switchgrass: This tall perennial sod-forming grass has as its natural habitat the Great Plains area. It grows to 5 ft (1.5 m) in height; it has short rhizomes which promote spreading. It is one of the easiest native grasses to establish, a common problem with the warm season perennial grasses. It produces well on droughty, infertile, eroded soil. Improved cultivars are available.

Within the last decade, switchgrass has been more commonly grown on Iowa and Missouri sites. When fertilized with nitrogen, yields have increased and are typified by those given in Table 5.18.

Table 5.18: Yield Potential of Switchgrass

| States | Cuts | .. Dry Matter... | | N Rate | Reference |
		MT/ha	tons/acre	(lb/acre)	
Iowa	2	6.3	2.8	0	Schaller and Murdock (1976)
Iowa	2	9.8	4.4	120	Schaller and Murdock (1976)
Iowa	2	9.6	4.3	240	Schaller and Murdock (1976)
Missouri	1 or 2	6.0	2.7	60	Matches (1976)
Missouri	3	11.2	5.0	?	Anderson et al (1976)

Source: DOE ALO/3729-1

Generally, it is expected that a species such as switchgrass could yield considerably better as rainfall increases. The grass has not, in practice, been grown for its total dry matter production, and in this respect, yields now being obtained when water and nitrogen are available must be carefully evaluated. Further, Rechentin (1956) has pointed out that the morphology of switchgrass suggests that it is not an ideal grazing plant, because there are only two to four short basal internodes, suggesting few basal buds available for recovery. On the other hand, allowing switchgrass to produce one crop, which could then be removed, is likely to be both advantageous to maximizing annual dry matter yield and maintaining the plant for the following year.

General Considerations — Climatic conditions suggest that rain-fed biomass production from grasses will be maximized when moving from West to East, more particularly from Northwest to Southeast in the North Central states or Corn Belt. This area is also the leader in row crop production, machine harvested forage for large ruminant livestock industries, and where improvement of long-term pasture offers great opportunities. For perennial grass use in the ruminant livestock industry, along with its harvested forage and pasture needs, this should be regarded as complementary with biomass production. Some of the same inputs which will make the beef industry profitable in the future will likely be of benefit economically to biomass production. Use of legumes for nitrogen needs of grasses, double cropping (Helsel, 1976), fostering proper land use to minimize environmental insults, interseeding and overseeding, devising new cropping systems over years, are but a few of the important ones for the decades ahead.

Grasses and Legumes for Warmer Regions: Abbreviated discussion of grasses and legumes for warmer regions are presented below. Some of the grasses used in the southern Corn Belt are also of importance in Kentucky, Tennessee, and Virginia; e.g., orchardgrass and tall fescue. Higher summer temperatures would tend to lower production on these. But the grasses and legumes of potential should be looked for in those adapted to the Midsouth (states as above plus northern areas of Mississippi, Alabama, Georgia, and extending into Arkansas, etc.), Lower South, or Deep South. Moisture is a plus factor in areas east of the Mississippi River, and thus, the humid South has several species which can be considered.

Annual Summer Grasses — (1) Sudangrass: Widely adapted, yields would be expected to be higher than in the North Central states, but it will take more fertilizer to do it.

(2) Pearl Millet: Not as extensively tested further north.

(3) Sorghum x Sudan Crosses and Forage Sorghum: Higher yields, maximized at one cut per season, could be oversown into a more permanent vegetation such as Johnsongrass or Bermudagrass sod, which might be used for grazing at other times.

(4) Cereals, Fall Sown and Barley: Adapted to cooler temperatures. Available early, can be doublecropped with summer row crop or other biomass producer.

(5) Annual Ryegrass: Possibilities for overseeding with off-season harvest.

(6) Lespedeza, Korean: Adapted to wide range of soils and in all southeastern states; responds well to fertility.

Winter Annual Legumes — (1) Crimson Clover: Adapted further north; overseeded in Bermudagrass sods will enhance production.

(2) Arrowleaf Clover: Adapted to lower south. Yield of 8,400 kg/ha (3.75 tons/acre) of dry matter reported in Alabama for rye and Yucchi arrowleaf clover (Hovelande, 1969); three harvests total in November, February, and April.

(3) Red Clover: Winter annual, as summer temperatures limit growth.

(4) Vetches, Clovers, Lupines, and Others: None seem to offer great promise for biomass production, unless by overseeding.

Perennial Grasses — (1) Perennial Ryegrass: Likes cooler temperatures; may winter kill; some yields exceeding 5 tons/acre (11 MT/ha) of dry matter, but with four cuts from May to August (Heath, 1973).

(2) Bermudagrass: Widely adapted; tops killed at freezing, can be oversown with winter annuals. One report at 100, 200, and 400 lb/acre of N had about 4, 6 and 8 tons/acre (9, 13.5 and 18 MT/ha) of dry matter produced (Heath, 1973).

(3) Johnsongrass: Good yields on any soil favorable for cotton or corn, but best adapted for heavy clay soils. In Mississippi about 5 tons/acre (11 MT/ha) of dry matter yield when 500 lb/acre (227 kg/ha) of N was applied (three to four cut total) (Watson, 1970).

Perennial Legumes — (1) Sweet Clover: Adapted for states west of the Mississippi; potential should be checked.

(2) Sericea Lespedeza: Not particularly high yielding.

(3) Crownvetch: Of more recent vintage; widely adapted, but primarily fits north of 35 degrees north, and east of 97 degrees longitude. Should definitely be looked at, maximizes yield into one cutting, needs to be looked at in combination with some southern grasses.

(4) Kudzu: Vigorous grower; could have weedy characteristics; would need to be monitored. It is a rapid grower and maintains itself.

(5) Cicer Milkvetch: Adapted in drier areas west of Mississippi—some promising yields more recently.

Overseeding Bermudagrass sods to give an off-season (for crops) yield is opening up in the South. These yields should be looked at more closely. Such a practice is commensurate with good land use; there are no-tillage contact herbicides to help in getting the overseeding done; and the whole concept of doublecropping or off season cropping would seem to be feasible. Overseeding for winter grazing in the South has not been widely accepted but, of course, dollar return from grazing livestock is not often easy to see. Cash cropping for biomass production should have a following, and it should and must be done as a sound economic and sound environmental alternative.

New Crops for Biomass Production

It would seem that a biomass-oriented agriculture might have a greatly expanded horizon relative to a food-fiber-forage-oriented agriculture. One might thus predict that little used grasses and grains might come to the forefront in biomass agriculture, as might adapted or genetically modified plant species. The present study failed, however, to bring to light more than a very few little-used plants which are sufficiently documented under U.S. conditions to merit serious consideration. Further, agricultural research has so accentuated improved food crop production that plant species modifications have usually not been aimed at biomass production.

The present study emphasized new plants (which, in fact, are likely to be quite old but are new because of only trival exploitation to the present) which show promise for specific situations.

The original list of new crops included weedy plants, plants grown in other countries which might be adapted to this country, and plants which are under study for specialized uses.

Plants which grow in natural systems and which often are termed weeds have been generally excluded from serious consideration; little yield data are available. What does exist would not indicate them to be superior to their nonweedy competitors, and little is known about plant culture, fertilizer response, moisture requirements, and other factors.

The sunflower is an exception to this generalization. Residue yields from sunflower seed production are about 2.5 tons/acre (5.6 MT/ha), high enough to be potentially attractive in special situations, but not higher than many competitive crops. Johnsongrass also is reasonably well documented, and yields are sufficiently attractive to merit its consideration, particularly on marginal lands.

Kudzu is a very vigorous legume, but limited yield data are not encouraging. The only plant which grows widely in natural systems, which has been proposed for serious consideration, is the ordinary cattail, which will be discussed later.

The selection process for new or underexploited plants involves the same basic considerations which have been applied in the past to selection of food, forage and fiber crops. The new crop must be adapted to a significantly large segment of the country, should, within limits of available knowledge, not be subject to uncontrollable pests or disease, should give attractive yields of biomass for the region in which it is adapted, and should not pose special or costly problems in growth or harvest. Additional criteria include adaptability to different types of marginal lands; i.e., lands with poor or erodible soils, droughty soils, or semiarid regions, and wet land, and nominal requirements for fertilization.

The results of preliminary economic analyses of biomass production indicate that costs for fertilizer and other production-enhancing practices pose a fairly high penalty on whole plant biomass crops, where costs of fertilization cannot be ascribed to grain costs.

In summary, the crops of particular interest are, first of all, producers of large volumes of fibrous raw materials. Other general requirements for a species to be considered as a potential source of cellulosic biomass are as follows:

    (a)   desirable chemical composition;

    (b)   wide range of climate and adaptation;

    (c)   ability, in semiarid regions, to produce well under conditions of moisture stress;

    (d)   low annual planting and cultivation costs;

    (e)   ease of harvesting; and

    (f)   stalk storability and minimal storage requirements after harvest.

Among those new or underexploited crops, which thus far appear to have a reasonable potential as producers of cellulosic biomass, are kenaf (*Hibiscus cannabinus*), roselle (*Hibiscus sabdariffa*), guayule (*Parthenium argentatum*), mariola (*Parthenium incanum*), giant reed (*Arundo donax*), and cattail (*Typha latifolia*).

Kenaf and Roselle: Kenaf breeding programs (Wilson, 1964), conducted first in Cuba and then in Florida during and after World War II, led to the development of fast growing, high yielding varieties. However, susceptibility of kenaf to root-knot nematodes and associated root diseases remains a problem.

Kenaf is cultivated widely in the tropics as a fiber plant and appears to have been cultivated originally for the purpose in western Africa. Next to cotton, it is the most widely cultivated fiber plant in the open country from Senegal to Nigeria (Dalziel). The leaves and flowers of kenaf are used as a vegetable and various plant parts are used in medicines and in various native rites.

Kenaf is a fast growing annual crop which generally grows on soils which produce good yields of cotton, soybeans, and corn. The climate adaptation is best in the southeast portion of the U.S., where under favorable conditions, field scale yields of 6 tons of dry matter per acre (13.4 MT/ha) may be grown, or up to 7.5 to 8.0 tons/acre (16.8 to 18.0 MT/ha) in northern Florida (Killinger, 1965). In the Midwest, yields appear to be highly dependent upon climatic factors, especially temperature and soil moisture. Kenaf will produce a fiber crop quickly (90 to 120 days) and requires little care during growth (White, 1970).

Seeding rates will vary depending mainly on location, row width, and the germination percentage of the seed. However, a plant population of 75,000 to 100,000 plants per acre is desirable, which requires 6 to 8 lb of seed per acre. Seed should be planted after the danger of a killing frost is over and when there is sufficient soil moisture. Dry matter yields have generally been better from wide [30 to 40 in (75 to 100 cm)] then from narrow row spacing [15 to 20 in (40 to 50 cm)].

Cultivation is required to control weeds in wider row spacing as no weed control chemicals are available for this crop.

According to Wilson, et al (1964), in dense stands, kenaf plants are largely unbranched and may grow to heights of 8 to 20 ft (2 to 6 m). The bark contains a soft bast fiber which is used in some countries for cordage and spinning. These bast fibers comprise 20 to 25% by weight, on a dry basis, of the stem.

The occurrence of serious disease has not been a problem with test plots. However, under intensive culture kenaf may become subject to some fo the serious diseases that affect cotton and okra. Undoubtedly, the most serious production problem with kenaf is its susceptibility to soil nematodes (Winchester, 1964). Nematode infestation is usually at its worst on light droughty soils of the Southeast and an apparently healthy crop suddenly may become unhealthy, defoliated, and unproductive.

The most satisfactory method of harvesting kenaf has been forage choppers for either the green (high moisture content) or air dried (standing crop killed by frost or chemicals) plants. Green material may be stored like silage or sugarcane bagasse, using bulk methods. Air dried materials can be stored in large piles or stacks.

Seed production costs are higher, and yields are lower in the U.S. than in tropical countries such as Haiti (Dryer, 1967). Most varieties of kenaf will not produce seed except in the Deep South of the U.S., but they will continue vegetative growth until frost. The climate of southern Florida is suitable for kenaf seed production (Jayner and Wilson, 1967).

Roselle (White, 1970) has not received much attention in the Western Hemisphere because of its slow growth rate. It requires about 180 days to produce a satisfactory yield of fiber. However, because of the resistance which can be demonstrated by roselle to rootknot nematodes, renewed interest has been shown in this crop.

Roselle was used in Africa as a food plant for several centuries (Haarer, 1956). However, roselle as a fiber plant developed elsewhere and the fiber types have been reintroduced into Africa only recently. The food plant roselle is called Florida cranberry in southern Florida, where it is used to a limited extent in the preparation of a sauce or jelly.

Because some varieties of roselle possess a high degree of genetic resistance to rootknot nematodes, it has been recommended that these species be grown in place of kenaf on nematode infested land in the U.S. Several disadvantages however exist. These are as follows:

    (a) slow growth rate of roselle increases the difficulty of weed control and prolongs the time a given land area is occupied; and

    (b) sparse seed production of fiber types of roselle in southern Florida.

The cultural, soil, and climatic requirements are very similar to those of kenaf as well as harvesting procedures. As a fiber crop, roselle is more difficult to handle during the ribboning process than kenaf. However, as a biomass crop, this disadvantage would be insignificant.

Guayule and Mariola: Guayule, a shrub which resembles the sagebrush, grows wild in north central Mexico and in the Big Bend Region of southwestern Texas. The plant usually attains a height of 2 to 3 ft (0.6 to 0.9 m) and has crooked, brittle branches. Its slender leaves are grayish-green and it produces inconspicuous flowers on short slender stems. The North American Indians discovered that the guayule plant contained rubber and extracted this by chewing (Whaley, 1948).

Because of the plant's resin content (which made it burn with a fierce, hot flame), it was used to fuel the Mexican adobe smelters in mining areas of northern Mexico. This extensive use resulted in the depletion of thousands of acres of guayule. It was also used as a fuel for the bread ovens of Mexican women in the northern provinces until it was established that the shrub had a more valuable use (Lloyd, 1975).

Where soil conditions are suitable for good growth of guayule, soil moisture is probably the most important factor affecting growth and rubber accumulation in the guayule plant.

Guayule was extensively studied during World War II in the Southwest for its potential to alleviate rubber/latex shortages (see references for McGinnies).

The usual procedure in guayule rubber production (Kelly, 1946) has been to sow the seed thickly in nursery beds, grow the plants to an arbitrary size, clip them within a few inches of the crown, and transplant them to fields at spacings ranging from 14 to 36 in (35 to 90 cm) in rows 24 to 36 in (60 to 90 cm) apart. It is usually more than a year under the most favorable conditions and sometimes several years under less favorable conditions before all the available area of fields planted in this way is fully occupied by plants.

Because guayule seed is so small, special seedbed preparation is required to grow this plant satisfactorily (Tingey and Clifford, 1947). Also, slow growth of the seedling allows the fast growing weeds to overtake them. Therefore, weed control is a problem in field seeded guayule.

Petroleum oil sprays at about 2 weeks after seeding gave almost complete killing of the weeds and no killing of the guayule. Air temperature was a critical factor and had to be between 70° and 80°F as a great many guayule plants were killed at 50° and 94°F.

Studies with guayule culture have been conducted on soils in the arid Southwest which are marginal because of certain excess quantities of salts (Kelly, 1945). The study included both alkaline and neutral soils, which were compared to soils free of salts. It was concluded that guayule production is not feasible on soils containing salts in excess of 0.3% in the 5 ft (1.5 m) profile.

Studies were conducted to determine the effect of irrigation on guayule production. In a study of the effects of soil moisture (Kelly, 1945) stresses on guayule nursery stock, it was found that the dry shrub weight increased with increasing soil moisture, but that the concentration of rubber and resin of the plants decreased at the same time.

In another study (Hunter and Kelly, 1946), soil moisture was varied under five different conditions:

- (a)   at or above field capacity;
- (b)   75% of field capacity;
- (c)   33% of field capacity;
- (d)   an initial heavy irrigation at the beginning of the trial; and
- (e)   no supplemental water at any time.

Shrub samples were taken with 6 to 8 in (15 to 20 cm) of root, defoliated, and oven dried at 100°C. Under the conditions of the test (e) with no supplemental irrigation on sandy loam soil, 1,300 lb/acre (1,450 kg/ha) of annual shrub yield was observed compared to 3,370 lb (3,770 kg/ha) on silty clay loam soil. Test (d) with one initial heavy application of supplemental water yielded 4,100 lb/acre (4,590 kg/ha) of oven dry shrub on sandy loam and 3,040 lb (3,400 kg/ha) on silty clay loam. Tests (a), (b), and (c) were fairly comparable and yielded approximately 8,500 to 10,000 lb/acre (9,500 to 11,000 kg/ha) dry shrub weight on sandy loam soil and approximately 7,000 to 7,500 lb/acre (7,800 to 8,400 kg/ha) dry shrub on the silty clay loam soils.

In studies conducted by Tingey and Clifford (1946) on Greenfield loam, unirrigated direct-seeded guayule yielded 1.7 tons/acre (3.8 MT/ha) of dried shrub and transplanted guayule yielded 2.4 tons/acre (5.4 MT/ha) at 10 in (25 cm) spacing. Direct-seeded, unirrigated at 20 in (50 cm) spacing yielded 1.1 tons/acre (2.5 MT/ha) of dry shrub, and 1.9 tons dried shrub per acre (4.3 MT/ha) for transplanted guayule. There were significant differences between the yield of shrub for the different irrigation treatments and yields were progressively higher for the greater amounts of water.

Cultivation of guayule plantations (McGinnies and Haase, 1975) is no different from that required for other row crops grown in the area. Cultivation early may be supplemented with oil sprays. Later cultivation may be continued until the plants became large enough to shade the ground.

Two methods of harvesting guayule are used (McGinnies and Haase, 1975). One is to undercut the plants to a depth of 7 to 9 in (18 to 23 cm), allow the plants to dry several days, and then rake and bale with a conventional baler. The second method is to harvest the aerial portion of the plant and leave the roots on 2 to 3 year old plants. (The roots of older plants will die instead of sprout.) This latter is a process called pollarding.

Mariola (McGinnies and Haase, 1975) is closely related to guayule and there is a tremendous introgression of mariola genes into guayule in its natural range. For rubber production alone, mariola is not of interest, but from the standpoint of total biomass and extending the range of hardiness, mariola is important. Mariola grows well up into New Mexico, and as far north as the

Grand Canyon of Arizona. In general, it ranges far to the north, far to the east, and far to the south of the natural range of guayule.

In general, the cultural practices for guayule apply to mariola; however, little work has been done specifically on mariola except from a plant breeding standpoint because there is no economic demand for this crop. No yield figures are available; however, the mariola shrub is larger than the guayule shrub, and yields should be equal to or be greater than for guayule.

Giant Reed: Giant reed is a tall erect perennial cane-like or reed-like grass which grows 6.7 to 26.3 ft (2 to 8 m) high (Perdue, 1958). It is one of the largest herbaceous grasses. In English-speaking countries, it is called one of several names—bamboo reed, Danubian reed, donax cane, Italian reed, Spanish reed, or Provence cane—but most commonly, giant reed.

The fleshy creeping rootstocks form compact masses from which arise fibrous roots that penetrate deeply into the soil. The culms reach a diameter of 1 to 4 cm and usually branch during the second year of growth. They are hollow with walls 2 to 7 mm thick, divided by nodes which vary in spacing from 12 to 30 cm. The outer tissue of the stem is of a siliceous nature, very hard and brittle with a smooth glossy surface that turns pale yellow when the culm is fully mature.

Rapid growth, a characteristic of giant reed, may occur at a rate of 0.3 to 0.7 m/week over a period of several months.

Young culms develop at approximately the full diameter of mature cane, but their walls increase in thickness after the initial growing season. The new growth is soft, very high in moisture and has little wind resistance.

Giant reed is native to the countries surrounding the Mediterranean Sea and has been introduced into almost all of the subtropical and warm temperature areas of the world. In the U.S., the plant has been cultivated successfully as far north as Washington, D.C., Virginia, and Missouri.

It has been used as an ornamental throughout the area south of this line, and in the northwestern part of the U.S. has been used along ditches for erosion control. There are abundant wild growths from this use along the Rio Grande River and other areas of the Southwest.

The soil requirements (Carnu, 1945; Wynd, 1948) are not rigid; it will flourish in all types of soils from heavy clays to loose sands and gravelly soils. In nature, giant reed is able to flourish in soils that are apparently very infertile. However, under cultivation, the plant is responsive to improved soil fertility and is particularly favored by abundant nitrogen.

Giant reed will survive extended periods of severe drought accompanied by low atmospheric humidity or periods of excessive moisture. The plant's ability to tolerate or grow well under conditions of apparent extreme drought is due to the development of coarse drought-resistance rhizomes and deeply penetrating roots that reach deep-seated sources of moisture. Giant reed can be seriously retarded by lack of moisture during its first year, but drought does not cause damage to stands 2 to 3 years old. The plant produces the most vigorous growth in well drained soils where abundant moisture is available, such as in areas that are suitable for the cultivation of corn (Romollo, 1952).

Giant reed is a perennial crop. New plantings are commonly started by root cuttings placed in rows 2 to 3 m apart and covered to a depth of about 10 cm. Under extreme drought conditions, it may be necessary to water the crop the first year and either chemicals or cultivation may be necessary for weed control. Established plantings require little attention other than periodic removal of large weeds.

Annual yields of dry matter vary widely. In India, 3.2 tons/acre (7.1 MT/ha) have been reported for wild stands (Raitt, 1913). An area not fully stocked yielded 5 tons (11.2 MT/ha) of dry reed per acre and a fully stocked area yielded 6 tons of dry cane per acre (13.4 MT/ha). An area in India that consisted of many persistent bases of old culms and the growth of a number of years yielded 43 tons (96 MT/ha) dry cane per acre.

In Argentina (Raitt, 1913), yields in infertile, partly fertile, and fertile soils were 4, 6, and 8 tons of dry matter, respectively, per acre (9, 13.4, and 18 MT/ha). In the U.S. along the Rio Grande (Gillespie, 1946), fairly good wild stands yield $8\frac{1}{3}$ tons of oven dry reed per acre (18.6 MT/ha) while poorer stands yield $5\frac{1}{4}$ tons of oven dry cane per acre (11.1 MT/ha). In Italy (Marinotti, 1941), average annual production under cultivation yielded 35 tons/acre (78 MT/ha) of green reed per acre and 13 tons (29 MT/ha) of dry cleaned cane for pulp production.

Cattail: Cattail is a hydrophyte (Snow, 1920) that is adapted to live partly in water and partly in air. This species of amphibious plant has extensive underground or creeping stems which are rooted in the mud and spread rapidly. Thus, while the roots and portions of the stems, as well as frequently part of the leaves, are under water, at least a portion, and often most of the shoot, is aerial. Because of its frequent occurrence at the water's edge, cattail has a wide range of adjustment and may grow for a time as mesophytes or partially submerged.

Owing to variations in the water content of marshes, the rhizome structure of cattail may be adapted to the immediate environment, that is, the prominence of mechanical and conductive tissue allows mesophytic growth, while the abundance of storage parenchyma allows hydrophytic growth. Since both mechanical and conductive tissues are well developed, the plants are able to grow erect without being supported by the water.

Research studies were conducted over a 3 year period at the University of Minnesota (Moss, 1977). Small cattail shoot segments were collected early in the spring of 1974 and planted in 4 x 4 ft (1.2 x 1.2 m) paddies. The paddies were fertilized similarly to field corn. In 1975 and 1976, the plots were sampled weekly to get seasonal patterns, and measurements were made of light penetration into the canopies, seasonal patterns of leaf area index, and individual leaf and canopy photosynthetic rates.

Cattails are plants and have rather ordinary leaf photosynthetic rates. In contrast, canopy photosynthetic rates were similar to maize during that part of the year when the leaf area index was high. The most striking result, however, was the high yearly biomass yields, as shown in Table 5.19. In all 3 years, the cattail biomass yields, including aerial and underground biomass, were nearly double those of nearby maize fields.

Table 5.19: Biomass Yields of Managed Cattails

| | | | Dry Matter Yield | | | |
| | ......Tops...... | | Roots and Rhizomes | | ......Total...... | |
Year	tons/acre	MT/ha	tons/acre	MT/ha	tons/acre	MT/ha
1974	3.7	8.3	13.0	29.1	16.7	37.4
1975	4.6	10.3	15.0	33.6	19.6	43.9
1976	7.6	17.0	14.0	31.4	21.6	48.4

Note: Maize–4.2 tons/acre (9.4 MT/ha).

Source: DOE ALO/3729-1

Miscellaneous Plant Studies

A limited amount of information was obtained in the present study of several other plants and is summarized in the remainder of this section.

Jojoba (*Simmondsia chinensis*) (Yermanos, 1974): The jojoba plant is a desert-oriented plant. It has been studied as a producer of oils. The plant requires several years to come into production. Annual yields of the beans are indicated to be in the ¼- to ¾-ton/acre (0.5 to 1.5 MT/ha) range, which places it in the marginal category for biomass production Rates of production of whole plant biomass are not available.

Bean harvest is presently labor intensive, though machinery for mechanical picking could surely be developed. Jojoba is attractive from the standpoint that it could utilize extensive acreages which are not suitable for crop production, but the relatively low yield of beans and the difficulties involved in establishing intensive standards are a deterrent to consideration of the plant for biomass.

Hemp (*Cannabis sativa*) (Robinson, 1943): Hemp has been grown for fiber in the U.S., primarily the North Central states. U.S. production has for many years been barely competitive with foreign sources of fiber and production is very limited. It requires rich salt or clay loam soils. Marshes have yielded rank growths, but the quality for fiber was low. Yield data indicate that hemp, as it is cultured for fiber, will not be a serious competitor to other plants for biomass production (2 to 2½ tons/acre (4.5 to 5.6 MT/ha) of air dry stalks). However, the indication of high yields of low quality fiber on marshy land is of potential interest.

Jute (*Corchorus capsularis*) or White Jute (*Corchorus olitorius tossa)*: Requires hot steaming climates and rich alluvial soils. It is grown primarily in Pakistan and India (Nelson, 1960). White jute is adapted to very wet (flooded) soils, tossa to higher ground and drained soils. Yields of total plant biomass are not available (fiber yields in Pakistan average 1,200 lb/acre (1.3 MT/ha). Jute could be considered only for very limited areas of the U.S., namely semitropical wetlands.

Flax (*Linum usitatissimum*) (Robinson, 1949): A plant which might be considered. Under southern Minnesota conditions in 1945 to 1946, a 100% stand produced 2.1 tons/acre (4.7 MT/ha) of oven dried straw and 0.65 tons/acre (1.5 MT/ha) of oven dried weed plants. In 1947, a 100% stand produced 2.0 tons/acre (4.5 MT/ha) of oven dried straw.

Kudzu (*Pueraria thunbergiana*) (Richardson, 1945): Kudzu has been used as a cover for depleted infertile soils, and under conditions in Alabama, was observed to yield 3.3 tons/acre (7.4 MT/ha) of green weight when not fertilized and 6.0 tons/acre (13.4 MT/ha) green weight when provided 65 to 130 lb of phosphorus per acre. Kudzu was originally introduced to this country as a promising legume, and has since become a serious pest.

Sunflower (*Helianthus annuus*): The sunflower, which is both a weed and a commercial plant, is perhaps deserving of serious consideration for biomass production. The yields of residue produced in sunflower seed production are of the order of 2 to 3 tons/acre (4.5 to 6.7 MT/ha). These yields are less than those reported by Drage, et al (1952) in test plots in Minnesota: 5 tons/acre (11.2 MT/ha) air dry forage with no manure and 8.8 tons/acre (19.7 MT/ha) air dry forage with 20 tons/acre (45 MT/ha) of manure.

The Buffalo Gourd (*Cucurbita foetidissima*): Adapted to Mexico and the southwestern U.S.; yields up to 1.1 tons/acre (2.5 MT/ha) of seed (National Academy of Sciences, 1975).

Calathea lutea: Yields 0.3 tons/acre (0.8 MT/ha) of wax in wetlands of Central and South America (National Academy of Sciences, 1975).

Guar (*Cyanopsis tetrogonoloba*): In semiarid areas (15 to 35 in; 40 to 90 cm rainfall) of the U.S., yields 0.3 tons/acre of beans (0.7 MT/ha) (Poats, 1960).

Aeschynomene scabra G. Don: A potential pulp source, gives estimated dry matter yields in Mexico of 11.4 MT/ha (White, 1971).

Sesbania exaltata (Raf) Cory (White, 1971): Also with pulp potential, is estimated to yield 17 MT/ha in Mexico.

C. juncea L. (White, 1971): Yields up to 15.7 MT/ha in Brazil.

Euphorbia lagascae: A herbaceous annual, which grows to 60 to 100 cm in height, yields up to 950 kg/ha in Kansas (White, 1971).

Urena lobata: A potential fiber plant, yielded nearly 5 tons/acre (11.2 MT/ha) of oven-dried stalks in Alabama (Ergle, 1944).

Oil- or Latex-producing Plants: The literature on oil- or latex-producing plants is extensive. This literature includes jojoba and guayule, which have been discussed earlier. Much of the literature has emphasized desert-oriented plants, in part because moisture stress in many plants favors higher concentrations of oils and latex in plant materials and seed, and in part because the semiarid regions of the country offer large available acreages which might be benefically used for oil/latex production, and are of little value for more conventional crops unless irrigation water is available.

It is difficult to arrive at definite conclusions regarding the technical and economic feasibility of utilizing such plants for biomass for two reasons:

(a)   the test or research results are somewhat inconclusive concerning the cultural practices required to obtain and maintain producing stands of candidate plants; and

(b)   the research conducted is lacking in descriptive detail about important data such as whole plants biomass yields.

The oil- or latex-producing plants are potentially important in that the oil/latex products are hydrocarbons of varying structures which might have relatively direct or indirect application as petrochemicals or petrochemical intermediates, and might also be considered in the context of crude oils from which fuels, e.g., gasoline or distillates, could be manufactured by relatively standard petroleum refining technology.

Not to be discounted is the possibility of direct combustion, with the oils and/or latexes providing more energy per unit weight than other cellulosic biomass materials such as corn stover or small-grain hay.

The actual potential of such plants under various cultural practices cannot be determined with confidence, however. On a yield basis alone, sorghums, corn, small grains, and hay crops appear generally to be considerably more productive than the oil- and latex-bearing plants.

The rubber tree, *Hevea*, is the best documented, and it has evolved from relatively low latex yields of less than 1,000 lb/acre (1.1 MT/ha) to 2,000 lb/acre (2.2 MT/ha) average, and to 5,000 lb/acre (5.6 MT/ha) for selected high-producing stands and cultural methods.

The evolvement of the rubber tree (*Hevea*) exemplifies the kind of progress which can come from plant breeding and cultural research and field experience. *Hevea* grows in warm humid climates where rainfall is not limiting.

A species of *Euphorbia* has recently been publicized as a fuel or gasoline plant. This plant is adapted to warm, low rainfall [5 to 10 in (13 to 25 cm) annually] areas. A *Science* (1976) release imputes to *Euphorbia lathyrus* yields of 10 to 50 barrels/acre/year.

Under arid to semiarid conditions (5 to 15 in of rainfall), latex yields from guayule have been well documented at about 1 to 3 barrels/acre (300 to 1,000 lb/acre, 0.3 to 1 MT/ha) annually. Yields of this magnitude might be interpreted to be representative of the present state of the art for desert-adapted latex- or oil-producing plants grown without irrigation.

Since there are large acreages of desert or semiarid land, modest yields of oils could nevertheless contribute significantly to energy needs. For example, 90 million acres (36 million hectares) of arid land (or about 5% of the area of the 48 states) producing a latex or oil at the rate of 1 barrel/acre/year would yield about 3% of present domestic oil production. At a 3 barrel/acre rate, nearly 10% of current oil production would result.

ANALYSIS OF BIOMASS PRODUCTION POTENTIAL

General

A data base and analytical procedures were derived to provide the basis for developing estimates of potential biomass availability under a large number of hypothetical scenarios. Many of the scenarios considered have no practical application but they serve to establish bounds or limits from which other more reasonable scenarios may be developed.

Land characteristics data and yield data for current agricultural crops and grasses are contained in the data base on a county basis. For the purpose of these analyses, counties are aggregated into LRAs and analysis results are displayed on that basis. For the hypothetical adaptation of underexploited crops (new crops) to selected LRAs, average values for all of the counties which comprise the LRA are assumed. These estimated yields for new crops may vary within each LRA in accordance with land characteristics.

The use of the LRA as the geographic unit of analysis is a compromise between the precise, albeit voluminous detail which would result if the county were used and the robust resolution which would result if the states or agricultural growing regions were used.

The LRA as defined by *Agricultural Handbook 296*, Soil Conservation Service, USDA, is based on geographic rather than geopolitical characteristics. A further compromise is made with geopolitical reality by modifying the geographic boundaries of the LRA to conform to county boundaries. As a result, some small LRAs are combined. The boundaries of the LRAs of the U.S. as modified to conform to county borders are shown in Figure 5.2.

Analysis Scenarios

Three types of scenarios have been analyzed in this study:

    (a)   theoretical maximum biomass production scenarios from current agricultural production;

    (b)   feasible biomass production scenarios, including new crops on marginal land; and

    (c)   alternative land use scenarios.

The assumptions, constraints, and methodologies associated with each of these scenario types is discussed in the paragraphs which follow.

Theoretical Maximum Biomass Production Scenarios from Current Agricultural Production: This first set of scenarios addresses current agricultural production of the selected grain and grass crops and shows the maximum biomass production potential for acres planted and harvested in the data base years (for each of the major source categories of crop residues, forages, and grasses, and also in various combinations of these categories).

Individual crops also have been analyzed to reveal geographic distribution of yields for each crop. The biomass yields which result from analysis of these scenarios are calculated from actual crop production data for the years 1971 through 1973 (or from 1969 COA data for forage crops) as discussed earlier.

Figure 5.2: Land Resource Areas of the United States

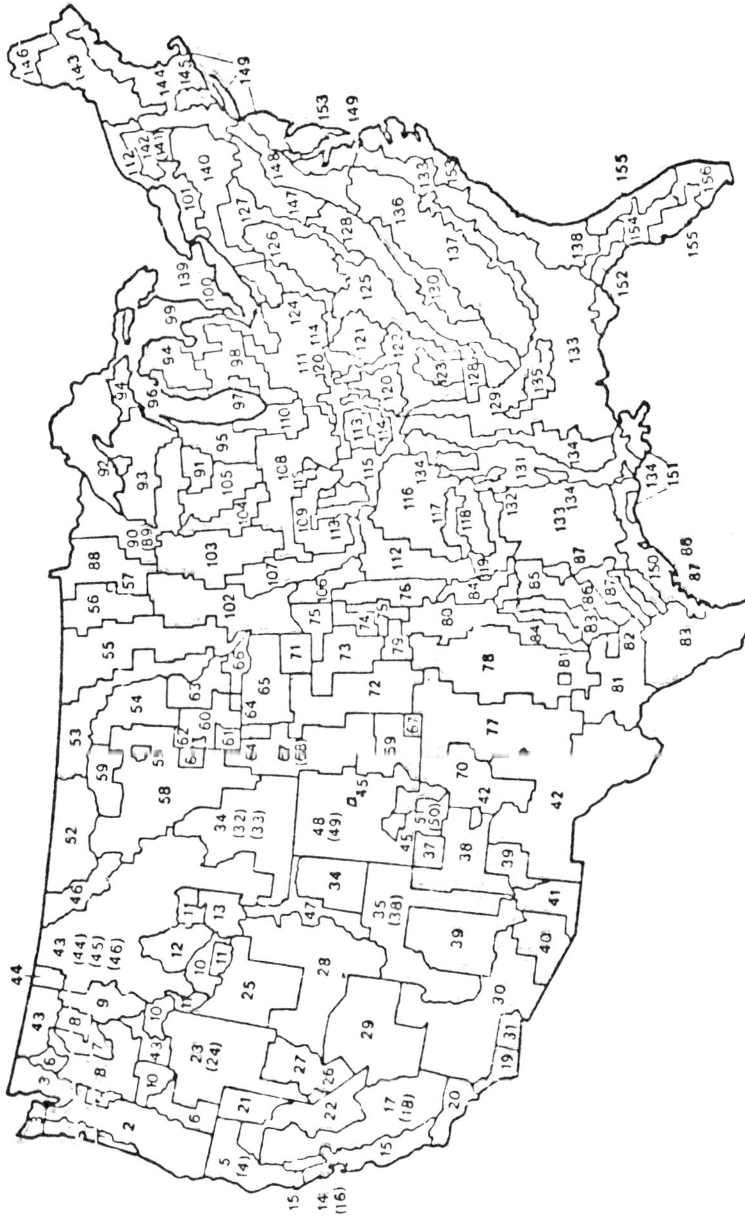

Source: DOE ALO/3729-1

Feasible Biomass Production Scenarios, Including New Crops on Marginal Land:
This second set of scenarios gives recognition to a number of factors relating to
soil conservation practices, competitive price of forages, and the high cost of
harvesting low density grasses.

In these scenarios, the acreage for each food crop is assigned a coefficient which
is indirectly proportional to the average need for residue to be left on the land
for erosion control. For example, a coefficient of 0.75 indicates that nationwide
an average of 25% (1.0 minus 0.75) of the available residue from that crop
would be left on the ground, and thus not be available as biomass for conversion
to fuels.

It must, of course, be admitted that this technique is robust; however, because
of the broad range of estimates of the requirements of different types of soils
for residue which must be left on the ground in accordance with the crop,
cultural practice, geographic location, and so forth, it does provide an average
reduction in biomass availability which addresses the effect of residue removal
on soil condition.

The current market price for forage ($50/ton or higher) raises questions as to
its use as an economical source of biomass. Even if the cost of biomass delivered to the conversion plant gate is $30/ton, the resultant fuel from biomass
would cost $4 to $6/million Btu, depending upon the conversion process used.
However, those forage crops or portions of crops which spoil and have no value
as animal feed could be sold as biomass. Approximately 10% of the annual forage crop yield falls into this category.

Since forage yields already take into consideration residue left on the land for
erosion control, the full 10% of the crop may be considered as potential biomass.

Biomass yields from the grasses grown on pasture and rangeland are substantial.
Indeed, this source exceeds all other categories. However, the yield density
from native grasses is low, averaging ⅔ ton/acre (1.49 MT/ha). Even the highest
density of grass yields rarely exceed 1½ tons/acre (3.4 MT/ha).

Fewer than 1% of the pasture and rangeland acreages attain this value of average
annual yield. The primary use for grasses grown on pasture and rangelands is the
grazing of animals, and thus these crops have competitive value as animal feed
without any cost of collection or harvest. The high cost of harvesting and collecting low density grasses suggests that the economics of its use as biomass for
fuel is questionable.

To avoid excess optimism regarding the potential availability of biomass from
agricultural production, grass crops were not included in this set of scenarios.

Scenarios of this second type do include the concept of bringing portions of
pastureland and rangeland into higher utilization, e.g., the production of biomass
from high yield species grown solely for their biomass content. The selection of
these high yield species (new crops) and of their probable adaptability was discussed earlier. A conservative goal of converting approximately 10% of current
grasslands to new crop biomass production was selected for one scenario. The

method of allocating this marginal land to the new crops was also discussed earlier, but bears repeating at this point. The fundamental assumption is that it should be possible to select a small fraction of the acreage in each LRA which is currently being utilized as pastureland, rangeland, federal rangeland, or other nonfarm land, which could be converted to biomass production without major land preparation effort. The selection process is aided by the designation of selected LCCs appropriate for each new crop and the specification of the fraction of that LCC/LUC combination expected to be made available for the cultivation of the new crop.

With the new crops on 10% marginal land approach, about 7% of grassland, etc. is diverted to new crops, since new crops were not planted in some LRAs. With the less empirical approach based on both LCC and land use, about 74 million acres (30 million hectares) (an average of 9.7% of grassland on a national basis) were planted to new crops.

Alternative Land Use Scenarios: The third type of scenario which was developed for analysis addresses possible changes in the use of a portion of the croplands of the U.S., possibly as a result of changing market conditions in the feeding of cattle, or perhaps as a result of national policy decisions which would emphasize production of biomass for fuel rather than grain for export.

The complex economic factors which might alter the demand for animal feed grains include food price controls, grain price subsidies, inflation, profitability of beef production, supply of imported beef, consumer demand, and a large number of other variables, all of which are beyond the scope of this study.

It is sufficient to speculate that beef growers, in efforts to reduce production costs, might develop feeding regimens which could reduce the demand for feed grains. If this should occur, a portion of the cropland acreage formerly needed for the production of animal feed grains would be vacated and thus be available for an alternate use. These scenarios examine the possible use of such vacated acreage for the production of biomass.

Two scenarios relating to changed feeding of cattle have been developed. One allows the feeding of grain only to animals over 700 lb (318 kg). The other sets 900 lb (408 kg) as the point at which animals are placed on grain feed, and places a further restriction on the feeding of grain to animals over 1,100 lb (499 kg).

The determination of the number of acres of cropland which would be vacated as a result of adoption of each of the reduced-grain-fed-to-cattle scenarios is based on USDA data for the years 1971 to 1973. These years are consistent with the crop production and land use data contained in the analytical data base. The principal grains fed to cattle on feed are corn, sorghum, oats, and barley.

From data on the tonnages of each of these grains fed to cattle and the grain consumption by cattle in each weight class, the fraction of acres of cropland in corn production which would be vacated as a result of a given cattle feeding scenario may be calculated. For example, mutliplying the percent of corn fed to cattle by the percent of grain consumed by cattle under 700 lb shows that a certain percentage of corn acreage would be vacated for the feed grain only to cattle-over-700-lb scenarios. Acreages vacated for other feed grains can also be calculated.

Analytically, this reduction in acreage devoted to food and feed grain production is treated as a fractional reduction applied uniformly to all LRAs which produce the crop. For biomass production, these vacated acres are planted in the highest yield species adapted to each LRA. The reduced-grain-fed-to-cattle scenarios also include biomass available from usable crop residues, spoiled forages, and new crops on 10% of marginal lands.

Reduced Exports Scenarios

The political and economical factors which might cause a marked change in U.S. policy regarding the export of grains are too problematical to be identified at this time. If, however, a significant change did occur in this export market demand, large numbers of acres would become idle.

A scenario was developed to investigate the potential for biomass production under these conditions. The scenario assumed a reduction in exports of 25%. Data from which to determine the number of acres devoted to the production of grain for export are obtained from USDA sources (USDA, 1976) for the years 1971 through 1973. These data are shown in Table 5.20 for the major export crops.

The export fraction is applied to the total acreage for each crop contained in the data base to determine the number of acres devoted to the production of grain for export. The hypothetical reduction of exports of corn by 25%, for example, would vacate one-fourth of 20% of the corn production acreage.

As in the case of the reduced-feed-to-cattle scenarios, the vacated acreage is assumed to be distributed among all LRAs on a proportional basis. The analytical program selects the highest yield crop species adapted to each LRA to be planted on the vacated croplands. Usable residues from grain crops, spoiled forages, and biomass from new crops planted on 10% marginal land are also included in the estimates of total biomass potential for the reduced exports scenarios.

Analysis Results

Current U.S. biomass production from grains and grasses is 1,258 million tons (1,141 million metric tons) annually as shown in Table 5.21. Food grains comprise 21% of this total but are not considered to be available as biomass for conversion to fuels. The relative contribution to biomass production of residues of grain crops, forages, and grasses is also shown in Table 5.21.

The geographic distribution of biomass from current residues, forages, and grasses as measured in yield density (tons per acre) is shown in Figure 5.3. The yield density is for acreage in grain and grass biomass production. Since the location of specific production tracts is not known, the average yield is attributed to the entire LRA. The geographic display of biomass production yields is useful in the identification of those LRAs with high production potential.

Table 5.20: Export Volumes of Grains from Agricultural Statistics

| | Millions of Tons | | | | | | | | Average |
| | Produced | | | | Exported | | | | Percent |
	1971	1972	1973	Average	1971	1972	1973	Average	Exported
Corn									
Tons	158	155	158	157	23	36	35	31	20
Metric tons	143	141	143	142	21	33	32	28	
Sorghum									
Tons	25	23	26	24	3.4	5.9	6.6	5.5	23
Metric tons	23	21	24	22	3.1	5.3	6.0	5.0	
Barley									
Tons	11	10	10	10	1.2	1.6	2.1	1.6	16
Metric tons	10	9	9	9	1.1	1.5	1.9	1.5	
Oats									
Tons	14	11	1	12	0.4	0.4	0.9	0.6	5
Metric tons	13	10	10	11	0.3	0.3	0.8	0.5	
Wheat									
Tons	49	46	51	49	19	36	34	30	61
Metric tons	44	42	46	44	17	33	31	27	
Rice									
Tons	4.3	4.3	5.6	4.7	2.8	2.7	2.5	2.7	57
Metric tons	3.9	3.9	5.1	4.3	2.5	2.5	2.3	2.5	
Rye									
Tons	1.38	0.82	0.074	0.76	0.05	0.27	0.75	0.36	46
Metric tons	1.25	0.74	0.067	0.69	0.044	0.25	0.68	0.33	
Soybeans									
Tons	35	38	46	40	18	14	16	14	36
Metric tons	32	34	42	36	12	13	15	13	

Table 5.21: Current U.S. Biomass Acreages and Production by Major Crop Classifications*

| | Grain Crops | | | | | | | | |
	Grain	% of Total	Residues	% of Total	Forages	% of Total	Grasses	% of Total	Total
Production area, 10^6									
Acres	203	20.0	203	–	50	4.9	760	75.1	1,013
Hectares	82	–	82	–	20	–	308	–	410
Weight, 10^6									
Tons	266	21.2	378	30.0	108	8.6	506	40.2	1,258
Metric tons	241	–	343	–	98	–	459	–	1,141

*Based on 1971-1973 for grain crops; 1969 census of agriculture data on forages; 1967 CNI acreages for pasture, range, and federal range (by differences).

Source: DOE-ALO/3729-1

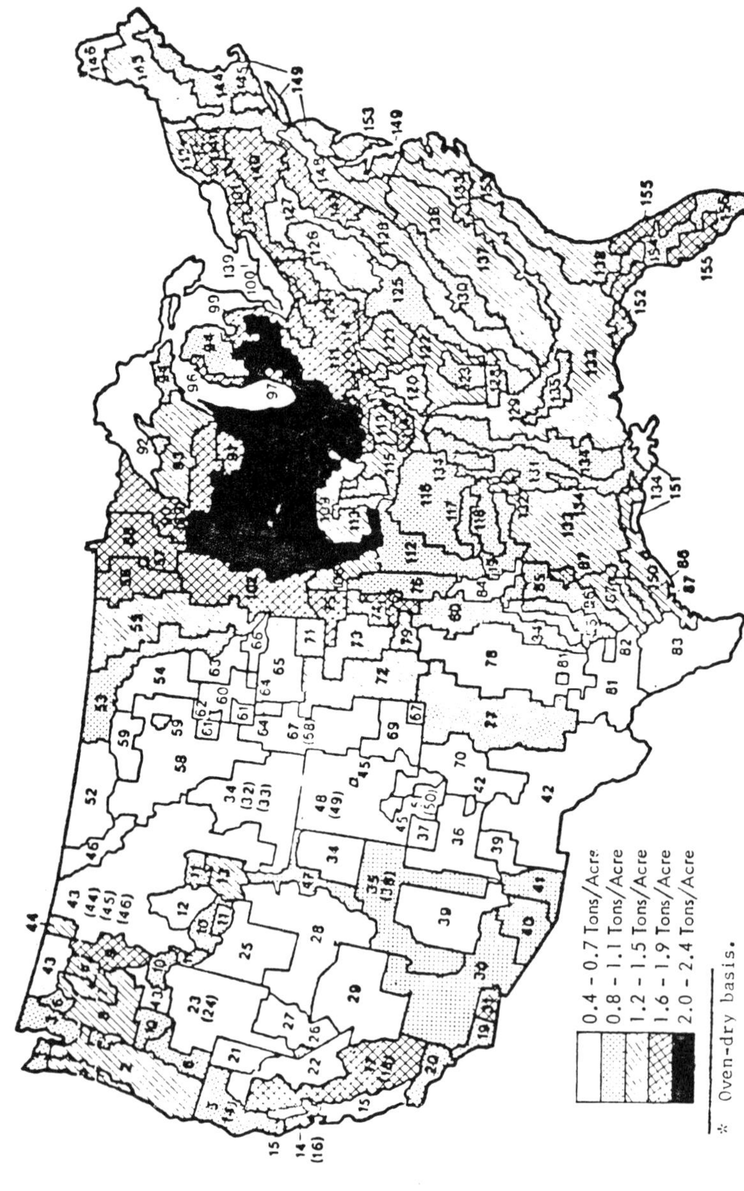

Figure 5.3: Geographic Distribution of Biomass Yields* from Current Residues, Forages, and Grasses by LRA

Source: DOE ALO/3729-1

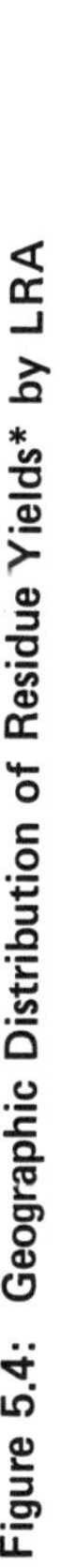

Figure 5.4: Geographic Distribution of Residue Yields* by LRA

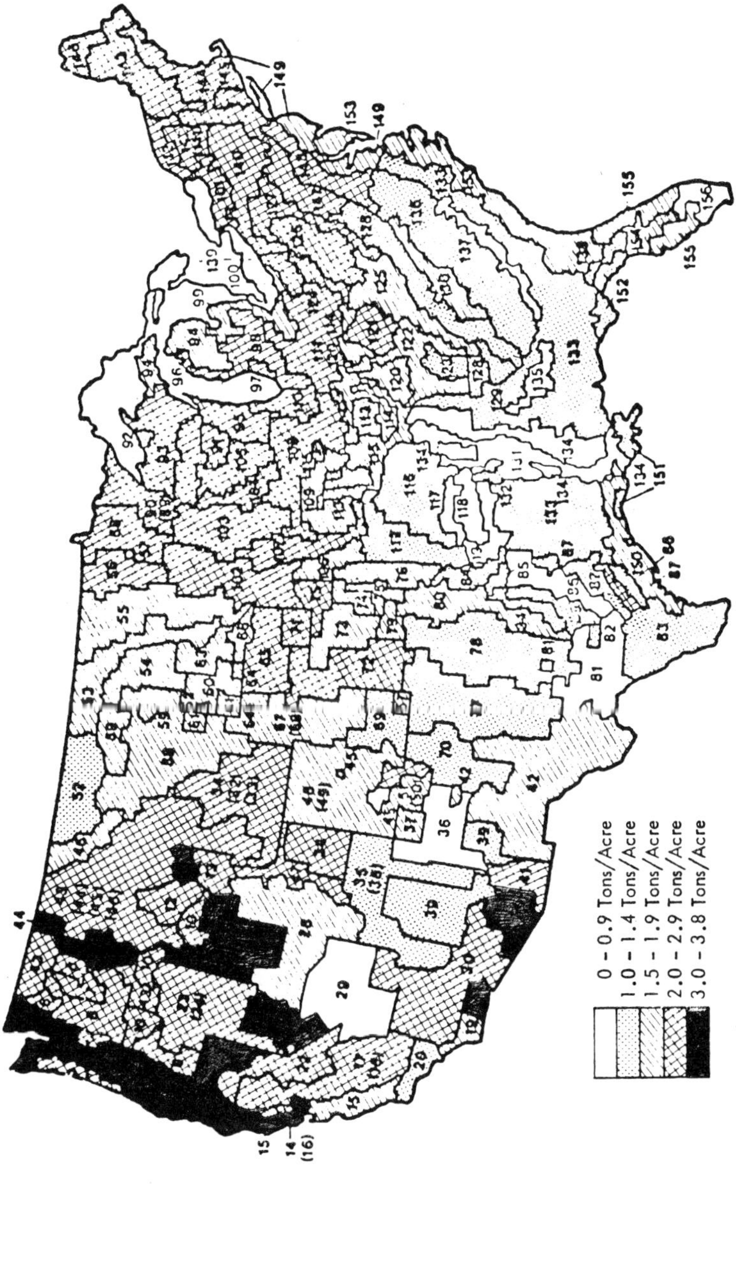

* Oven-dry basis

Source: DOE ALO/3729-1

Table 5.22: Current U.S. Grain Crop Residue Yields*

	Production Area, 10^6		Yield per Unit Area		Total Yield, 10^6	
	Acres	Hectares	Tons/Acre	Metric Tons/Hectare	Tons	Metric Tons
Corn	62	25.1	2.8	6.27	173	156.9
Wheat	49	19.8	1.8	4.03	88	79.8
Soybeans	48	19.4	0.7	1.57	34	30.8
Sorghum grain	15	6.1	1.0	2.24	15	13.6
Oats	14	5.7	2.3	5.15	32	29.0
Barley	10	4.0	2.4	5.38	25	22.7
Rice straw	1.9	0.77	2.6	5.82	5.0	4.5
Peanut hay	1.5	0.61	1.3	2.91	1.9	1.7
Rye	1.2	0.49	1.4	3.14	1.7	1.5
Sunflower	0.6	0.24	2.6	5.82	1.5	1.4
Safflower	0.2	0.08	0.9	2.02	0.2	0.16
Average	—	—	1.8	4.03	—	—
Total	203.4	82.3	—	—	377	342

*Based on 1971 to 1973 data.

Table 5.23: Production of Biomass from Forage Crops*

	Production Area, 10^6		Yield per Unit Area		Total Yield, 10^6	
	Acres	Hectares	Tons/Acre	Metric Tons/Hectare	Tons	Metric Tons
Alfalfa	19.2	7.8	2.5	5.6	48	44
Other hay	10	4.0	1.1	2.5	11	10
Clover	8.8	3.6	1.6	3.6	14	13
Corn silage	6.1	2.5	3.6	8.1	22	20
Small grain hay	1.9	0.77	1.4	3.1	2.7	2.5
Grass silage	1.2	0.48	1.9	4.3	2.3	2.1
Sorghum hay	0.86	0.35	2.1	4.7	1.8	1.6
Sorghum silage	0.73	0.30	3.4	7.6	2.5	2.3
Green chop hay	0.58	0.23	1.9	4.3	1.1	1.0
Sorghum grazed	0.5	0.20	2.2	4.9	1.1	1.0
Corn green chop	0.33	0.13	3.6	8.1	1.2	1.1
Total	50	20	—	—	108	99

*Based on 1969 data

Source: DOE ALO/3729-1

Table 5.24: Biomass Production from Five Assumed Crop Scenarios*

		Corn	Wheat	Soybeans	Grain Sorghum	Oats	Barley	Rice	Rye	Peanuts, Sunflower, Safflower	Total Grain Crops	Forage	New Crops	Total
I	Current crop residues plus new crops, 10% marginal land													
	Acres	62	49	48	15	14	10	1.9	1.2	2.3	203	50	74	327
	Hectares	25.1	19.8	19.4	6.1	5.7	4.0	0.77	0.49	0.93	82	20	30	132
	Biomass - tons	130	66	17	13	24	19	4.5	1.3	3.3	—	11	168	457
	MT	118	60	15.4	12	22	17	4.1	1.2	3.0	—	10	152	415
II	Grain to cattle over 700 lb add new crops on vacated cropland													
	Acres	59.5	49	48	13	13	9.3	1.9	1.2	2.3	197	50	80	327
	Hectares	24.1	19.8	19.4	5.3	5.3	3.8	0.77	0.49	0.93	80	20	32	132
	Biomass - tons	125	66	17	11	22	17.5	4.5	1.3	3.3	—	11	194	473
	MT	113	60	15.4	10	20	16	4.1	1.2	3.0	—	10	179	429
III	Grain to cattle 900–1,100 lb add new crops on vacated land													
	Acres	56	49	48	10	12	8.3	1.9	1.2	2.3	188	50	89	327
	Hectares	22.7	19.8	19.4	4.0	4.9	3.4	0.77	0.49	0.93	76	20	36	132
	Biomass - tons	118	66	17	9	20	15.5	4.5	1.3	3.3	—	11	231	497
	MT	107	60	15.4	8	18	14	4.1	1.2	3.0	—	10	210	451
IV	Reduce grain exports 25%, add new crops on vacated land													
	Acres	59	42	44	14	14	9.6	1.6	1.1	2.3	187	50	90	327
	Hectares	23.9	17	17.8	5.7	5.7	3.9	0.65	0.45	0.93	76	20	36	132
	Biomass - tons	123	56	17	12	24	18	3.9	1.1	3.3	—	11	243	512
	MT	112	51	15.4	11	22	16	5.5	1.0	3.0	—	10	220	464
V	Combine III and IV													
	Acres	53	42	44	9.5	12	7.9	1.6	1.1	2.3	173	50	104	327
	Hectares	21.4	17	17.8	3.8	4.9	3.2	0.65	0.45	0.93	70	20	42	132
	Biomass - tons	111	56	16	8	20	14.8	3.9	1.1	3.3	—	11	307	552
	MT	101	51	15	7	18	13.4	3.5	1.0	3.0	—	10	279	501

*All values given are 10^6.

Source: DOE ALO/3729-1

Table 5.25: Potential Biomass Availability for Selected Land Use Scenarios

Scenario	Total Biomass of ... Oven-Dry Tons ...		... Potential Energy	
	Millions	Metric Tons	10^{15} Btu*	10^9 Gigajoules
Usable residues and spoiled forages	289	262	4.2	4.4
Usable residues, spoiled forages, and new crops	457	415	6.7	7.1
Only feed grain to cattle >700 lb	473	429	6.9	7.3
Only feed grain to cattle >900<1,100 lb	498	452	7.3	7.7
Reduce grain exports 25%	511	463	7.5	7.9
Reduce grain exports 25% and only feed grain to cattle >900<1,100 lb	552	501	8.1	8.5

*Assuming an average lower heating value of 14.6 x 10^6 Btu/oven-dry ton (17 gigajoules/MT).

Source: DOE ALO/3729-1

Focusing on only the residues from grain crops presents a different geographic distribution pattern. This is shown in Figure 5.4. It should be noted that only 1% of the U.S. grain production acreage is included in those LRAs with the highest average yields (3.0 to 3.8 tons/acre) (6.7 to 8.5 MT/ha). Nearly half (48%) of the U.S. production acreage falls in the 2.0 to 2.9 ton/acre (4.5 to 6.5 MT/ha) stratum. These percentages are not shown in Figure 5.4, but are derived from the analysis results obtained in this study.

The relative contribution of the residue from each grain food crop to total U.S. residue yields is shown in Table 5.22. The dominance of corn and wheat in the production of biomass from grain crop residue is evident. It should also be noted that oats produce approximately the same amount of residue biomass as soybeans from less than one-third of the acreage. Rice and sunflower produce residues nearly as dense as that of corn, but differences in the market demand for corn, rice and sunflower seeds is reflected in the acreage devoted to each.

The relative contribution of forage crops to total biomass production is shown in Table 5.23. As noted earlier, forages are grown for their value as animal feed, but it is of interest to this study to note the high yield densities available from the forages, particularly corn silage and sorghum silage. These high yield forages frequently are the selected high yield crop to be planted on the vacated land in the alternate land use scenarios.

Table 5.24 displays the results of feasible biomass production scenarios and alternate land use scenarios, identifying the sources contributing to the total biomass production under each scenario.

A summary of potential biomass availability and the associated energy potential for selected analysis scenarios is shown in Table 5.25. The biomass available is shown in oven-dry tons, and the energy potential is calculated using a lower heating value for biomass of 14,600,000 Btu/dry ton (17 gigajoules/MT).

The energy potential from usable residues and spoiled forages alone is significant when it is considered that it is equivalent to the Btu contained in 2 million barrels (318,000 m^3) of oil a day for a full year. The introduction of under-exploited plant species (new crops) on marginal land (primarily pasture and rangeland in LCCs I through IV) increases the energy potential to 6.7 quadrillion Btu (7.07 x 10^9 gigajoules) or the energy equivalent of over 3 million barrels (477,000 m^3) of oil a day for a year. The results of the analysis of other alternate land use scenarios in Table 5.25 indicate even greater potential.

FINDINGS

The major findings which bear on the question of the potential for production of biomass from grain and grass sources are listed below.

(a) The food component of current grain production is not appropriate as biomass for conversion to fuels for both economic and moral reasons. This conclusion does not necessarily apply to periods of overproduction or to food crop acreages added for deliberate production of grains for energy.

(b) Current grain crop residues offer a high potential for providing a major contribution to biomass production for conversion to fuels: the equivalent of 4 quadrillion Btu (4.2 x 10^9 gigajoules) from this source alone, even when residue requirements for soil conservation needs are taken into account. Corn is the most important residue crop, accounting for nearly half of the biomass from residue production. Wheat and soybeans follow, although the majority of soybean residues need to be returned to the land. Alternate uses for grain crop residues, such as fodder for cattle or perceived need for residues to remain on the land to provide nutrients and improved tilth may serve to reduce the availability of these residues.

(c) Forage production is intended as feed for animals to provide food for or serve man. The economic value of these current forage crops as feed places them out of reach as a source of biomass for conversion to fuel. The approximate 10% of the annual crop of forages which becomes spoiled could have salvage value as biomass. The high yield of corn silage [3.6 tons/acre (8 MT/ha), average] suggests a possible use for corn as whole plant biomass when in excess of market demands for forage or grain.

(d) Native grasses grown on pasture and rangeland account for an impressive 500 million tons (450 million metric tons) of biomass annually. However, the low density of grasses [0.75 ton/acre (1.7 MT/ha), average] indicates that the high cost of harvesting native grasses would make its use as biomass for conversion to fuels uneconomic. The large quantity of acreage devoted to pasture and rangeland

[760 million acres (308 million hectares)] makes it an attractive resource for possible upgrading to a more productive use.

(e)   Underexploited species (new crops) which could be incorporated into the U.S. agricultural production system, largely on underutilized land, have a high biomass production potential. The conservative yield and adaptation estimates used in this study project biomass yields from a mix of new crops to be 60% of that of usable residues, when only 10% of pasture and rangeland is converted to new crop production. Giant reed, a drought-resistant perennial, is particularly promising as is cattail, a species suited to wetlands including peat bogs. Yield and adaptability data on new crops are limited and must be used with caution.

(f)   The availability of underutilized (marginal) land for conversion to the production of biomass from high-yield crops is sufficient to warrant serious consideration of integrating new crops on marginal land into the U.S. agricultural system, or of expanding the acreages currently under production with the more promising current grain, hay, and forage crops.

This study has made robust projections of this potential, but ownership patterns of land, and compatibility of the new crop cultural requirements with local agricultural production practices and equipment will be important factors influencing specific cases.

About 250 million acres (101 million hectares) of grass or idle lands are in LCCs I through IV and an additional 300 million acres (121 million hectares) of grass or idle land are in LCCs V through VIII. The potential of such resources is limited by problems inherent in the land, by climate and rainfall, by the availability of adapted species, and by current production uses.

Nevertheless, this land is a candidate for introduction of new crops and extended use of current crops for biomass production, with careful consideration being given both to the specific modes of use and to possible conflicts with current use. A reasonable goal would appear to consist of utilization of about 75 million acres (30 million hectares) for biomass production.

CONCLUSIONS

A number of significant conclusions may be drawn from the findings of this study.

(a)   The biomass production potential from current grain and grass sources of U.S. agriculture is of such a magnitude to warrant serious consideration of this resource as part of the U.S. energy supply system.

(b) Underexploited crop species (new crops) adapted to the contiguous U.S., and suitable for planting on noncrop land (underutilized or marginal land) show promise as whole plant biomass crops.

(c) Large acreages of underutilized (marginal) land are available for conversion to the production of biomass from underexploited crop species (new crops) or expanded use of current crops.

REFERENCES

Adigun, E.O. (1969) "Dry Matter Accumulation and In Vitro Digestibility of Pasture-Type Sorghums as Affected by Planting Dates", unpublished Master of Science thesis, Iowa State University, Ames, Iowa.

Alich, J.L. (August 1976) *An Evaluation of the Use of Agricultural Residues as an Energy Feedstock*, Stanford Research Institute.

Anderson, B., Mitchell, M., and Matches, A.G. (1976) "Management of Warm-Season Grasses", Research Report (Special) No. 192, Southwest Missouri Center, University of Missouri.

Atchison, J.E. (1973) "Non-Wood Plant Fiber Pulping–Progress Report No. 4", CA Report No. 52, Technical Association of the Pulp and Paper Industry, Atlanta, Georgia, pp 69–88.

Auckland, A.K. (1974) "Inheritance and Interrelationships of Canopy Characteristics and Specific Leaf Weight in Soybeans", Ph.D. thesis, University of Minnesota.

Austin, M.F. (1972) *Land Resource Regions and Major Land Resource Areas Agriculture Handbook 296*, USDA Soil Conservation Service, Government Printing Office.

Baker, D.N. (1965) "Effects of Certain Environmental Factors on Net Assimilation in Cotton", *Crop Sci.* 5:53–56.

Benedict, H.M., and Kiofchek, A.W. (1947) *J. Amer. Soc. Agron.*, Vol 39.

Bhatt, G.M. (1976) "Variation of Harvest Index in Several Wheat Crosses", *Euphytica*, 25:41–60.

Blakely, B.D., and Williams, R.E. (1974) "Our Grazing Land Resources", *Grasslands of the United States*, Iowa State University Press, Ames, Iowa, Chapter 2.

Burns, J.C., and Wedin, W.F. (1964) "Yield and Chemical Composition of Sudangrass and Forage Sorghum Under Three Systems of Summer Management for Late Fall in situ Utilization", *J. Agron.*, 56:457–460.

Bradshaw, A.D. (1965) "Evolutionary Significance of Phenotypic Plasticity in Plants", *Advances in Genetics*, 13.115–155.

Clark, T.F. (1965) "Plant Fibers in the Paper Industry", *Econ. Bot.*, 19:394–405.

Cooper, J.P., (Ed.) (1975) *Photosynthesis and Productivity in Different Environments*, Cambridge University Press, London.

Carnu, C. (1945) "L'*Arundo donax* (Canne de Provence ou Roseau a Quenquille)", *Petit Casablancais*, 31(12221):3–4 and 31(12222):4.

Dalziel, J.M., *The Useful Plants of West Tropical Africa*, Crown Agents for the Colonies, London.

Decker, A.M., Jung, G.A., Washko, J.B., Wolf, D.D., and Wright, M.J. (1967) *Management and Productivity of Perennial Grasses in the Northeast: I. Reed Canarygrass*, West Virginia Agricultural Experiment Station Research Publication No. 550T.

Drage, B.L., Thompson, M.J., and Caldwell, A.C. (1952) "Long-Time Effect of Applying Barnyard Manure at Varied Rates on Crop Yield and Some Chemical Constituents of the Soil", *Agronomy Journal*, Vol. 44.

Dryer, J.F. (1967) "Kenaf Seed Varieties", *Proceedings: First Conference on Kenaf for Pulp Processing*, Gainesville, Florida.

Ergle, D.R., Robinson, B.B., and Dempsey, J.M. (1944) "Malvaceous Bast Fiber Studies", *Journal of the American Society of Agronomy*, pp. 113–126.

Etherington, J.R. (1975) *Environment and Plant Ecology*, John Wiley & Sons, New York.

Fox, C. (1975) "Energy Capture by Plants", M.S. thesis, University of Minnesota.

Gillespie, W.F., Goss, J.J., Dyer, J.K., Ratliff, F.C., and Reed, A.E. (1946) "*Arundo donax* Cane", Paper presented at Annual Meeting, TAPPI, New York City.

Haarer, A.E. (1956) "The Roselle Family", *Fibres Eng. Chem.*, 17:105–107.

Hanway, J.J., and Russell, W.A. (1969) "Dry Matter Accumulation in Corn (*Zea mays* L.) Plants: Comparison Among Single-Cross Hybrids" *Agron. J.*, 61:947–951.

Hanway, J.J. and Weber, C.R. (1971) "Dry Matter Accumulation in Eight Soybean (*Glycine max* L. Merrill) Varieties", *Agron. J.*, 63:227–230.

Hardman, L.L. (1970) "The Effects of Some Environmental Conditions on Flower Production and Pod Set in Soybean Variety Hark", Ph.D. thesis, University of Minnesota.

Harlan, J.R. (1959) "Plant Exploration and the Search for Superior Germ Plasm for Grasslands" *Grasslands*, American Association for the Advancement of Science Publication 53, Washington, D.C., Chapter 1.

Helsel, Z. (1976) "Double Cropping Experiment Harvests First Crop as Forage" *Crops and Soils*, 28(8):24.

Heath, M.F., Metcalf, D.S., and Barnes, R.E. (1973) *Forages*, 3rd edition, Iowa State University Press, Ames, Iowa.

Hovelande, C.F., E.L. Carden, G.A. Buckauren, E.M. Evan, W.B. Anthony, E.L. Macon, and H.E. Burgess (1969).

Hunter, A.S., and Kelly, O.J. (1946) "The Growth and Rubber Content as Affected by Variations in Soil Moisture Stresses" *J. Amer. Soc. Agron.*, Vol. 38.

Jackobs, J.A. (1950) "The Influence of Spring-Clipping Interval between Cuttings, and Date of Last Cutting Alfalfa on Yields in the Yakima Valley" *Agronomy Journal*, Vol. unknown.

Joyner, J.F., and Wilson, F.D. (1967) "Effects of Row and Plant Spacing and Time of Planting on Seed Yield of Kenaf" *Econ. Bot.* 2(1).

Kelley, O.J., Hunter, A.S., and Hobbs, C.H. (1945) "The Effect of Moisture Stress on Nursery Grown Guayule with Respect to the Amount and Type of Growth and Growth Response on Transplanting" *J. Amer. Soc. Agron.*, Vol. 37.

Kelly, O.J., Haise, H., Markham, R., and Hunter, H.S. (1946) *J. Amer. Soc. Agron.*, Vol. 38.

Killinger, G.B. (1965) "Kenaf, a Potential Paper-Pulp Crop for Florida", Florida Agricultural Experiment Station, Sunshine State Agricultural Research Report, 10(2):4–5.

Kipps, M.S. (1970) *Production of Field Crops*, 6th edition, McGraw-Hill Book Company, New York.

Lawn, R.J., and Brun, W.A. (1974a) "Symbiotic Nitrogen Fixation in Soybeans. I. Effect of Photosynthetic Source-Sink Manipulation" *Crop Sci.* 14:11–16.

Lawn, R.J., and Brun, W.A. (1974b) "Symbiotic Nitrogen Fixation in Soybeans. III. Effect of Supplemental Nitrogen and Intervarietal Grafting" *Crop Sci.*, 14:22–25.

Leith, H. (1968) "The Measurement of Caloric Values of Biological Material and the Determination of Ecological Efficiency" *Functioning of Terrestrial Ecosystems in the Primary Production Level*, UNESCO, Paris.

Lavine, L. (1969) *Biology of the Gene*, C.V. Marley Company, St. Louis.

Litzenberger, S.C., (Ed.) (November 1974) *Guide for Field Crops in the Tropics and the Subtropics*, Office of Agriculture, Technical Assistance Bureau, Agency for International Development, Washington, D.C.

Lloyd, F.E. (1975) "Guayule–A Rubber Plant of the Chihuahua Desert" *International Conference on Utilization of Guayule*, Washington, D.C., p. 3.

Long, R.W. (1974) "Future of Rangelands in the United States" *J. Range Mgt.*, 27:253.

Loomis, R.S., and Gerakis, P.A. (1975) "Productivity of Agricultural Ecosystems" *Photosynthesis and Productivity in Different Environments*, Cambridge University Press, London.

McGinnies, W.G., and Haase, E.F. (1975) An International Conference on the Utilization of Guayule.

Marinotti, F. (1941) "L'utilizzazione Della Canna Gentile *Arundo donax* per la Produzione Autarchica di Cellulosa Nobile per Raion" *La Chimica*, 8:349–355.

Matches, A.G. (1976) "Growing Legumes with Warm-Season Grasses" Research Report (Special) No. 192, Southwest Missouri Center, University of Missouri, pp. 32–33.

Metcalf, C.R. (1967) "Distribution of Latex in the Plant Kingdom" *Econ. Bot.*, 21(2):115–125.

Morrison, F.B. (1951) *Feeds and Feeding*, Morrison Publishing Company, Ithaca, New York.

Mors, W.B., and Rizzini, C.T. (1966) *Useful Plants of Brazil*, Holden-Day Inc., San Francisco, California, pp. 116–117; 139–140.

Moss, D.N. (March 2, 1977) "An Agronomist's View of Biomass Production", paper presented at Biomass—Cash Crop for the Future Conference, Kansas City, Missouri.

Moss, D.N. (1976) "Studies on Increasing Photosynthesis in Crop Plants" *CO₂ Metabolism and Plant Productivity*, University Park Press, Baltimore.

Moss, D.N., Musgrave, R.B., and Lemon, E.R. (1961) "Photosynthesis Under Field Conditions. III. Some Effects of Carbon Dioxide, Temperature, and Soil Moisture on Photosynthesis, Respiration, and Transpiration of Corn" *Crop Sci.*, 1:83–87.

National Academy of Sciences, (1977).

National Academy of Sciences (1976) *Making Aquatic Weeds Useful: Some Perspectives for Developing Countries*, Washington, D.C.

National Academy of Sciences (1975) *Underexploited Tropical Plants with Promising Economic Value*, Washington, D.C.

National Academy of Sciences (1971) *Nutrient Requirements of Poultry*, Washington, D.C.

National Academy of Sciences (1970) *Nutrient Requirements of Beef Cattle*, Washington, D.C.

Nelson, E.G. (1960) *Jute*, United States Department of Agriculture, Crop Research Division, Beltsville, Maryland.

Perdue, R.E., Jr. (1958) "*Arundo donax*—Source of Musical Reeds and Industrial Cellulose" *Econ. Bot.*, Vol. 12.

Perry, L.F., Jr., and Olson, R.A. (1975) "Yield and Quality of Corn and Grain Sorghum Grain and Residues as Influenced by N Fertilization" *Agron. J.*, 67:816–818.

Poats, F.J. (1960) "Guar, A Summer Row Crop for the Southwest" *Econ. Bot.*, 14(3).

Puckridge, D.W., and Ratkowsky, D.A., (1971) "Photosynthesis of Wheat Under Field Conditions. IV. The Influence of Density and Leaf Area Index on the Response to Radiation" *Aust. J. Agric. Res.*, 22:11–20.

Raitt, W. (1913) "Report on the Investigation of Savannah Grasses as Material for the Production of Paper Pulp" *Ind. For. Rec.*, 5(3):74–116.

Rechentin, C.A. (1956) "Elementary Morphology of Grass Growth and How It Affects Utilization" *J. Range Mgt.*, 9:167–170.

Retzer, J.L., and Mogen, C.A. "The Salt Tolerance of Guayule" *J. Amer. Soc. Agron.*, Vol. 38.

Richardson, E.C. (1945) "The Effect of Fertilizer on Stand and Yield of Kudzu on Depleted Soils" *J. Agron*, 37:763–770.

Robertson, E.E. (Date unknown) *Perpetually Renewable Biomass Prospects*, The Biomass Energy Institute, Inc., Winnipeg, Canada, pp. 157–179.

Robinson, B.B., and Wright, A.H. (1940) "Hemp Its Production and Use as a Fiber Crop", U.S. Department of Agriculture Bulletin, Bureau of Plant Industry.

Robinson, R.G. (1949) "The Effect of Flax Stand on Yields of Flaxseed, Flaxstraw, and Weeds" *J. Agron.*, 4:483–484.

Schaller, F.W. and Murdock, S.J. (1976) "Switchgrass Production. Annual Progress Report (1975), Shelby-Grundy Experimental Farm", Iowa State University, pp. 5–6.

Science (October 1976) 194:46.

Sims, H.J. (1963) "Changes in the Hay Production and Harvest Index of Australian Oat Varieties" *Austr. J. Exp. Agric. Anim. Husb.*, 3:198–202.

Singh, I.D. and Stoskopf, N.C. (1971) "Harvest Index in Cereals" *Agron. J.*, 63:224–226.

Snow, L.M. (1920) "Contributions to the Knowledge of the Diaphragms of Water Plants" *Bot. Gaz.*, 58:495–517 (1914), pp. 297–317.

Tingey, D.C., and Clifford, E.D. (1947) "Comparative Yields of Rubber from Seeding Guayule Directly in the Field and Transplanted Nursery Stock" *J. Amer. Soc. Agron*, Vol. 38.

U.S. Bureau of Census (1972) *Census of Agriculture, 1969. Area Reports*, U.S. Government Printing Office, Washington, D.C., Vol. I.

U.S. Department of Agriculture (1969) *Agricultural Statistics*, U.S. Government Printing Office, Washington, D.C.

U.S. Department of Agriculture (1970) *Agricultural Statistics*, U.S. Government Printing Office, Washington, D.C.

U.S. Department of Agriculture (1971) *Agricultural Statistics*, U.S. Government Printing Office, Washington, D.C.

U.S. Department of Agriculture (1972) *Agricultural Statistics*, U.S. Government Printing Office, Washington, D.C.

U.S. Department of Agriculture (1974) *Agricultural Statistics*, U.S. Government Printing Office, Washington, D.C.

U.S. Department of Agriculture (1975) *Agricultural Statistics*, U.S. Government Printing Office, Washington, D.C.

U.S. Department of Agriculture, (1976) *Agricultural Statistics*, U.S. Government Printing Office, Washington, D.C.

U.S. Department of Agriculture (1969) *Conservation Needs Inventory of 1967.*

U.S. Department of Agriculture (1941) *The Yearbook of Agriculture.*

U.S. Department of Agriculture (1950-1951) *The Yearbook of Agriculture.*

U.S. Department of Agriculture (1953) *The Yearbook of Agriculture.*

U.S. Department of Agriculture (1957) *The Yearbook of Agriculture.*

U.S. Department of Agriculture (1961) *The Yearbook of Agriculture.*

U.S. Department of Agriculture (1970) *The Yearbook of Agriculture.*

Watson, V.H., Ward, C.Y., and Thurmon, W., (1970) "Response of Johnson Grass Sward to Various Levels of Nutrient and Management" *Proceedings of Association of Southern Agricultural Workers*, Vol. 67.

Weaver, J.E., and Clements, F.E. (1938) *Plant Ecology*, McGraw-Hill Book Co., Inc., New York and London.

Wedin, W.F., Hodgson, J.J., and Jacobson, N.L. (1975) "Utilizing Plant and Animal Resources in Producing Human Food" *J. Animal Sci.*, 41:667–686.

Wedin, W.F. (1974) "Fertilization of Cool-Season Grasses" *Forage Fertilization*, American Society of Agronomy, Madison, Wisconsin, Chapter 5, pp. 95–118.

Wedin, W.F. (1970) "Digestible Dry Matter, Crude Protein and Dry Matter Yields of Grazing-Type Sorghum Cultivars as Affected by Harvest Frequency" *J. Agron,* 62:359–363.

Wedin, W.F., and Vetter, R.L. (1970) "Pastures for Beef Production in the Western Corn Belt, USA" *Proceedings: 11th International Grassland Congress*, Brisbane, Australia, pp. 842–845.

Whaley, W.G. (April 1948) "Rubber—The Primary Source for American Production" *Econ. Bot.*, 2:99.

White, G.A., Clark, T.F., Craigmiles J.P., Mitchell, R.L., Robinson, R.G. Whiteley, E.L., and Lessman, K.J. (1974) "Agronomic and Chemical Evaluation of Selected Sorghums as Sources of Pulp" *Econ. Bot.*, 28:136–144.

White, G.A., Adamson, W.C., Whiteley, E.L., and Massey, J.H. (1971) "Emergence of Kenaf Seedlings as Affected by Seed Fungicides" *Agron. J.*, Vol 63.

White, G.A. (1970) "Cultural and Harvesting Methods for Kenaf—An Annual Crop Source of Pulp in the Southwest", Agricultural Research Service, United States Department of Agriculture, Washington, D.C., Report 113, p. 38.

Wilson, F.D., and Menzel, M.Y. (1964) "Kenaf (*Hibiscus cannabinus*), Roselle (*Hibiscus sabdariffa*)" *Econ. Bot.*, 18(1):80–91.

Winchester, J.A. (1964) "Kenaf Nematodes", *Proceedings: Second International Kenaf Conference*, Palm Beach, Florida, pp. 188–189.

Woods, J.E., Hafenrichter, A.L., Schwendiman, J.L., and Law, A.G. (1953) "The Effect of Grasses on Yield of Forage and Production of Roots by Alfalfa-Grass Mixtures with Special Reference to Soil Conservation" *Agron. J.*, 45:590–595.

Wynd, F.L., Steinbauer, G.P., and Diaz, N.R. (1948) "*Arundo donax* as a Forage Grass in Sandy Soils" *Lloydia*, 11:181–184.

Yermanos, D.M. (1974) "Agronomic Survey of Jojoba in California" *Econ. Bot.*, 28:160–174.

AVAILABILITY OF BIOMASS
WOOD

The information in this chapter is based on *Silvicultural Biomass Farms, Volume I—Summary,* prepared by R.E. Inman of The Mitre Corporation, Metrek Division, for the Energy Research and Development Administration under ERDA Contract E(49-18)-2081, May 1977.

There are several options for procuring silvicultural biomass for conversion to energy. These include wood and bark mill residues which accumulate at primary wood manufacturing plants, logging residues which are traditionally left in the woods following timber harvests, noncommercial or surplus stands of timber, and thinnings from commercial timber stands. These biomass sources exist today; mill residues, in fact, are already being utilized as an energy feedstock by the forest industry. The expansion of the use of this resource constitutes a viable potential whose effect can be felt in the near-term as well as in future years.

The availability and utility of the forest residue resource has been the subject of a number of recent studies. The use of such existing sources of biomass for energy production in the near-term can be expected to provide a stimulus for the development of reliable supplies for the long-term. The major emphasis of this report has been an analysis of the feasibility of formal silvicultural biomass farms, whose potential will not be significantly tapped within the near-term because of the developmental research needed to bring the concept to a commercial status.

The concept of silvicultural energy farms accommodates the purposeful production of energy feedstock as a primary product. Although it is visionary at this time, this concept can be applied directly to the national energy predicament as a whole. It is not susceptible to compromise by associated interests or associated primary wood products, nor severely by location. It is, on the other hand, highly susceptible to manipulation, and to augmentation through the practice of agricultural technology. As such, it has a substantial, long-term potential as a renewable and reliable source of energy.

225

THE SILVICULTURAL ENERGY FARM

The energy farm, as envisioned in this study, would consist of plantings of selected, rapidly growing tree species at close spacings. The crop would be harvested at appropriate intervals, or rotations, with the succeeding crops arising by coppicing (sprouting from stumps). Since most conifers do not coppice, plantings would be restricted in most cases to selected hardwood species. Coppicing would preclude the need to replant after each harvest. Intensive crop management would be practiced, including fertilization, irrigation and weed control. In this regard, energy crop production would be similar to field crop production rather than conventional forestry.

There are two major factors which will control the extent to which the energy farm concept will impact upon the nation's energy supply. The first is biomass productivity, or the yield that can be produced per unit time-space. Productivity levels vary with the species planted, the cultural practices used, and the conditions specified by the farm site, including climate, weather, and soil characteristics. All of these factors may be controlled or modified to some extent by man.

Promising species can be selected through experimentation and improved by genetic recombination. Cultural practices may also be selected and optimized through experimentation to provide cost-effective approaches to energy crop management. Crop management practices will vary, however, depending upon the species and the site. Site-specific factors can be either modified by cultural practices such as irrigation, fertilization, and liming, or accommodated by appropriate species selection, or both. Hence, biomass productivity is a technological factor that can be, within certain biological limits, molded and fashioned by man to best fit the needs of the concept. Success in optimizing biomass yield will largely be a function of the time and research dollars spent on the question.

The second controlling factor is land availability. Based upon current productivity levels of short rotation tree crops, the production of one quad of energy (10^{15} Btu) in the form of biomass would require from 5 to 12 million acres of land. The farming of energy crops, therefore, will require large amounts of land to make a significant national impact on energy supply.

The prospects of acquiring land for biomass production are restricted by current competing land uses in those regions where large biomass farms might conceivably be located. As the demands for food and fiber products increase in the future, competing land uses may become even more restrictive if these increased demands are not accompanied by proportional increases in food/fiber production efficiency. Hence, the land availability factor differs from that of biomass productivity. It cannot be confronted directly by developing energy crop technology (although increasing biomass productivity would afford more effective use of energy farm land) and is heavily dependent upon socioeconomic factors only tangentially related to energy farming.

If the energy farm concept is to attain a respectable status, the land required for its purpose must either be obtained from lower valued land of minor use in agriculture and forestry, or the value of energy products must increase to the point where biomass farming can successfully compete for land with high value uses.

BIOMASS SPECIES AND PRODUCTIVITY

The plantation culture of forest trees is well established in the United States.
Trees are grown in such plantations to provide raw material for a wide variety
of products, including lumber, plywood, pulp and paper, and many other minor
products. They are generally grown widely spaced, much as they occur in natural stands. Further, they are grown to sizes large enough for the manufacture
of solid wood products, such as lumber and plywood and large enough also for
the use of conventional single-tree harvesting methods.

The traditional practice of silviculture requires relatively long growing periods,
or rotations. These rotations generally range from 30 to 80 years or more.
Trees can be grown using much shorter rotations provided that size and form
are not important considerations. The use of short-rotations cultures has been
advocated for producing wood fiber. A modification of this concept is suggested as a means to provide energy feedstock on a commercial scale.

For sheer biomass production, short-rotation management appears to offer several advantages vis-a-vis conventional management practices:

> Higher yields per unit land area.
>
> Lower land requirements for a given biomass output.
>
> Shorter time span from initial investment in stand estab-
> lishment to positive cash flow from harvestable crop.
>
> Increased labor efficiency through mechanization.
>
> Increased harvesting efficiency through the use of methods
> similar to those used in agriculture.
>
> The capacity of most short-rotation candidate species to
> regenerate by coppicing.
>
> Ability to take advantage of cultural and genetic advances
> quickly.

Indications from initial short-rotation research are that short-rotation biomass
yields can far exceed those of conventional forest crops. Average annual biomass increments of several conventional forest crops range from about 1.3 to
2.7 dry ton equivalents (DTE) per acre. Current data indicate that short-rotation biomass yields of 4 to 8 DTE per acre and more are currently possible.
It should be noted, however, that the short-rotation management necessary to
glean such high yields from a given land area is generally more intensive and
thus more costly than forest management programs currently in use.

Rapid establishment of a vigorous stand will be essential for the efficient production of biomass with short-rotation forest crops. The key to short-rotation production is early utilization of the growth capacity of the site. This is accomplished by utilizing much higher planting densities than those used for conventional forest crops. Most experimental work with short-rotation forest crops
has utilized stand densities ranging from 1,200 to 77,000 stems per acre. In
contrast, conventional forest crops are generally planted at densities ranging
from 600 to 2,000 stems per acre.

Tree species which constitute conventional forest crops are generally chosen on the basis of desirable wood properties. Other criteria, such as rapid growth and ease of establishment, are also important, but they remain secondary to wood properties. In the selection of tree species for short-rotation biomass production the single most important factor is the ability to grow rapidly. Other factors such as ease of establishment and regeneration are important, but remain secondary to rapid growth.

Candidate biomass species may be chosen primarily on the basis of the following criteria:

(1)   Rapid Juvenile Growth—Either by way of genetic propensity or stimulative silviculture, trees grown in an intensive mode must get off to a rapid start and avoid a lag period after field establishment.

(2)   Ease of Establishment and Regeneration—Hardwoods are particularly advantageous since stands can be established with cuttings, and regeneration after harvesting can be achieved by coppicing.

(3)   Freedom from Major Insect and Fungal Pests—Genetic diversity in susceptibility to pests does exist and can be exploited to build in resistance to attack. Pests can also adapt, however, and the threat of epidemic pest outbreak in monocultures is real. Thus, selection for resistance, while at the same time maintaining genetic diversity, is a goal.

Softwoods (conifers) generally appear to offer less promise than hardwoods. Conifers in general do not satisfy short-rotation criteria very well, although exceptions occur. Most conifers do not exhibit rapid juvenile growth, starting somewhat slowly compared to hardwoods. In addition, true coppice reproduction of conifers does not occur except in the case of the redwoods.

On the positive side, nursery and field planting techniques for conifers are well established, so stand establishment and regeneration should not present a problem from a technical standpoint. Furthermore, conifers are generally less site demanding than hardwoods, and could be established on both moist and dry upland sites, where medium heavy to sandy soils predominate.

In contrast to the situation with conifers, many hardwoods fit the short-rotation criteria quite well. Most hardwood species coppice well. Many can be established by cuttings rather than seedlings and exhibit rapid early growth. On the negative side, most hardwoods do not exhibit great site adaptability. The full productive potential of hardwoods can usually be achieved only on sites which are inherently fertile, well-drained, well-aerated, and adequately supplied with moisture throughout the growing season. Intensive management of hardwoods, however, could be expected to reduce the significance of this characteristic.

A number of species and species groups may be viewed as candidates for use in silvicultural biomass farms on the basis of the above criteria. The ranges of these species are indicated in Table 6.1. All of the species listed have exhibited the rapid juvenile growth necessary for short-rotation biomass production, with the exception of the aspens *(Populus tremuloides* and *Populus grandidentata)*. In the *Populus* group, the cottonwoods and the poplars, including hybrid poplars, are attractive from a productivity standpoint. Furthermore, all of these species, with the exception of tulip-poplar, are at least moderately adaptable to varying site conditions. Tulip-poplar, in contrast, has fairly exacting site requirements as far as soil and moisture are concerned.

Table 6.1: Ranges of Candidate Biomass Species

Species	Range
American sycamore	All states east of the Great Plains except Minnesota.
Eucalyptus spp	Generally frost-free areas of the southeast and California.
Loblolly pine	Coastal Plain and Piedmont from Delaware and Central Maryland south to Central Florida and west to Eastern Texas.
Populus spp	
Eastern cottonwood	Southern Quebec and Ontario to Southeastern North Dakota, south to Western Kansas, Western Oklahoma, Southern Texas, Northwestern Florida, and Georgia.
Black cottonwood	South along the Pacific Coast from Kodiak Island and Southeastern Alaska to mountains in Southern California. Eastward into Southwestern Alberta, South-Central Montana, Central Idaho, Northern Utah and Nevada.
Sweetgum	Connecticut southward throughout the East to Central Florida and Eastern Texas. It is found as far west as Missouri, Arkansas and Oklahoma, and north to Southern Illinois.
Tulip-poplar	Throughout the Eastern United States from Southern New England west to Michigan and south to Central Florida and Louisiana.
Red alder	Confined to the Pacific Coast region from Southeastern Alaska south through Washington, Northern Idaho, and Western Oregon to Santa Barbara, California.

Source: ERDA E(49-18)-2081

An indication of the soil and moisture requirements for the best growth of each species is found in Table 6.2, along with remarks on adaptability to other sites.

Table 6.2: Site Requirements for Selected Species

Species	Preferred Site(s)	Remarks
American sycamore	Alluvial soils along streams and in bottomlands.	Tolerant of wide fluctuations in ground-water levels, but generally requires a good supply of ground-water.
Eucalyptus spp	(Over 500 species in genus with a wide variety of site requirements.)	A few species are somewhat frost resistant.

(continued)

Table 6.2: (continued)

Species	Preferred Site(s)	Remarks
Loblolly pine	Soils with poor surface drainage, a deep surface layer, and a firm subsoil.	Grows on a wide variety of soils, from the flat, poorly-drained, ground-water podsols of the Coastal Plain to the old residual soils of the upper Piedmont.
Populus spp Eastern cottonwood	Moist, well-drained, fine sandy loams or silts, with an abundant and continuous supply of moisture throughout the growing season.	Will survive on infertile sands and fairly stiff clays. Does not grow well more than 15 or 20 feet above stream levels.
Black cottonwood	Deep, alluvial soils at low elevations and moist, sandy, gravelly, or loam soils at higher elevations. Requires abundant moisture.	Grows well on some upland soils if sufficient moisture is available.
Sweetgum	Rich, moist, alluvial clay and loam soils of river bottoms.	Very tolerant of different soils and sites.
Tulip-poplar	Moderately moist, well-drained, loose-textured soils.	Exacting in soil and moisture requirements. Rarely grows in very dry or very wet situations.
Red alder	Deep, well-drained loams or loamy sands of alluvial origin.	Grows on soils varying from gravel or sand to clay. Soils are not a serious limitation except as they affect availability of soil moisture.

Source: ERDA E(49-18)-2081

The biomass productivity of a short-rotation silvicultural farm, using a given species on a given site, will be a function of stand density and management intensity. Generally, the more dense a stand, the sooner crown closure and thus full utilization of the growth potential of a site is achieved. The biomass productivities of several of the species listed above are presented in Table 6.3, along with corresponding rotation lengths. It must be kept in mind that intensive application of management practices are required to obtain the short-rotation biomass yields shown.

Table 6.3: Biomass Productivity of Short Rotation Species

Species	Estimated Annual Biomass Production (DTE/acre/yr)	Estimated Rotation Age (yr)
American sycamore	6	4
Eucalyptus	12	6
Loblolly pine	5	10
Populas spp		
Eastern cottonwood	5	9
Black cottonwood	6	4
Hybrid poplars	10	8–10
Red alder	9	4

Source: ERDA E(49-18)-2081

The yields reported in Table 6.3 represent current yield potential under intensive management conditions. It is generally recognized at this point that research in this area has only scratched the surface of biomass yield potential, especially as

regards the energy farm concept. Very little screening has been initiated to
designate species or selections within species that have high yield capacities.
Essentially no work has been performed on developing preferred high-yielding
hybrids specifically adapted for selected sites.

It is entirely reasonable to presume that current yields demonstrated by the preliminary efforts could be essentially doubled if research were to be directed
toward this end. Future biomass yields, therefore, are expected to attain levels
of 15 to 20 DTE per acre per year with the development of improved species
and selections and advanced cultural methods. At such levels of productivity,
one quad of biomass energy could be produced on 3 to 4 million acres of land.

LAND SUITABILITY AND AVAILABILITY

Due to the importance of land to the biomass farming concept, the subject of
land suitability and availability warrants special consideration. An analysis of
potentially available land and its potential impact on energy feedstock production was conducted. A four-step approach was used.

(1) Suitability criteria for potential silvicultural biomass
farm sites were established.

(2) Regional biomass yields on intensively managed silvicultural farms were estimated.

(3) Land potentially available for biomass production in
the United States was identified, according to various
land availability scenarios.

(4) The quantity of energy feedstock which would be
produced was estimated.

The criteria selected for designating suitable sites were: (a) at least 25 inches
of precipitation annually to sustain estimated yield, (b) arable land, and (c) slope
equal to or less than 30% (17 degrees) to allow mechanized crop management.

Arable land was defined as that in Soil Conservation Service (SCS) capability
classes I-IV and National Forest System land in U.S. Forest Service (USFS) commercial forest site classes I-IV. Land in SCS classes V-VIII was excluded. These
classes include most of the land with steep slope. Nearly all of the land which
was included, therefore, had generally acceptable slope for farming operations.

Current and future biomass yields from representative biomass candidate species
which might be established in areas meeting the three land suitability criteria
were projected. These were used to estimate energy production in nine of the
ten USDA Farm Production Regions. The Mountain Region was not included
because it did not meet suitability criteria. The national average anticipated
yield from intensive silviculture is approximately eight dry tons per acre per
year in the near-term and fifteen dry tons in the future (Figure 6.1).

Four land availability scenarios were developed to identify potentially usable
land in counties which receive 25 inches of precipitation annually. The 1967
SCS "Conservation Needs Inventory" was used as the data base for this survey.

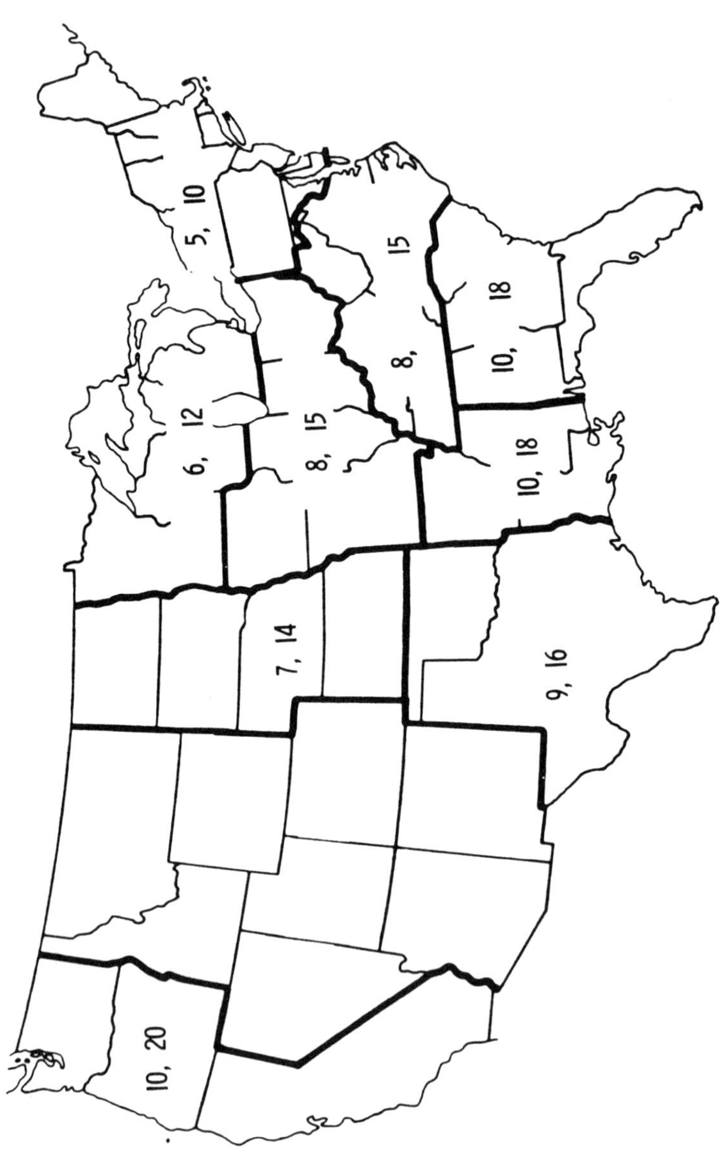

Figure 6.1: Estimated Current and Future Biomass Yields in Nine Farm Production Regions (dry tons per acre per year)

Source: ERDA E(49-18)-2081

The four successively inclusive scenarios which were explored were:

(1) SCS Class IV; permanent pasture, forest and range.

(2) SCS Classes I-IV; permanent pasture, forest and range.

(3) SCS Classes I-IV; permanent pasture, forest, range, rotation hay and pasture, hayland and open land formerly cropped.

(4) SCS Classes I-IV; permanent pasture, forest, range and all cropland except that in conservation use only. This included cropland in the following categories: field crops; rotation hay and pasture; hayland; temporarily idle cropland; orchards, vineyards and bush fruit.

Two additional scenarios were developed on a state and regional basis to identify potentially available land which is limited by wetness or which is part of the National Forest System. These consisted of:

(5) SCS Classes II_w-IV_w; permanent pasture, forest and range (w indicates wetness problem).

(6) USFS Classes I-III; and I-IV National Forest System land.

The most likely sources of land for silvicultural biomass plantations are included in Scenarios (2) and (3). These consist of 270 and 320 million acres, respectively (Figures 6.2 and 6.3). This land is under relatively less pressure than prime cropland, but is nonetheless arable and suitable for biomass farming. Nearly half of the acreage in these categories is in the southern United States. More than 10% is in the Lake States. The latter is particularly attractive because the agricultural potential of the region is limited by climate but the intensive culture of trees would not be as severely limited.

The Northern Plains show relatively little potential because of moisture limitations. The Appalachians and Northeast are limited by several factors including slope, land ownership patterns, climate and urbanization. Nevertheless, one-fourth of the potentially available land exists in these two regions.

The potential of the Corn Belt for biomass production is limited by intensive agriculture in the region. The only category in which the Pacific Region demonstrates potential is in the National Forest System Scenario (6). More than one-third of the United States Forest System land is in the Pacific Region. The only other sizable source of USFS land is in the Lake States, most of which is in the marginal USFS Class IV.

Total land areas and the energy yield which could be obtained from 10% of the acreage in each scenario are summarized in Table 6.4. These data indicate that between 4 and 7 quads of feedstock might eventually be obtained from Scenario (2) (Figure 6.4). Scenario (3) could yield between 4 and 8 quads (Figure 6.5). If 10% of the suitable USFS land could be added to this base, an additional quad could be produced. Though it is unlikely that prime cropland will be used for energy production, this land has enormous potential and an additional 6 quads could be produced if 10% of this acreage could be made available.

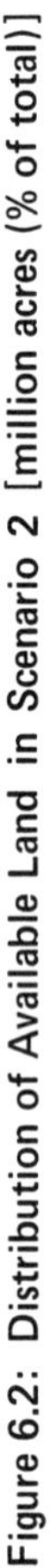

Figure 6.2: Distribution of Available Land in Scenario 2 [million acres (% of total)]

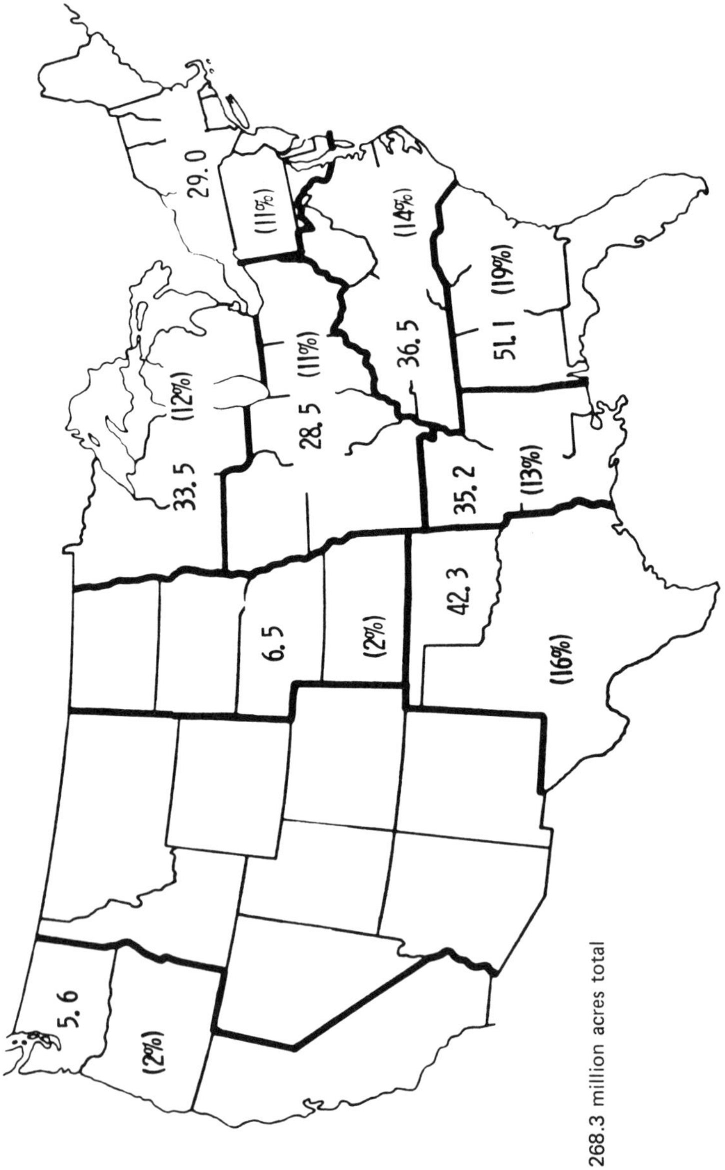

268.3 million acres total

Source: ERDA E(49-18)-2081

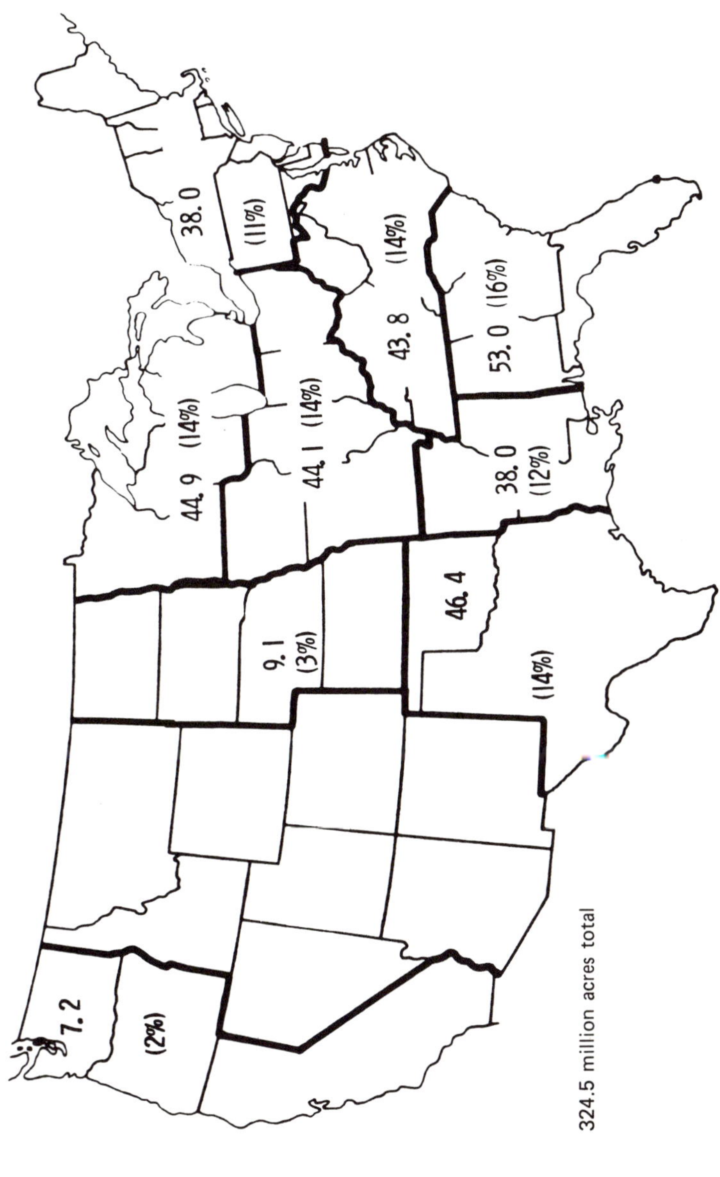

Figure 6.3: Distribution of Available Land in Scenario 3 [million acres (% of total)]

Source: ERDA E(49-18)-2081

Figure 6.4: Current and Future Energy Yield (Quads) Using 10% of the Land in Scenario 2

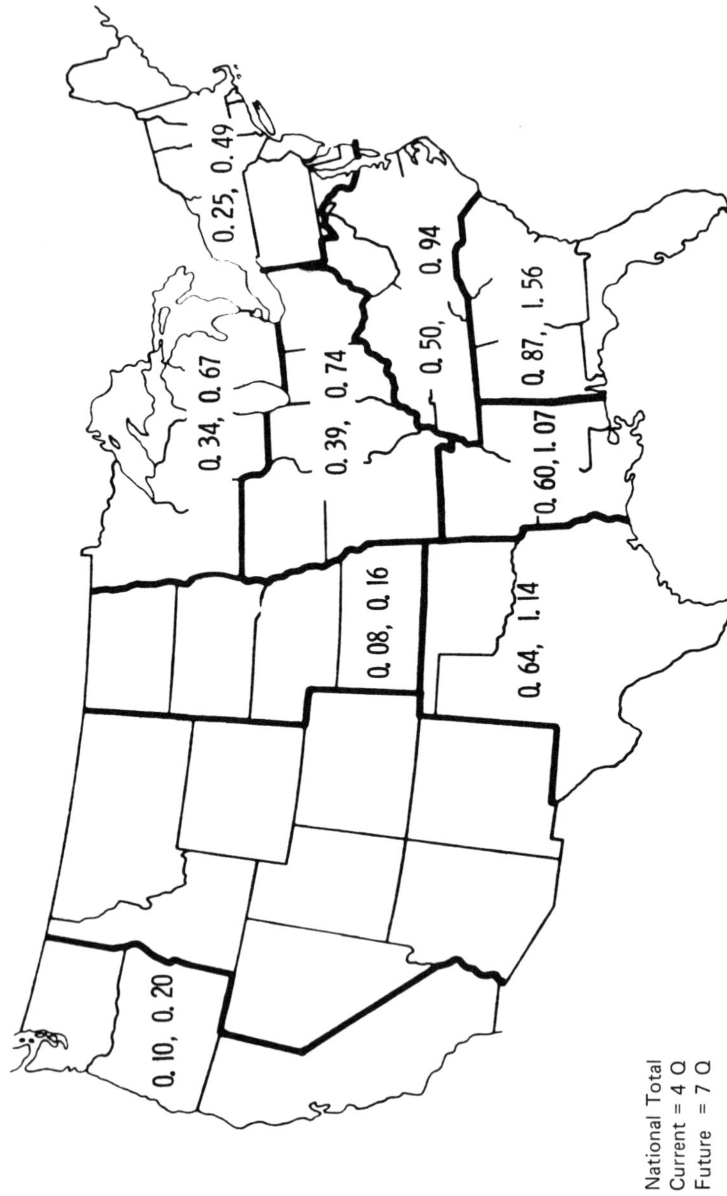

Source: ERDA E(49-18)-2081

Figure 6.5: Current and Future Energy Yield (Quads) Using 10% of the Land in Scenario 3

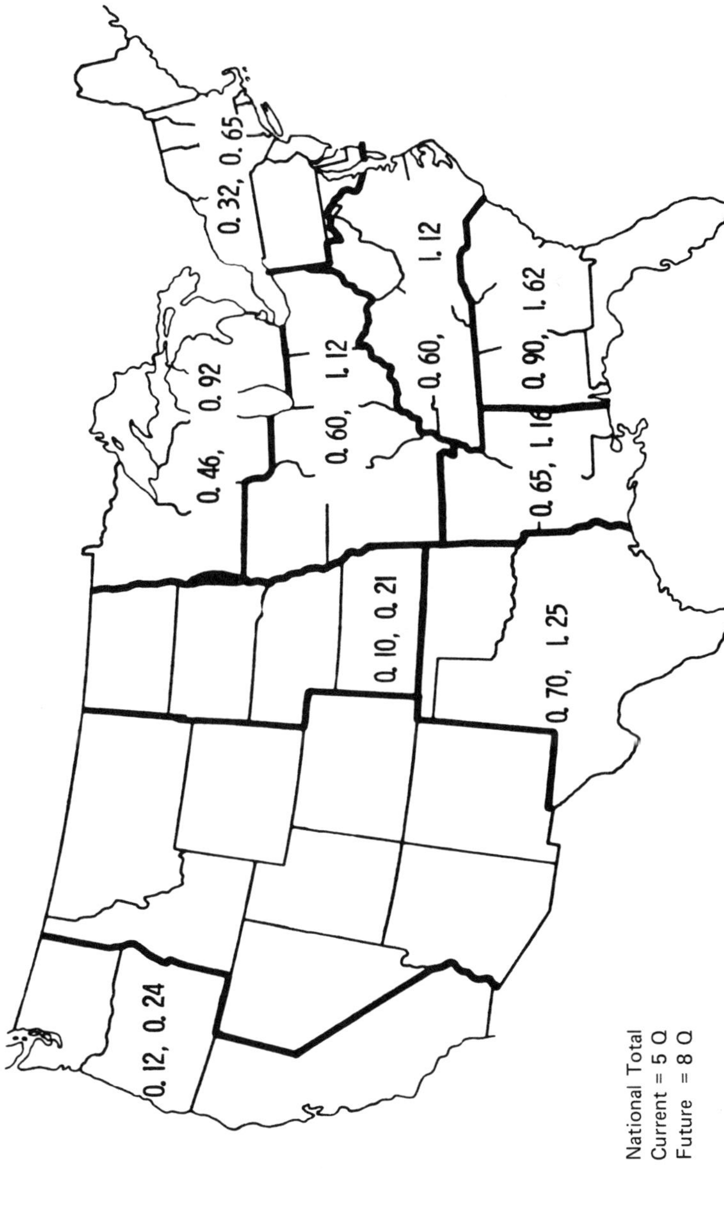

Source: ERDA E(49-18)-2081

Table 6.4: Land Area and Projected Energy Yield from 10% of the Area in Six Scenarios

Scenario	Total Land Area (million acres)	Energy Yield from 10% of Land [quads (10^{15} Btu)] *	
		Current	Future
1	81	1.1	2.1
2	268	3.8	7.0
3	323	4.5	8.3
4	489	6.7	13.7
5	90	1.3	2.3
6A**	21	0.3	0.5
6B***	51	0.7	1.4

*Assumptions: A current average yield of 8 dry tons per acre per
year and a future yield of 15 dry tons per acre per year,
8,500 Btu per dry pound
**USFS Classes I-III National Forest System Land
***USFS Classes I-IV National Forest System Land

Source: ERDA E(49-18)-2081

Land use patterns change constantly. One hundred million acres of SCS inventoried land changed use between 1967 and 1975. Economics, national needs and legal and institutional considerations will determine how much land eventually may be made available for biomass farming. It may be necessary to establish the first farms on large land holdings such as those of timber companies and the federal government. This would overcome many of the barriers associated with acquiring large quantities of land from many small owners.

SITE CHARACTERIZATIONS FOR SITE-SPECIFIC PRODUCTION ANALYSES

It became apparent early in the study that important differences relating to all phases of biomass production exist between different locations. For the purposes of production cost analyses, therefore, ten sites were selected representing a variety of climate, topographic and land-use situations. Six of the ten sites were representative of "preferred" site conditions or of locations where it is believed that plantings might reasonably be placed in the future. These were designated largely on the basis of current land-use patterns and private ownership. These sites were in the states of Washington, Wisconsin, Missouri, Louisiana, Georgia, and New Hampshire/Vermont and did not involve the use of prime agricultural land, public land, or swampland.

Four additional sites were chosen to broaden the analytical base by including other land-use options. Sites in Illinois and California, respectively, represented the option of using prime agricultural land for biomass production. Crop production at the California site is 100% dependent on irrigation; the Illinois site is not irrigated. The Florida site was chosen to represent the option of using wetlands (53% of the Levy County site would require drainage), and the Mississippi site was selected to entertain the notion of using national forest land (DeSoto National Forest) for biomass production. Each of the ten sites is characterized briefly below. The tree species selected for each site is representative of the types of trees which may be grown in these regions. Actual species selected should await site-specific field research.

Washington Site: Represents an area of high forest productivity and logging activity; low level of land availability; suitable rainfall; terrain averages 12% slope; red alder selected as representative biomass candidate.

Wisconsin Site: Represents an area of low agricultural and forestry value; land relatively inexpensive and available; suitable rainfall; slope averages 3 to 4%; relatively low production potential; hybrid poplars selected as representative biomass candidates.

Missouri Site: Represents an area of relatively low agricultural value; moderate land values and availability; suitable rainfall but droughty soils; moderate production potential; slope averages 4 to 5%; hybrid poplars selected as representative biomass candidates.

Louisiana Site: Represents an area of low agricultural and forestry values; land considered "marginal"; moderate land values; availability relatively high; suitable rainfall; slope averages 3 to 4%; *Eucalyptus* spp selected as representative biomass candidates.

Georgia Site: Represents an area of low agricultural value and high commercial forestry value; moderate land values and availability; suitable rainfall; moderate production potential; slope averages 5 to 6%; sycamore selected as representative biomass candidate.

Vermont/New Hampshire Site: Represents an area of low agricultural activity; wood availability greatly exceeds consumption; moderate land prices and availability; suitable rainfall; terrain rocky; slope averages 15 to 16%; winter snow cover averages 3 to 4 feet; relatively low production potential; hybrid poplars selected as representative biomass candidates.

California Site: Represents an area of intense agricultural activity; crop production totally dependent on irrigation; high land values and low availability; very high production potential; slope averages less than 1%; *Eucalyptus* spp selected as representative biomass candidates.

Illinois Site: Represents an area of intense agricultural activity; very high land prices and low availability; suitable rainfall; moderate production potential; slope averages 2 to 5%; hybrid poplars selected as representative biomass candidates.

Mississippi Site: Represents federally-owned land (DeSoto National Forest); current prospects low for energy farms; low-moderate land charges; suitable rainfall; high production potential; slope averages 4 to 5%; *Eucalyptus* spp selected as representative biomass candidates.

Florida Site: Represents a wetland area with low agricultural and forestry activity; low land availability potential; moderate land values; land use requires drainage; suitable rainfall; slope averages less than 1%; high production potential; *Eucalyptus* spp selected as representative biomass candidates.

Table 6.5: Major Characteristics of Ten Sites Chosen for Production Analyses

. . . . . .Site Location		Average Percent Slope	Usable Acres (%)	. . Vegetation . . .		. . . . . Current Scenario				. . . . . .Future Scenario		
				Percent Wooded	Percent Open	MAI*	Acres Planted**	Acres Acquired***	MAI*	Acres Planted**	Acres Acquired***	
State	County											
Preferred												
Wisconsin	Rusk	4	73	69	31	5	50,000	71,000	10	25,000	36,250	
Missouri	Polk	5	89	38	62	7	35,715	38,930	14	17,860	20,000	
Louisiana	Sabine	3	81	80	20	12	20,835	23,540	20	12,500	14,625	
Georgia	Hancock	6	85	86	14	8	31,250	34,375	16	15,625	17,815	
New England	Windsor (VT)	16	71	68	32	5	50,000	64,500	10	25,000	33,000	
Washington	Lewis	12	55	76	24	10	25,000	33,000	20	12,500	17,250	
Agricultural												
Illinois	McLean	2	89	6	95	8	31,250	35,000	15	16,670	19,165	
California	Kern	<1	96	0	100	13	19,230	21,345	22	11,365	13,070	
Swampland												
Florida	Levy	<1	85	85	15	12	20,835	23,125	20	12,500	14,375	
National Forest												
Mississippi	Perry	4	94	91	9	12	20,835	22,710	20	12,500	14,125	

*Mean annual increment.

**Acreage required for annual production of 250,000 DTE.

***Included planted acreage, areas for storage, irrigation lanes and roads, and unusable acreage with lease or purchase of usable land.

Source: ERDA-E(49-18)-2081

Each of the ten sites was defined in regard to those characteristics that would influence the logistics and economics of biomass production. These include the proportion of the total area composed of purchasable (leaseable) acreage, wooded acreage, and open acreage; the average slope and surface character of the usable acreage; the length of the growing season (frost-free period) and other climatic characteristics; the estimated level of productivity obtainable during a 6-year rotating period, and the cost of land, labor, fuel, and other materials required for plantation operation. Some of these characteristics are included in Table 6.5.

No attempt was made in this study to assign species to each site, although representative species were designated. An attempt was made, however, to estimate the annual level of biomass productivity that could be attained if the representative species were used, since this is the primary factor determining the economics of biomass production. These site-specific designations of productivity (mean annual increment—MAI) were made on the basis of results of field tests and observations. Estimates were also offered for future productivity levels.

The cost of land at the ten study sites was calculated on the basis of the weighted average of values for wooded land and open land at the respective sites. Current cost among the six preferred sites ranged from a low of $150 per acre at the Wisconsin site to a high of $430 per acre at the Washington site. Among the four alternative sites, cost of prime, nonirrigated agricultural land at the Illinois site was valued at $2,860 per acre, while irrigated farmland at the California site was valued at $1,400 per acre. Land at the Florida swampland site averaged $280 per acre. National Forest System land (DeSoto National Forest, Mississippi) was leased for $20 per acre/year, on the basis of the USFS receipts for sawtimber and roundwood traditionally removed from that land.

ENERGY FARM DESIGN

There is not yet in existence a silvicultural energy farm; therefore, the approach used to analyze and evaluate this concept was to estimate biomass production costs. This is deemed a logical approach in view of the lack of precedent, for it attempts to answer the leading question of how much it is going to cost, by visualizing, defining and analyzing as many of the pertinent parameters as can currently be identified. The major parameters identified include:

- Species composition—the types of tree species likely to comprise the biomass crop
- Productivity—the average yield attainable per acre-year
- Production—the total quantity of biomass produced per farm-year
- Farm size—a function of production and productivity, coupled with the ability to acquire and manage land
- Land acquisition—the method by and extent to which land may be acquired, its suitability for biomass production, and its cost
- Farm configuration—the spatial relationship between planted acreage and the land surrounding it
- Crop management—practices that might be used to optimize productivity or to ensure a desired level of production

- Operations specifications—the equipment, labor, and material required for operations, and their costs
- Site location—the effect of site-specific conditions and characteristics on each parameter
- Time frame—the effect of anticipated or potential technological developments on productivity, operations and costs.

As additional research on short-rotation crops is performed, species other than those listed in Tables 6.1, 6.2, and 6.3 may emerge as suitable candidates. The use of *Eucalyptus* would be confined to relatively frost-free sites, although further screening holds promise of extending its range in the United States. The choice of a species or species array must be made on a case-by-case basis, however, for it is highly doubtful that a wide range of conditions could satisfy the requirements of one species or one species array. The respective levels of productivity selected as representative for the ten study sites have been presented in Table 6.5.

A conceptual production design (CPD) was formulated for use as a vehicle in estimating production costs. This was necessary due to the lack of actual experience in formal biomass production. The CPD was formulated on the basis of current knowledge regarding the short-rotation culture of coppicing hardwood species and discussions with field research experts and equipment manufacturers, and accommodated the major parameters listed previously. The major features of the CPD are summarized in Table 6.6.

Table 6.6: Summary of the Conceptual Design for Silvicultural Biomass Production

Feature or Operation	General Characteristics
Annual production for one farm	500,000 green tons per year 250,000 DTE/year $4{,}250 \times 10^9$ Btu/year 235 farms produce 1 quad annually

Productivity (DTE/acre/year)	Site- and species-dependent
	   Range*: Current — 5 to 13
	                Future — 10 to 22

Land acquisition	Lease or purchase Cost of land is site-dependent 10% farm density
Land clearing/preparation	Costs depend on vegetative cover
Planting	Seedlings or cuttings 4' x 4' spacing (2,725 plants per acre) Plantings staggered over first rotation period
Rotation period	6 years each; 30-year lifetime per planting 5 rotations (1 plant crop; 4 coppice crops)
Irrigation	Automatic sprinkler (traveller) at all sites except California site

(continued)

Table 6.6: (continued)

Feature or Operation	General Characteristics
Irrigation	First three years per rotation One acre-foot water per acre-year Flood irrigation used at California site; 3.5 acre-feet per acre-year, applied every year
Fertilization	Annual N, P, K application (60% recovery efficiency) Lime applied once per rotation; no lime applied at California site Ground application; first year per rotation Air application: second-sixth year per rotation Rate of application depends on productivity level
Pest control	Mechanical weed control; first year per rotation No insect control No disease control
Harvesting	Self-propelled harvester Tractor/wagon teams (green-haul) Season: winter (3 to 4 months) Chipped biomass stored in field storage areas
Transportation	Semitrailers Front-end loaders Season: year-round
Roads	454 miles of work roads per farm Used mainly during harvest season and for crop management operations
Miscellaneous operations	Planning Supervision Field support/supply

*See Table 6.5, MAI

Source: ERDA E(49-18)-2081

A level of production of 500,000 green tons per year was assigned to each of the ten study sites. This amount of biomass is equivalent to 250,000 dry tons per year, assuming a moisture content of 50% on a green weight basis.

"Production" is the term applied to the total quantity produced on the farm each year; "productivity" refers to the average quantity produced per acre per year over a single rotation and is synonymous to mean annual increment (MAI). A production level of 250,000 DTE yields $4,250 \times 10^9$ Btu per year, assuming an average higher heating value of 8,500 Btu per dry pound, and would be approximately sufficient to fire a 50 megawatt power plant. At this rate, 235 such farms would be required to contribute one quad of energy each year to the nation's supply.

The CPD accommodated the acquisition of land by either long-term lease or direct purchase. It was assumed that no more than 10% of the usable acres at any given site could be realistically acquired for planting, and that a portion of the nonusable land would also have to be acquired in any lease or purchase contract. Planted acreage was calculated by dividing production by productivity; the total acreage acquired, however, included planted acreage, additional land needed for roads, irrigation lanes and storage, and a quantity of nonusable land which varied between sites according to the ratio of purchasable land to usable land. Only the planted and supporting acreage was assumed to be cleared and prepared for farming. Costs for this operation varied between sites according to the proportion of wooded land to open land, and were based on published land clearing costs.

The basis for biomass production was the intensive culture of closely spaced, rapidly growing hardwoods under short-rotation conditions. Seedlings or cuttings were assumed to be planted at 4' x 4' spacings, a planting density of 2,725 plants per acre. A rotation period of six years and a productive lifetime of 30 years were assumed allowing the harvest of one plant crop and four coppice crops from each planting before replanting was required.

No actual field data are available, however, regarding the maximum number of coppice crops that can realistically be obtained from a single planting. The same level of productivity was assumed for the coppice crops as for the plant crop, although field research has indicated that the productivity of coppice crops is usually greater than that of the plant crop, at least through the first two coppicings.

Crop management involved the application of supplemental irrigation water during the first three years of each rotation, the annual application of nitrogen, phosphorus and potassium fertilizers at rates sufficient to prevent nutrient mining at a 60% nutrient recovery efficiency, the application of lime during the first year of each rotation, and mechanical weed control practiced twice during the first year of each rotation. (See Table 6.6 for variations from these crop management practices for the California site.)

No provision was made for insect and pest control, since the potential cost effectiveness of this practice in biomass production is undetermined at this time. Fertilizer application is accomplished by aircraft during the greatest portion of each rotation, since the presence of the biomass crop would impede operation of ground equipment.

Harvesting the biomass crop at the end of each rotation is accomplished by a conceptualized self-propelled harvester, which cuts the tree stems near the ground-line, chips the biomass, and delivers the chips to a wagon towed behind the harvester. Although such a machine is not yet in production, its development is considered imperative to the farming concept. Preliminary design of a similar harvester has been completed by the USFS. Its major advantage would be reduced costs and the reduction in field traffic that occurs with conventional forest harvesting methods. Any amount of field traffic is expected to be detrimental to the development and productive capacity of successive coppice crops over short-rotation conditions. The use of a self-propelled harvester would restrict field traffic to a minimum. A network of work roads would support the harvesting and crop management operations.

The harvested biomass is shuttled to field storage areas to await transportation to the conversion facility, which is presumably located near the center of the planted area. The means of transportation is by truck, which was concluded to be the most convenient and least expensive method of transport currently available.

BIOMASS PRODUCTION COSTS

A computerized economic model incorporated the various quantifications of the CPD and site characterizations in the estimation of production costs. The model was fashioned after the accounting procedures used by industrial corporations. It accommodated a debt/equity ratio of 50:50, interest on debt at 9% per annum, a general inflation rate of 5% (fuel, and N-P fertilizer escalated at 7%), a 5.3% rate of escalation for biomass price, and income tax rate of 48%, and a discounted cash flow rate of return of 10%. The model was comprised of a group of functional models which computed the capital and operation/maintenance costs for each of the farm's component activities and operations, and combined this information to provide an overall financial analysis for the energy farm.

A summary of the production costs estimated for each of the ten study sites is presented in Table 6.7. Since in each case the "cost" value listed includes a discounted cash flow (DCF) rate of return of 10%, this value actually represents the price at which the biomass should be sold to provide such a return. A considerable variation in price exists between sites for both current and future scenarios, ranging from $1.21 per MMBtu at the Louisiana site to a high of $2.47 at the Illinois location at current productivity levels, and from about $1.00 to $1.57 per MMBtu at levels projected for the future. Capital costs comprised only 10% of the annualized total production costs, indicating that biomass production is not a capital intensive operation.

Table 6.7: Summary of Total Production Costs

	Production Costs (1976 $/MMBtu)	
Site	**Current**	**Future**
Wisconsin	1.78	1.18
Missouri	1.49	1.03
Louisiana	1.21	1.00
Georgia	1.37	0.99
New England	1.96	1.30
Washington	1.43	1.03
Illinois	2.47	1.57
California	2.00	1.47
Florida	1.24	1.03
Mississippi	1.25	1.03

Source: ERDA E(49-18)-2081

A breakdown of production costs by operation is presented in Table 6.8 for the low-yield Wisconsin site and the high-yield Louisiana site. The primary cost items are those involved with intensive crop management, essentially irrigation and fertilization costs. Fertilization and irrigation account for 36% of the total current costs at the Wisconsin site and almost 40% at the Louisiana site. These

Table 6.8: Production Costs by Cost Category – Wisconsin and Louisiana Sites

Cost Category	. . . . 1976 $/DTE (% Total Costs)			
	. . . . Wisconsin		. . . . Louisiana	
	Current	Future	Current	Future
Planning, supervision field support	1.74 (5.7)	1.63 (8.1)	1.74 (8.5)	1.63 (9.6)
Land lease	2.09 (6.9)	0.98 (4.9)	1.06 (5.2)	0.62 (3.7)
Land preparation	1.26 (4.2)	0.59 (3.0)	0.60 (2.9)	0.34 (2.0)
Roads	0.16 (0.5)	0.10 (0.5)	0.07 (0.3)	0.05 (0.3)
Planting	0.60 (2.0)	0.22 (1.1)	0.53 (2.6)	0.20 (1.2)
Irrigation	6.04 (19.9)	2.82 (14.1)	2.52 (12.3)	1.41 (8.3)
Fertilization	5.00 (16.5)	5.02 (25.0)	5.45 (26.8)	6.30 (37.1)
Weed control	0.17 (0.6)	0.08 (0.4)	0.10 (0.5)	0.05 (0.3)
Harvesting	1.37 (4.6)	1.00 (5.0)	0.93 (4.6)	0.94 (5.5)
Tractors/wagons	0.68 (2.3)	0.79 (4.0)	0.51 (2.5)	0.50 (3.0)
Loading	0.45 (1.5)	0.45 (2.3)	0.39 (1.9)	0.39 (2.3)
Transportation	3.11 (10.3)	2.45 (12.2)	1.89 (9.3)	1.56 (9.2)
Interest	1.15 (3.8)	0.55 (2.7)	0.69 (3.4)	0.38 (2.2)
Taxes	3.44 (11.4)	1.83 (9.2)	2.10 (10.3)	1.35 (7.9)
Return to investment	3.23 (10.7)	1.72 (8.0)	1.97 (9.7)	1.27 (7.5)
Salvage value	(0.15) (0.5)	(0.09) (0.4)	(0.08) (0.4)	(0.06) (0.4)
Totals	30.33	20.14	20.47	16.94

Source: ERDA E(49-18)-2081

costs increase in proportion to total costs in future scenarios, because of the continued high costs of fertilization, which are governed mostly by the static annual level of production and fertilization rate per ton of biomass produced.

Higher fertilization costs for the Louisiana site are due to the calculation of fertilizer needs based on harvesting *Eucalyptus* spp, which would include foliage as well as stems and branches. Since foliage contains greater proportions of nutrients than stems and branches, higher application rates of fertilizer were required to satisfy the requirements for nutrient replacement.

The third highest operational cost item is transportation, which comprises roughly 10% of the total costs at both the Louisiana and Wisconsin sites. Land leasing costs comprise 7% and 5% of the current costs at the Wisconsin and Louisiana sites, respectively, and decrease proportionately in the future scenarios as land requirements decrease.

The major factor influencing production costs is the level of productivity. As productivity increases, costs decrease. The effect of other site-specific variables as a rule was secondary to the effect of productivity. In certain extreme cases, however, such as the Illinois site, the difference in one variable (land cost) may be sufficient to override the effect of productivity (Table 6.7).

Further evidence of the sensitivity of production costs to productivity is revealed by the analysis that, at the high-acreage, low-yield site (Wisconsin), the productivity sensitive costs comprise 52% of total costs excluding transportation and fixed costs, whereas, at the low-acreage, high-productivity site (Louisiana), productivity sensitive costs are but 35% of the total considered. Hence, costs at low-yield sites are dominated by those which are most sensitive to productivity, while nonsensitive cost components dominate costs at high-yield sites (see Table 6.8).

Difference in productivity (DTE/acre-year) is the main reason for cost variations between current and future scenarios at the same site, and between comparable scenarios at different sites. Costs in many of the listed categories depend largely upon the number of acres required to produce 250,000 DTE annually. This, in turn, depends upon productivity; the higher the productivity the less acreage required to produce a given tonnage of biomass. Hence, changes in productivity are reflected in all cost items which are determined on the basis of acreage. These cost categories include irrigation, land lease, land clearing and preparation, planting, and weed control.

Cost categories which are largely independent of acreage considerations include fertilization, planning, supervision, field support, roads, loading, and harvesting. Transportation may be considered as falling between the two cost groups, for while trucking costs depend to some extent upon farm acreage, inasmuch as it affects the average hauling distance, the amount of biomass that has to be hauled is the primary cost determinant. Moreover, certain elements of fertilization costs are controlled by acreage, including the cost of lime and cost of ground application of lime and fertilizer during the first year of each rotation. Since average annual productivity varies with the age of the crop at the time it is harvested, production costs can also be expected to vary with the rotation period. This effect is depicted in Figure 6.6, which presents an analysis of annualized production costs at the Wisconsin site for hybrid poplar grown at

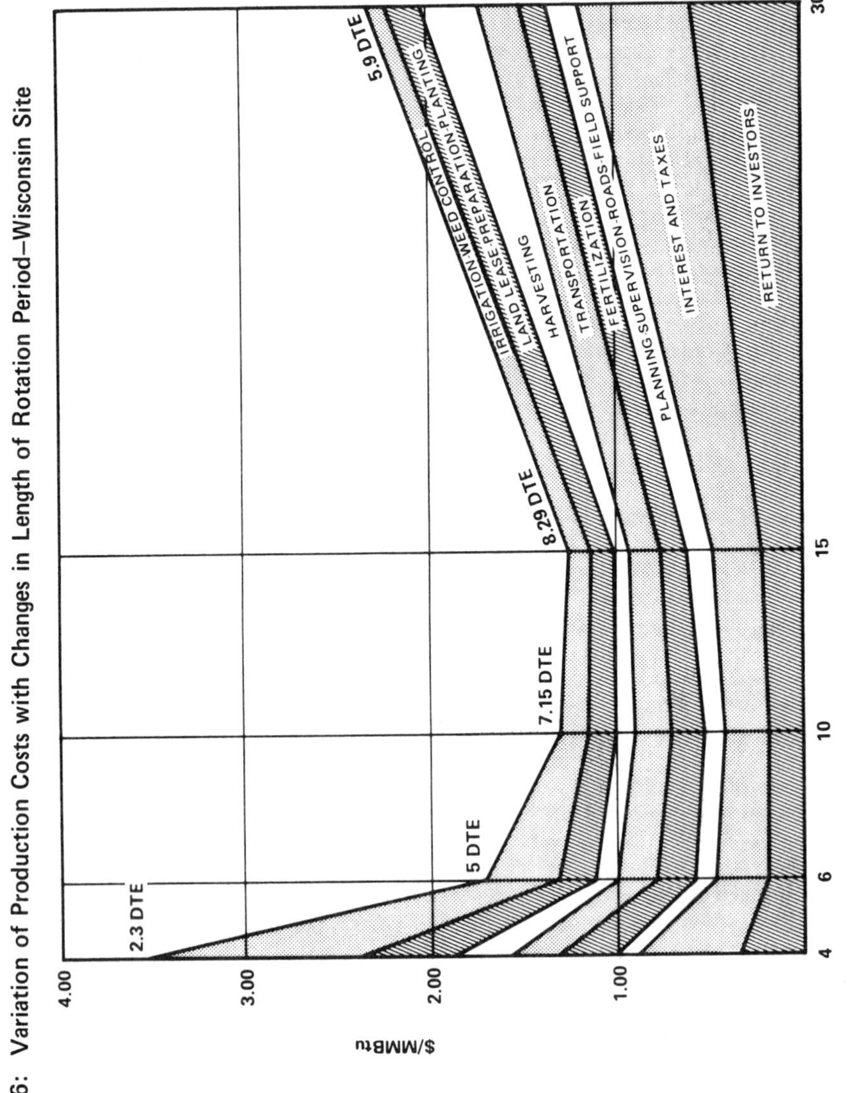

Source: ERDA E(49-18)-2081

rotations of 4, 6, 10, 15 and 30 years, based upon a 30-year planting life. The productivity of the first crop was assumed for all coppice crops.

Costs decrease inversely to productivity from the 4-year rotation to the 15-year rotation, largely due to the progressively decreasing prominence of the productivity dependent cost factors, irrigation, land lease charges, land clearing and preparation, and planting. Total production costs for the 15-year scenario were estimated to be $1.25 per MMBtu, compared with $1.31, $1.71, and $3.46 for the 10-year, 6-year, and 4-year rotations, respectively. Costs for the 30-year rotation are higher than those for the 6-year rotation in spite of its higher productivity level, both because of the effect of discounting, and increased harvesting costs associated with the use of conventional harvesting equipment rather than the conceptualized self-propelled harvester assumed for the other rotations.

The results of the analysis cannot be construed as an indication that a 15-year rotation is the "best" of all possible rotation schemes. Other factors than cost considerations must be taken into account and weighed collectively in the choice of rotation period for a given site and species. This cost analysis is based only on the "plant crop" yield, and ignores the effect of a rotation length (tree age) on coppice yields. As a general rule, older trees are not expected to coppice as strongly as younger trees, due to progressively increasing bark thickness with age which restricts the formation of adventitious buds.

Although the productivity of the first rotation may be higher the longer it is extended, the same may not hold for the coppice crops which are to follow. If coppice crops are to be considered, therefore, long-term production costs may be benefited by using rotations shorter than 15 years. Another factor which will surely play a role in consideration of rotation period selection, is that associated with obtaining and maintaining sufficient financing for plantation operations during the first rotation income hiatus.

The longer the rotation period the longer the initial returns are delayed, and the less attractive the proposition becomes as a new business venture. The lack of income during the first period also compromises the ability of management to pay interest on the debt-finance portion of the investment—the longer the period the greater the interest burden. Although a rotation period of around 15 years may be optimum on the basis of first-period production costs alone, considerations of coppice crop productivity and financing would favor shorter rotations.

Obviously, the selection of rotation periods must also consider the variations in site conditions and the species selected as crop components, for it can be expected that the optimum rotation length for managing one species at a given site may not be optimum for another site/species combination. A considerable amount of long-term (up to 20 years) field research will be required in this area before the proper designations can be made. The results of production costs analyses permit the following conclusions to be drawn:

(1) Biomass may be produced under intensive management for between $1.20 and $2.00 per MMBtu at preferred sites. These prices are generally competitive with the costs of fuel oil on the basis of higher heating values. Future production costs at the more highly productive sites may approach $1.00 per MMBtu if higher productivity levels can be developed through an organized field research program.

(2)   Production costs vary from site to site depending largely upon the level of productivity assumed, the higher the productivity the lower the total costs of production. Costs are less sensitive to changes in productivity at highly productive sites than at low productive sites.

(3)   Biomass production is not a capital intensive operation. Capital costs comprise only from 9% to 11% of the total costs.

(4)   Major production costs occur in the management of the biomass crop. Irrigation and fertilization costs alone may comprise up to 40% of total production costs.

(5)   Biomass transportation costs comprise roughly 10% of the total costs and are the third most expensive cost item.

(6)   The cost of land, whether leased or purchased, comprises 5% to 7% of the total production costs at the two sites analyzed.

(7)   The sensitivity of total costs to individual cost items varies according to the productivity level (needed acreage). Costs at low productive sites are most sensitive to those cost items which, in turn, are sensitive to the level of productivity; i.e., irrigation and land costs. Costs at highly productive sites are most sensitive to productivity-independent cost items, such as fertilization.

(8)   The south and southeast regions of the country appear to be the best suited to biomass production, both from the standpoint of high productivity levels and lower costs of production. These regions also have the greatest potential in terms of land suitability and availability.

(9)   The use of prime agricultural land is discouraged by the relatively high costs of production due to high land costs.

ENERGY BUDGET FOR BIOMASS PRODUCTION

The solar energy captured by the biomass farm is from 10 to 15 times the "primary" energy consumed during its operation. The energy budget of the process roughly parallels the dollar budget, i.e., the higher the energy consumption the greater the cost per unit biomass produced. Energy consumption per unit yield is also sensitive to productivity; assuming the same level of production, the higher the productivity the lower the amount of energy consumed per unit of biomass produced. Hence, the energy gained through the farming process is greater at the higher yielding sites. A breakdown of energy consumption by operation is shown in Table 6.9 for the current scenarios at the Wisconsin and Louisiana sites. The energy required to manufacture farming equipment is not included in this accounting.

Table 6.9: Energy Consumption and Balances—Current Scenarios

Operation	Material*	..Energy Consumed (10^9 Btu/yr)..	
		Wisconsin	Louisiana
Supervision	Gasoline	1.67	1.67
Field supply	Diesel/gasoline	0.53	0.53
Harvesters	Diesel	9.43	7.87
Tractor haul	Diesel	5.45	5.49

(continued)

Table 6.9: (continued)

| | | . .Energy Consumed (10^9 Btu/yr). . | |
Operation	Material*	Wisconsin	Louisiana
Loading	Diesel	3.32	3.29
Transportation	Diesel	28.83	17.32
Irrigation move	Diesel	1.30	0.65
Irrigation pumping	Diesel	223.63	111.83
Manufacture	Urea	99.36	102.32
Manufacture	P_2O_5	8.03	8.17
Manufacture	K_2O	15.42	17.08
Ground operations	Diesel	1.16	0.06
Aircraft operations	Gasoline	0.15	0.16
Fertilizer transport	Diesel (rail)	0.98	0.64
Total energy consumption		399.26	277.62
Total energy yield		4,250	4,250
Net energy yield		3,851	3,972
Net energy efficiency		0.906	0.935
Energy in:energy out		1:10.6	1:15.3

*Energy content: 124,000 Btu/gal of gasoline; 138,690 Btu/gal of diesel fuel; 27,730 Btu/lb N of urea; 6,019 Btu/lb of P_2O_5; 4,158 Btu/lb of K_2O; and 800 ton-miles per gal for rail.

Source: ERDA E(49-18)-2081

UTILITY OF FORESTRY RESIDUES

Mill Residues

Wood and bark residues are generated in mills which process timber into primary products, such as lumber and plywood. They are also generated, along with some foliage residues, in the growing and harvesting of timber.

Mill residues can provide a ready source of energy, particularly for the forest products industry. They are also a source of raw material for a number of products, such as wood pulp and particleboard.

The heating value of mill residues depends on the species of wood involved, the relative contents of wood and bark, and moisture content. Heating values of representative North American commercial wood species range from about 7,900 to 9,700 Btu/lb, on a moisture-free basis. The heating values of bark are generally somewhat higher than for wood of the same species. Moisture content of mill residues may range from 2 to 8% for sander dust to 75% for green bark. The combustion efficiency of oven-dry wood is estimated at about 82%. Raising the moisture content to 50% reduces combustion efficiency to about 68%.

Wood and bark residues have negligible sulfur content, unlike most coals and many heavy fuel oils. Thus, combustion does not present a sulfur oxide emission problem. Furthermore, wood and bark have relatively low ash contents, compared to coal. Wood and bark have high oxygen contents, which means that less oxygen has to be supplied from air for combustion.

Mill residues are generated in very large amounts, owing to the fact that less than half of the volume of timber (including bark) processed into lumber and plywood becomes a finished product. Total mill residues generated in the U.S. in 1970 are estimated at about 86 million dry ton equivalents (DTE), including about 67 million DTE of wood and 19 million DTE of bark. This figure includes only residues generated in the manufacture of lumber, plywood and miscellaneous wood products, such as cooperage, shingles, pilings and posts. Approximately 78% of total residues were generated in the lumber industry, 15% in the plywood industry, and the remaining 7% in the manufacture of miscellaneous wood products.

Some representative prices of mill residues are presented in Table 6.10. Prices of wood and bark residues fluctuate widely and exhibit considerable variability with respect to location. Further, there are no centralized sources of price information, as there are for primary wood products.

Table 6.10: Some Representative Prices Paid for Wood and Bark Mill Residues ($/MMBtu)

	West Coast	South	Maine
Chips			
Softwood			
1975	1.47–2.21	–	–
1976	2.35	1.47–1.65	2.35
Hardwood			
1976	–	1.06–1.18	–
Shavings			
1975	0.22–0.32	–	–
1976	–	0.44	–
Sawdust			
1975	0.15	–	–
1976	–	0.09–0.12	0.29
Bark			
1975	0.14	–	–
1976	0.34	0.06	–

Source: ERDA E(49-18)-2081

Some mill residues now being left unused are available at zero or nominal cost. In some cases they may be enough of a nuisance that producers would actually be willing to subsidize their removal to some extent. Industry sources indicate this is actually happening for some residues in some locations.

Estimates of future residue generation through 2020 have been made, based on projections of primary product output in each of the three industries mentioned above. Total mill residue output expected in key future years is as follows:

	Projections in Millions of DTE (quads)	
	Low	High
1980	98 (1.67)	124 (2.1)
2000	88 (1.5)	128 (2.2)
2020	76 (1.3)	143 (2.4)

The high projection indicates a 66% increase in residue production over 1970 levels by 2020, while the low projection indicates a 12% decrease. These estimates are based on alternative high and low projections of primary product output which are based in turn on alternative high and low projections of socio-economic variables such as population growth and income growth. Both projections assume a relative decline in lumber production, which is a relatively heavy producer of residues, in favor of plywood production and residue-based panel products.

Approximately 75% of all wood residues generated by the lumber, plywood and miscellaneous wood products industries in 1970 were used for some purpose. Approximately 56% of all wood residues generated were used for nonenergy products, primarily for wood pulp. The remainder of the used wood residues, about 13 million DTE or 19% of total wood residues, were either used as fuel where produced or sold as fuel.

Approximately 11.6 million DTE, or some 60%, of bark residues were put to some use in 1970. It is likely that most of this amount was used for energy production.

Demands for mill residues are apt to increase rapidly, both for energy and non-energy products, in the forest products industry itself. The allocation of such residues among uses in the long-term will depend on the relative prices of residue-based forest products and alternative fuels. Mill residues are not likely in any case to be available for energy uses outside the forest products industry, except in relatively limited local situations. However, cogeneration of electricity is a real possibility. There are a number of forest products plants in the country that occasionally sell power to the local grid.

Forest Residues

Forest residues represent a potentially large source of biomass for energy production. The term "forest residues" has been interpreted to include logging residues, intermediate cuttings, understory removal and annual mortality. The last of these, annual mortality, refers to trees killed by natural agents. Such trees are generally widely dispersed, hence making it likely that their collection for any purpose would be noneconomic in most instances.

Intermediate cuttings involve the removal of small or inferior trees from a stand for the sake of stand improvement. Understory removal refers to the removal of shade-tolerant shrubs and trees growing beneath the canopy of an older commercial forest. Both of these operations could provide biomass for energy production, but only in relatively limited circumstances.

Logging residues represent the only widely and readily available source of forest residues. These are essentially the "leftovers" of commercial logging operations and are generally regarded as consisting of the aboveground portions of trees not removed in such logging operations, that is, branches, foliage, and pieces of stems too small or too defective for conversion into conventional wood products.

Forest residues could conceivably be collected and reduced to fuel sizes by chipping or by processing through a hammer mill similar to those used for reducing mill residues to fuel. The physical characteristics of forest residues which would be of importance with respect to energy production would be similar in most respects to those of mill residues.

Total aboveground logging residues generated in the U.S. in 1970 are estimated at about 83 million DTE. Stump-root systems left as logging residues in 1970 are estimated at 104 million DTE.

On a per-acre basis, logging residues are generally heaviest in the Pacific Northwest, where much of the timber harvested comes from stands of large, overmature, old-growth timber. Trees in these stands are often rotten and subject to breakage on falling, with broken or rotten pieces being left on site. Aboveground logging residues in the Pacific Northwest in 1970 averaged 24 DTE per acre of logged ground compared to a national average of 9.1 DTE per acre.

Forest residues generally have no positive value to forest landowners at the present time, and it may be assumed that they are currently available at no cost.

Costs of collecting, reducing and transporting forest residues are currently estimated at about $23 per DTE in the south to between $44 and $61 per DTE in the Pacific Northwest.

Estimates of future logging residue generation through 2020 have been made based on projections of timber harvest levels. Total logging residue outputs expected in key future years are as follows:

| | Projections in Millions of DTE (quads) | |
	Low	High
1980	83 (1.4)	110 (1.9)
2000	81 (1.4)	148 (2.5)
2020	107 (1.8)	195 (3.3)

The high projection indicates a 135% increase by 2020, while the low projection indicates a 29% increase. These estimates are based on alternative high and low harvest estimates, which correspond in turn to high and low projections of primary product outputs. Utilization of forest residues in the United States at the present time is negligible.

During periods of high pulp chip prices in recent years, there have been instances of larger-sized residues being salvaged and chipped in the Pacific Northwest. However, this has generally taken place in the national forests, where timber cutting contracts require that logging residues be collected and piled. Thus, collection costs are subsidized, with the user paying only for chipping and transportation.

Another important area of residue utilization is really a reduction in residue production. This is occasioned by whole-tree chipping, wherein entire trees, foliage and all, are reduced to pulp chips by mobile chippers. The ability and willingness of pulp mill operators to accept chips of this kind is limited, but is increasing and will probably continue to do so as fiber becomes more expensive. Obviously, logging residues are limited where whole-tree chipping is practiced.

Utilization of forest residues, particularly aboveground logging residues, is likely to increase during the projection period, for both energy and nonenergy uses. Forest residues, properly processed, can be put to many of the same uses as mill residues. The admixture of wood and bark is a problem in many non-energy uses, but separation of bark from chipped residues is a subject of current research, and significant progress has been made.

If the value of fiber for wood pulp and reconstituted building materials increases, and the value of residue biomass for energy increases, it is expected that manufacturers will look to much of what is now being left as residues on the forest floor. This will manifest itself both in changes in utilization standards and in increased efforts to collect residues after primary harvesting operations.

As in the case of mill residues, the allocation of forest residues as among uses in the long-term future will depend on the relative future prices of residue-based forest products and alternative fuels.

Two surveys of residue fuel availability and cost were carried out, one in the northern New England area and one in the southern Arkansas area.

The New England study found that generally adequate supplies of mill residues were available at attractive prices ($0.63 per 10^6 Btu) compared to fuel oil. It also concluded that the future outlook for residue fuel was favorable, with little competition for available supplies likely to develop. At the same time, it may become feasible to tap new sources of supply, including logging residues and low-grade chips from whole-tree chipping.

The Arkansas study found, on the other hand, that investment in increased residue fuel handling capacity would not be justified, based on expected costs of the increased supply. Costs of additional residue fuel in the area were estimated at about $10 per green ton, making it somewhat less expensive than oil at $12 per barrel. However, it was concluded that much higher residue fuel prices were likely in the future, due to expected competition for residues for both energy and nonenergy uses.

The two surveys emphasize the regional differences in residue markets and the necessity for considering the feasibility of using wood and bark residues for energy production on a case-by-case basis.

SUMMARY

Major potential sources of wood/bark biomass for conversion to useful energy products include under-utilized standing forests, logging residues, mill residues, precommercial thinnings from commercial forests, and silvicultural biomass farms. Biomass farms represent a substantial long-term potential for contributing to the nation's energy supply. Other sources represent both near-term and long-term potential in varying degrees. Under-utilized standing forests, in particular, warrant consideration for near-term use.

A silvicultural energy farm would involve the intensive management of a densely planted energy crop under short rotations, utilizing selected coppicing species. Major technical problems are in the areas of biomass harvesting and storage, and in increasing biomass productivity.

The two major factors which will influence the practicability of biomass farming are biomass productivity and the availability of suitable land.

Biomass productivity under close-spaced, short-rotation conditions is estimated to range from 5 to 13 DTE per acre-year with current technology, depending upon species and site selection. It is anticipated that these yields could be essentially doubled within 25 years by a concerted research effort on species selection and improvement, and energy crop management.

If 10% of the arable land currently used for private forest, pasture, range, and hayland were to be used for biomass production, up to 4.5 quads of energy could be produced annually at current yield levels, and 8.3 quads at anticipated future yield levels. The production of one quad of energy would require the use of 5.9 million acres at a productivity level of 10 DTE per acre-year.

In terms of potential availability of suitable land, the Southeast, Delta States, and Lake States regions of the country are most likely locations for biomass farms.

ETHANOL PRODUCTION TECHNOLOGY

The material in this chapter is based on *The Report of the Alcohol Fuels Policy Review,* DOE Report PE-0012 prepared by the U.S. Department of Energy, June 1979; *Survey of Alcohol Fuel Technology. Volume I,* NSF Report 760009, prepared by B. Baratz, R. Oulette, W. Park and B. Stokes of Mitre Corporation for the National Science Foundation, November 1975; and "Researchers Foment Better Ways to Ferment," *Chemical Week,* August 8, 1979.

Commercial production of ethanol by fermentation is based on the conversion of sugars with six carbon atoms, or C_6 sugars to ethanol by yeast. Fermentation is the decomposition of organic compounds into simpler compounds through the use of enzymes, the enzymes acting as catalysts to the process:

$$C_{12}H_{22}O_{11} \; + \; H_2O \; \xrightarrow{\text{invertase}} \; 2C_6H_{12}O_6$$
$$\text{sucrose} \hspace{5cm} \text{invert sugar}$$

$$C_6H_{12}O_6 \; \xrightarrow{\text{zymase}} \; 2C_2H_5OH \; + \; 2CO_2$$
$$\hspace{4cm} \text{ethanol}$$

The fermentation process takes about 36 to 48 hours, with the resulting fluid containing 6 to 12% ethanol. These fluids are then distilled to draw off the ethanol. In the distillation process, some impurities (aldehydes, esters, and fusel oil) distill with the ethanol. Moreover, this commercial grade of alcohol contains about 95% ethanol (190 proof). Further distillation cannot increase this percentage as the composition is a constant boiling point mixture of alcohol and water. Anhydrous ethanol (200 proof) is obtained by the addition of benzene to commercial grade ethanol and further distillation.

Until the end of World War II, the fermentation of molasses, fruits, and grains was the main source of ethanol in the United States. Fermentation is still the only source for ethanol produced for human consumption. Today, however, approximately 80% of all industrial grade (nonbeverage) ethanol in the United States is produced synthetically by the direct hydration of ethylene. In 1977, approximately 202.9 million gallons of ethanol from ethylene, and 50 million gallons of ethanol from fermentation of sugars were produced for the industrial market. Production of beverage-grade ethanol that year was about 30 million gallons (1).

Since ethylene is produced from ethane and propane, components of natural gas, and from naphtha, industrial alcohol production in the United States is based primarily on domestic nonrenewable or imported resources.

More advanced technology for ethanol production is in a relatively early stage of research and development. This newer technology pertains not only to improvement in conventional processes, but particularly to utilization of agricultural and forest residues as raw materials for alcohol production. It is estimated that these technologies should develop to commercial feasibility levels within the next five to ten years.

FERMENTATION

Although batch fermentation has been the conventional method of operation in industrial and beverage fermentation alcohol plants, there have been some developmental as well as full scale test operations of continuous fermentation.

Although this should lead to faster fermentation and reduction of volumetric capacity requirements, uncertainties concerning maintenance of sterile conditions has led to the requirement for redundant equipment to permit periodic switchover for cleaning and sterilization. This, in effect, nullifies the potentially substantial reduction in investment for continuous fermentation. It is possible that with improved techniques for continuous fermentation, and maintenance of sterile conditions, the redundancy requirement could be eliminated. In any event, continuous fermentation should also lead to reduced energy requirements, and increased overall fermentation efficiency.

The use of multiple batch fermentation vessels sometimes has been called continuous fermentation. This multibatch operation has the advantage of speeding up the fermentation process. However, it also requires increased capital investment for duplicate equipment. At the present time, there are no true continuous fermentation systems commercially operating in the United States.

Research and development work is now proceeding on continuous fermentation systems operating under vacuum, with the concept that the heat generated by fermentation will be sufficient to distill off the alcohol. This should result in reducing the need for fermenter cooling water, accelerating fermentation, and improving fermentation yields; in effect reducing the overall investment and operating costs of fermentation. However, it is unlikely that such operation can achieve complete recovery of alcohol from the fermented beer (95% by conventional beer still stripping).

It must also be taken into account that the use of vacuum vapors from the fermenter in the distillation system requires energy either to compress the vapors to a level where conventional cooling water can be used in the alcohol condensers, or to provide refrigeration for the condensing operation. No comprehensive technical and economic evaluation appears to have been made for such a system. It is possible that there could be some gain by such operations, but the potential savings may be counterbalanced by increased costs of construction of fermenters suitable for vacuum operation, the cost and energy consumption of vapor compression or refrigeration facilities, and the potential decrease in alcohol recovery.

Of greater promise are techniques for recovering and recycling yeast (once it can be separated from grain solids). Also promising is the development of high yeast concentration continuous fermentation systems such as the countercurrent flow fermentation tower. In these systems, the yeast concentration is maintained at a high level; fermentation time can be shortened to a few hours instead of a few days. This requires a higher rate of localized heat removal from the fermenters, which entails mechanical complexities. However, further research along this line offers more economic promise than the vacuum fermentation scheme.

Bacterial Fermentation

While yeast is the standard microorganism used to ferment carbohydrates into ethanol, many other organisms can do the job as well. At Australia's University of South Wales and at the U.S. Department of Agriculture's Northern Regional Research Laboratory (Peoria, Ill.), researchers are experimenting with *Zymomonas mobilis,* the bacteria strain used to make tequila.

The chief of USDA's Peoria fermentation laboratory says a researcher there has found that *Zymomonas mobilis* achieves 6 to 8% ethanol yield. This is not impressive compared with the 10 to 12% yields with conventional yeast; but the Peoria group is fascinated by *Zymomonas* because of its rapid growth at high temperatures.

Yeast won't grow and thrive above 20°C, but *Zymomonas mobilis* can withstand substantially higher temperatures. And as the temperature goes up, so does the fermentation rate.

One example of what USDA hopes to achieve is fermentation to 12% ethanol at 37°C in 40 hours. To move closer to such a goal, Peoria researchers are using mutation and selection techniques to develop *Zymomonas* strains that will give higher yields.

At the Massachusetts Institute of Technology, the Department of Nutrition is experimenting with *Clostridium thermocellum.* An anaerobic, thermophilic organism, it thrives at elevated temperatures, typically 55° to 60°C, only in the absence of oxygen. This bacteria strain has the ability to hydrolyze cellulose into glucose sugar and simultaneously ferment the glucose to ethanol. (Most other schemes for converting green-plant material into ethanol require some form of pretreatment, mild acid hydrolysis of the glucose, for example.)

It is thought that this simple one-step approach will represent an important alternative to those involving pretreatment and then fermentation. No yields are revealed, except to say "they need to be improved." The MIT group has used corn stover as its starting material, but any agricultural or municipal waste could conceivably be used.

Mold Research

At Purdue University, a group is working with *Rhizopus* and related mold species. *Rhizopus* organisms are used in fermentation of certain types of Chinese wines, where the alcohol content approaches 18%. The researchers have employed the mold in a column-type fermentation reactor, instead of the standard tank-type reactors that require constant stirring.

Because the molds can be fashioned easily into pellets, they can be inserted into the column reactors, simplifying the fermentation process. It involves nothing more than pouring a sugar-containing solution in the top of the reactor and collecting alcohol out of the bottom. Yields are about one-half pound of ethanol per pound of sugar feed.

The group has found a way of maintaining the viability of the mold once it is inside the column. The column itself, could eventually be employed in a continuous fermentation operation, cutting down fermentation time.

DISTILLATION

There is continuing discussion and debate concerning the necessity for production of anhydrous alcohol (199 proof—99.5 volume %) for addition to gasoline for gasohol use. Claims are made that conventional spirits (190-192 proof) can be utilized. Although this statement is essentially true, such mixtures can lead to considerable difficulties, particularly with seasonal variations, not only in cold climates but also in subtropical, humid climates, due to changes in moisture content of the mixture affecting adversely the relative solubilities of alcohol, gasoline and water (1% moisture content is sufficient to cause separation of alcohol and gasoline).

These changes can lead to separation of the alcohol component from the major gasoline component, leading to stratification in storage tanks and in motor car fuel tanks. Such stratification leads to the feeding of two different types of fuel to the carburetor, which cannot be readily adjusted for such a change. Operating practice with motor fuel alcohol blends with gasoline in several countries, such as France, Cuba, and Brazil in past years, has led to the conclusion that it is safest to proceed with use of anhydrous alcohol for such blends.

The major argument against production of anhydrous alcohol has usually been based on the high total energy consumption of first producing alcohol spirits (about 190 proof), and then using additional energy to dehydrate the spirits to the anhydrous grade (200 proof). Although this has been a viable argument in the past, current technology, employing a dual-pressure level distillation system, permits the manufacture of anhydrous alcohol with no additional energy input over that required for the alcohol spirits. Furthermore, with increased economic justification for heat recovery, and application of novel but proved techniques in this area, energy consumption for spirits production can be reduced substantially below what was previously considered a required level.

Additional developments in design and operation should permit additional improvements in the distillation area, with potential reductions in investment and further moderate reductions in energy requirements.

Research is continuing on alternatives to the distillation method of separating alcohol from water. Methods such as liquid-liquid extraction, adsorption agents, reverse osmosis and ultrafiltration are being developed on a laboratory scale. However, the economics of these systems is uncertain. The research has not yet faced the real problems of suspended solid contaminants in the fermented beer which would adversely affect extraction solvents, adsorbents, and membranes in reverse osmosis and ultrafiltration cells. Therefore, research should be directed toward economical systems of pretreatment with minimal energy use, to give a highly clarified feed to the separation system.

Even when such separation systems are developed, it is unlikely that if they removed the alcohol to the very high degree of recovery required economically, they can produce high concentrations without a final distillation step. Thus, economics must eventually be evaluated on the basis of combined primary separation and secondary high concentration distillation systems for spirits, and dehydrating units for motor fuel alcohol.

CELLULOSE CONVERSION

There is a very major research and development effort nationwide and abroad, in industrial, government, and university laboratories, to develop improved technology for conversion of cellulosic wastes to sugars, which, in turn, can be fermented to alcohol. The effort in the United States is justified on the basis of potential utilization of hundreds of millions of tons per year of agricultural and forest residues.

Technical development of processes utilizing either acid or enzyme hydrolysis, or a combination thereof is expected to lead to reasonable sugar yields from cellulose with practical investment and operating costs within the next five to ten years. One of the keys to commercialization will be the development of economic methods of collecting the raw materials and storing them.

Development of cellulose-to-alcohol processes has not proceeded far enough to justify immediate investment in full-scale plants, but technical and economic justification may be achievable in the next five to ten years. Therefore, consideration should be given in the design of alcohol facilities to process agricultural and forest residues.

The major specific technologies being developed in this area have advantages and disadvantages, which can only be evaluated in comparison with each other, and against the economics of conventional production of alcohol from grain or sugar cane. Ongoing research should then be directed toward more economical means of overcoming deficiencies. Several of the major processes being developed, and comments on these technologies are as follows.

Dartmouth College and Bureau of Solid Waste Management

An acid hydrolysis technique was developed by Porteous (2) with continued research by Converse, Grethlein, and their associates at Dartmouth (3)(4) and also by Chapman (5) from the Bureau of Solid Waste Management. The process operates at much higher than normal temperatures, reducing the cooking time to a matter of seconds. Also, under these circumstances the process obtains high sugar yields, in the region of 50%. In contrast, the historical experience with acid hydrolysis in a batch reactor shows run times of several hours and shows production of a large number of unwanted by-products. It is reported that the fermentation yield from these acid hydrolysis procedures compares very favorably with grain fermentation.

U.S. Army—Natick Laboratories

This laboratory first identified the fungal enzyme complex which hydrolyzes cellulose to glucose. In recent years, they have continued their research on improving enzyme production by fungus mutation, and developing basic knowledge concerning the three-enzyme system necessary to convert cellulose to glucose. However, the system developed at Natick has not been effective on lignin-containing materials, such as forest and agricultural residues. Essentially, only purified cellulose has given reasonable glucose yields, but such waste raw materials are limited in amount (pulp and paper mill wastes) or are expensive (primary pulp products).

More recently, Natick has given attention to replacement of the pretreatment method involving reduction of raw materials to fine particle size (below 200 mesh), with a roller mill flaking system, which makes lignocellulosic materials more accessible to enzymes for hydrolysis. However, there is no indication that this method is practical or economical on a large scale.

University of California at Berkeley (Wilke)

Realizing the difficulty and expense of preparing the raw materials by Natick technology, Wilke has investigated dilute acid pretreatment, which removes hemicelluloses and opens up the cellulose for enzyme hydrolysis. The economics of this process has not been thoroughly evaluated. This may be fairly expensive, as the acid treatment involves large corrosion resistant vessels or a very large continuous countercurrent reactor. Also, the acidic solution resulting from this pretreatment must be neutralized, and either concentrated for sale as feed, or other products must be made from it.

Wilke has also been investigating improved fermentation technology, including a high yeast concentration system, as well as a high sugar concentration system. However, the relatively dilute sugar solutions obtainable by acid and/or enzyme hydrolysis of cellulose preclude such fermentation improvement, without preconcentrating the sugar solutions. This would, in turn, take considerable energy.

University of Pennsylvania (Humphrey) and General Electric Company

The joint research of the University of Pennsylvania and the General Electric Company relates to dilute acid prehydrolysis utilizing sulfur dioxide and water as a hydrolyzing agent, prior to enzyme treatment. This has some advantages, in that after hydrolysis, a major part of the SO_2 can be stripped from the liquor, recovered and recycled. This would minimize neutralization costs and waste disposal. However, this system is also very corrosive and would require expensive alloy equipment for the pretreatment steps. The resultant prehydrolyzed materials would have to be washed or neutralized to increase pH to a level suitable for enzyme hydrolysis.

An alternate technology, being researched at the University of Pennsylvania, utilizes aqueous butanol as a delignifying agent, to yield cellulose suitable for enzyme hydrolysis. This process, in turn, requires separation of dissolved lignin (and pentoses) from the aqueous butanol extract. This is a complicated and difficult separation step. Solvent losses could be substantial and could make the process uneconomic.

The fermentation media being studied by General Electric yield substantial amounts of acetic acid as well as ethanol. This would be undesirable from a fuel standpoint, because of the corrosive nature of acetic acid. However, some microorganisms can convert pentoses to alcohol, which offers some promise. This line of research should be extended.

Purdue (Tsao)

Three different techniques have been investigated in the Purdue program. The recent evaluation report of the Purdue process by the Arthur G. McKee Co. for DOE (6) sheds some light on the current preferred approach.

One of the Tsao routes starts with dilute sulfuric acid pretreatment, which yields a pentose solution. Lime is added to the solution as a neutralizing agent. This would create a substantial scaling problem. The step for squeezing excess dilute acid solution from the lignocellulosic residue is carried out on conventional pulp mill equipment, such as screw presses. Such presses are not designed for handling hot dilute acid. They would have to be redesigned and built with expensive alloy material, as considerable heat is developed, as well as pressure stresses, which would adversely affect the resistance of conventional alloys.

Tsao has advocated converting the pentose solution to butanediol. Such a product, however, is anticipated to have a limited market.

In the next step of the process, the lignocellulosic residue from the pretreatment operation is treated with 70% sulfuric acid for dissolution of the cellulose. The acid cellulose solution, containing suspended lignin, is then treated with methanol to precipitate the cellulose. The separation of suspended cellulose and lignin solids from the methanol-sulfuric acid mixture is indicated as utilizing belt filter and other conventional pulp mill equipment.

None of this equipment is designed for use with 70% sulfuric acid. Wherever heat is generated, or the acid can be subject to slight dilution, as in startup and shutdown operations, excessive corrosion will occur. The materials of construction indicated would not withstand partially diluted sulfuric acid.

The problem is particularly difficult in the next step; evaporation to separate methanol from the 70% sulfuric acid. Corrosion would be a major problem at the elevated evaporator temperatures. Standard or special alloys may be unsuitable. Nonmetallic materials, which are difficult to maintain, may be required. There would also be a problem of methanol degradation by reaction with the hot sulfuric acid. Such reactions probably will produce dimethyl ether, dimethyl sulfate, dimethyl sulfoxide and other lesser products. It is unlikely that methanol recovery could achieve the 99% level assumed by the McKee report. Further, the by-product sugars outweigh the primary glucose sugar product by 70%.

At the present time, there still appear to be technical and economic questions about the Purdue/Tsao process.

Gulf Oil Chemicals Company

Gulf Oil Chemicals' cellulose alcohol technology appears to be the most advanced from a technical and economic standpoint. Their research has led to development of enzyme mutants which permit hydrolysis of cellulose to sugars in the presence of lignin, with quite satisfactory yields. This is done with only moderate pretreatment of raw materials, which may consist of municipal solid waste (cellulose fraction), pulp mill wastes (primary sludge and digester rejects), and agricultural and forest residues.

Enzyme production has been accelerated in a continuous process, as compared to long term batch processes of other investigators (one to two weeks). The enzymes need not be extracted from the enzyme broth, but can be used directly in the next stage.

This next stage is simultaneous saccharification and fermentation. This reduces time requirements radically, as compared to other processes. Most other processes have separate steps for saccharification, which takes on the order of one week, and fermentation, which may be the order of two days.

It is notable that this process does not utilize any acid or solvents, thereby minimizing corrosion and eliminating solvent recovery problems. However, the investment base and related operating charges are substantial. Gulf is operating a one ton per day pilot plant. The demonstration plant may indicate ways and means of reducing these cost elements.

REFERENCES

(1) U.S. Department of the Treasury, Bureau of Alcohol, Tobacco, and Firearms, *Summary Statistics,* FY (1976).

(2) Porteous, A., *A Profitable Means of Municipal Refuse Disposal,* Thayer School of Engineering, Dartmouth College, Hanover, N.H. (November 1966).

(3) Converse, A.O., Grethlein, H.E., Karandikar, S., and Kuhrtz, S., *Acid Hydrolysis of Cellulose in Refuse to Sugar and Its Fermentation to Alcohol,* Thayer School of Engineering, Dartmouth College, Hanover, N.H., NTIS No. PB-221 239 (June 1973).

(4) Grethlein, H.E., "The Acid Hydrolysis of Refuse," presented at the NSF Special Seminar on Cellulose as a Chemical and Energy Resource, University of California, Berkeley (June 25–27, 1974).

(5) Chapman, R.A., *Acid Hydrolysis of Cellulose in Municipal Refuse,* Report RC-02-68-11, Public Health Service, Department of Health, Education, and Welfare (1970).

(6) McKee Engineers and Contractors (Chicago), *Preliminary Engineering and Cost Analysis of Purdue/Tsao Cellulose Hydrolysis (Solvent) Process,* U.S. Department of Energy Fuels from Biomass Branch, PO No. EJ-78-X-01-4641, McKee Contract GC 4967 (August 1978).

ADDENDUM–SEPARATION PROCESSES

In an article by F.F. Hartline, "Lowering the Cost of Alcohol," *Science,* Volume 206, October 5, 1979 (copyright 1979 by the American Association for the

Advancement of Science), technologies for separating alcohol from water solutions are discussed which may significantly lower the cost of producing ethyl alcohol from fermented grain, sugar crops or cellulose. This is possible because these new processes use less energy in recovering the alcohol than does the traditional method of distillation, which was developed primarily to produce beverage-grade alcohol. If laboratory estimates of the energy savings are borne out on an industrial scale, alcohol may become competitive with gasoline.

Traditional ways of producing and recovering beverage-grade ethyl alcohol require 1.4 to 1.6 times more energy than would be liberated if the alcohol were burned as a fuel, and the cost of growing and gathering the raw biomass is additional. Including the energy required to remove the last of the water, a step omitted by beverage makers but essential if the alcohol is to be used in gasohol (90% gasoline, 10% alcohol), the distillation steps alone require a little more energy than the fuel value of the alcohol. The new processes might halve the total energy cost of alcohol just by making the purification steps more efficient.

M. Ladisch and K. Dyck, researchers at Purdue University announced that the energy used in recovering dry alcohol from fermentation broths, called beers, can be cut to about 10% of the fuel value of the alcohol if cellulose or other dry plant materials are used to absorb some of the water (*Science,* August 31, 1979, p 898).

The Purdue technique for separating alcohol from water solutions is a variant of a much older process. It has been known for years that alcohol can be dried with desiccants such as calcium oxide, which reacts with the water present, forming calcium hydroxide.

Ladisch and Dyck's approach is to dry the vapors from an alcohol-water solution that has already been concentrated by distillation to perhaps 85% alcohol. It was found that desiccants such as cellulose and cornstarch provide greater energy savings than calcium oxide, because less energy is needed to dry cellulose or cornstarch than to drive the water off calcium hydroxide. If cracked corn is used as the desiccant it does not even have to be redried, but can be fermented wet to make more alcohol.

Another technique for separating alcohol from water solutions is a version of gas chromatography. B. Miller of the Textile Research Institute in Princeton says that textile yarns such as rayon retard the movement of water vapor, but allow organic vapors to travel freely. Miller and his coworkers have developed a continuous process for separating alcohol from water based on this principle. An endless loop of yarn fibers is pulled slowly through a tube into which alcohol-water vapors are introduced. The water is removed by the yarn (which is then dried by heating), and pure alcohol vapor is recovered from the other end of the tube. However, the energy requirements for this gas chromatograph have yet to be determined.

Alcohol dissolves readily in some liquids that do not mix with water to any great extent. By exploiting this solubility difference, alcohol can be recovered from aqueous solutions by solvent extraction. One version of the process uses a so-called critical fluid, a gas that has been compressed to the point where the distinction between gas and liquid disappears. At 50 to 80 times atmospheric pressure, carbon dioxide becomes a critical fluid, and can be used to extract alcohol from fermentation beer that has been filtered to remove solids, says R. de Filippi of Arthur D. Little, Inc. Details of how alcohol is recovered from the carbon dioxide are proprietary, but de Filippi estimates the energy cost of the whole extraction process "conservatively" at 40 to 60% of the alcohol fuel value. One advantage of this scheme is that carbon dioxide is a by-product of the fermentation process, so the solvent costs are minimal, an important consideration since some of the solvent inevitably escapes in the alcohol recovery process.

Several other laboratories have different solvent extraction schemes under development. For example, the University of Pennsylvania and General Electric are working together on a process using dibutyl phthalate, a water-immiscible solvent for alcohols. This solvent has a much higher boiling point than fuel alcohols, so the alcohols can be driven off easily in a single distillation step and solvent losses are minimal.

Initially the researchers bubbled the dibutyl phthalate through the fermentation liquid, but emulsions of beer and solvent formed. The Pennsylvania group is now exploring several different ways to circumvent this problem. One method they have tried with some success is to keep the solvent separate from the beer with a thin membrane.

H. Gregor, of Columbia University, is optimistic that membranes, without additional solvents, can be used for alcohol purification. He calculates that with membranes, the energy cost of recovering pure alcohol from fermentation beer could be reduced to about 0.6% of the alcohol fuel value. This corresponds to an energy expenditure of only about four times the theoretical minimum imposed by the second law of thermodynamics.

Alcohol could be separated from water with membranes that are comparatively impermeable to alcohol but allow the water to pass through. Such membranes already exist, says R. Riley of UOP Incorporated's Fluid Systems Division in San Diego, but the pressures that would be needed to drive an industrial alcohol separation process are too high to be practical.

Another approach is to use membranes that are permeable to alcohol but not water. Gregor reports that he has such a membrane, one that passes alcohol about twice as fast as water and permits reasonable fluxes of alcohol. But more work is needed to find membranes suitable for commercial use, where high selectivity and high alcohol flux are essential for the overall economics of the separation process.

Zeolites, molecular sieves, selectively absorb molecules smaller than a certain size. These porous rocks might be useful in alcohol recovery, notes G. Baughman of the Colorado School of Mines Research Institute.

A promising candidate is clinoptilolite, a naturally occurring zeolite with holes that water, but not alcohol, should fit into. Clinoptilolite has yet to be tested for its selectivity in absorbing water, but if it works, the relative abundance of clinoptilolite could make its use economically attractive at least in some parts of the country. According to J. Gelo of Union Carbide Corporation's Linde Division, researchers there are exploring the use of synthetic zeolites for the final purification step. Gelo says that such a process can compare favorably with distillation in terms of cost and energy consumption.

With such a wide variety of techniques under investigation, it is unclear which, if any, will displace distillation for the recovery of alcohol or other fermentation products. In the end, choices may be made on the basis of fringe benefits or capital investment costs that affect the economics of the whole process, rather than on the basis of energy efficiency of the purification step alone.

NEBRASKA GASOHOL PROGRAM

The information in this chapter is based on "A General History of the Nebraska Grain Alcohol and Gasohol Program," by C.R. Fricke of the Agricultural Products Industrial Utilization Committee, presented at the Symposium on Utilization of Alternate Fuels for Transportation, University of Santa Clara, Santa Clara, California, June 19-23, 1978, and furnished by the National Gasohol Commission.

BACKGROUND

The history of Nebraska involvement in blending agriculturally derived ethyl alcohol and gasoline has been extensive. The first widespread evidences of this idea date back to the 1930s when an alcohol-blended fuel called Agrol was sold and distributed in Nebraska and surrounding states. Another alcohol-blended fuel called Alky-Gas was distributed in Nebraska and surrounding states in the late 1930s and early 1940s.

Generally, there was a great deal of interest at that time expressed by several states in this Midwest region in grain alcohol-blended fuels. Three consecutive national conferences were held in Dearborn, Michigan, sponsored by Henry Ford, in 1935, 1936, and 1938 to discuss the widespread use and development of grain alcohol-blended fuels. Several interest groups from Nebraska participated in these meetings.

During World War II, three alcohol plants were established in the Midwest with the assistance of the Defense Plant Corporation. One of these plants was built in Omaha, Nebraska to assist in the manufacturing of fuel extenders and synthetic rubber for the war effort in the early 1940s. This plant was later scrapped. During the late 1950s and 1960s, there were several efforts to promote the use of grain alcohol in automotive fuels. The primary thrust behind the movements, as in the past, was to utilize surplus Nebraska agricultural crops.

FORMATION OF AGRICULTURAL PRODUCTS INDUSTRIAL UTILIZATION COMMITTEE

As grain surpluses mounted and export markets were troublesome in 1968, the idea gained considerable strength in Nebraska. At this time, the Nebraska Wheat Growers became quite active in promoting the concept of using wheat and other feed grains for conversion into ethyl alcohol. In 1971, several farm senators in the Nebraska legislature obtained legislation providing for the establishment of the Agricultural Products Industrial Utilization Committee to administer the Nebraska Grain Alcohol Program. The Committee's primary responsibilities are:

(1) Establish procedures and processes to manufacture and market agricultural ethyl alcohol-blended fuels;

(2) Establish procedures to make the blended fuel marketable by private enterprise;

(3) Analyze the marketing process and testing of marketing procedures to assure acceptance in the private marketplace of such blended fuels and by-products resulting from its manufacture;

(4) Cooperate with private industry to establish privately owned agricultural ethyl alcohol manufacturing plants in Nebraska to supply demand for such product;

(5) Sponsor research and development of uses for by-products resulting from the manufacture of agricultural ethyl alcohol in order to enhance economic feasibility for alcohol; and

(6) Confirm the acceptability of the utilization of an alcohol-gasoline blend as an automotive fuel.

The ultimate goal of the committee is to obtain the construction and completion of one or more ethyl alcohol plants in the State of Nebraska.

Funds to carry out these broadly outlined activities are obtained by withholding one-eighth cent from the gasoline tax refund which is otherwise retained by users of gasoline for off-highway purposes. The revenue generated by this tax measure has amounted to an average of $82,000 annually. Obviously, the majority of these funds is derived from Nebraska's farmers. Legislation also provided for a reduction in the state gasoline tax of three cents per gallon on the first ten million gallons of Gasohol sold each year.

The Committee's first projects and studies began in 1972. Due to the small budget and structure of the Committee, its administrative staff is small. Therefore, the Committee awards grants for the necessary research and studies elsewhere to public institutions or private entities. One of the first things that the Committee did was to begin collecting all the available information on the subject of blending alcohol and gasoline.

The Committee gained the extensive cooperation in this regard of the Department of Chemical Engineering at the University of Nebraska at Lincoln. Much of the cooperation was generated by Dr. William Scheller, the chairman of the department. Dr. Scheller is credited with the origination of the tradename, "Gasohol," in the early days of the Committee's existence. The Committee then initiated several small projects investigating the technical aspects of Gasohol.

One such project was conducted in 1972. This project was a small scale road test with half-ton pickup trucks using Gasohol and regular gasoline.

TWO MILLION MILE ROAD TEST

In 1973, the Committee reviewed the results of this project. They were encouraged enough to have a large scale automotive fleet demonstration test designed by the University of Nebraska at Lincoln. This road test was designed with the cooperation of the Nebraska Department of Roads and the Department of Chemical Engineering at U.N.L.

Dr. Scheller was designated as the principal investigator in this road test. Dr. Scheller recommended that 200 proof anhydrous ethanol be used for the alcohol in Gasohol. Using this type of alcohol would avoid any water problems and provide good starting in cold weather. So, the official definition of Gasohol became: 10% agriculturally derived (200 proof) anhydrous ethanol and 90% unleaded gasoline. The alcohol for this project and remaining projects was obtained from a Georgia-Pacific Corporation plant in Bellingham, Washington. Georgia-Pacific fermented wood wastes into anhydrous ethanol. This alcohol is chemically identical to alcohol fermented from grain. This alcohol satisfied the Committee's project needs especially since there was no other major fermenter of anhydrous ethanol closely available at that time.

The alcohol was shipped to a Farmland Industries Cooperative Refinery at Phillipsburg, Kansas where it was stored and blended into Gasohol. The Gasohol was shipped to three Department of Roads test stations spread strategically across the State of Nebraska. The Department of Roads furnished 45 vehicles to be fueled with Gasohol, unleaded gasoline, and regular gasoline. The road test began in December of 1974 and was completed in October of 1977. This project was the primary and most comprehensive that the Committee had ever undertaken.

The preliminary results of the Gasohol Two Million Mile Road Test are encouraging. Consumption of Gasohol was about 5% less than for unleaded gasoline. No unusual engine wear or carbon buildup was found. Drivers reported that they experienced no problems of starting, vapor lock, or drivability. The Committee obtained the cooperation of the U.S. Department of Energy on exhaust emission tests at the Bartlesville Research Lab. These tests show that Gasohol emitted one-third less carbon monoxide than unleaded gasoline on the Nebraska road test vehicles. Due to the scale and results of this road test, it gained considerable national recognition.

CONSUMER ACCEPTANCE TEST

While the road test was in progress, the Committee increased its pace of activities. One of the most far-reaching Committee projects took place in 1975. An experiment entitled, "The Holdrege Gasohol Consumer Acceptance Test and Marketing Survey" was organized and implemented during that year. After the first positive reports of the road tests were obtained, the Committee decided that the public acceptance of Gasohol should be tested. Arrangements were made with the Holdrege Cooperative Service Station at Holdrege, Nebraska to sell Gasohol. The cooperation of Farmland Industries, Inc., of Kansas City, Missouri was easily obtained for this experiment since they were already storing the alcohol and blending it with their unleaded gasoline for the road test.

This experiment proved to be extremely successful. What was to be a project spread over a period of nine months to one year turned out to last only 2½ months. The Committee sold approximately 93,000 gallons of Gasohol. The supply of Gasohol was insufficient to meet the demand of consumers. Customers reported increased mileage and performance with Gasohol. The Committee published the sales and consumer acceptance data in December of 1975 after all of the information had been compiled and computed. The Committee concluded from this project that a market for Gasohol definitely existed in Nebraska.

FOOD AND FUEL FOR THE FUTURE

The Committee dispelled the misconception in 1976 that food would be taken away from starving people if Gasohol becomes widely distributed. This was done through the auspices of a National Science Foundation grant to the University of Nebraska at Lincoln for research on the extraction of protein from the distillers dried grains of ethyl alcohol fermentation. This extracted protein, commonly called a protein isolate, can be used in human food. This development reveals the prospect that the world's supply of protein can be increased.

This research indicates that when one makes ethyl alcohol from grain, recovers half of the protein from the distiller's by-products, and feeds the remaining residual grain to cattle, 50% more protein is available for human consumption than if the original whole grain had been fed directly to cattle. This research is the reason that the Committee adopted as its motto, "Food and Fuel for the Future," for its Grain Alcohol Program.

FEASIBILITY STUDIES

During the last two years the Committee has undertaken many other significant projects related to Gasohol and agricultural product fermentation. In the latter part of 1976, the Committee obtained the services of an outside engineering consulting firm (Stone & Webster Engineering Corporation) to conduct a preliminary economic feasibility study based on a coal-fired ethyl alcohol fermentation plant producing 20 million gallons per year. The study, released in January 1977, revealed that there was a profitable return on the capital investment of approximately $22 million. The Committee conducted other research on the by-products of ethanol fermentation, such as feeding trials, with promising results.

In August of 1977, the Committee voted, in the midst of much controversy, to undertake a comprehensive economic Gasohol feasibility study recommended by Governor J. James Exon. Governor Exon obtained $30,000 from the U.S. Department of Energy and $30,000 from the Old West Regional Commission to finance the feasibility study. An outside consulting firm was selected to conduct the study. The study was completed in July 1978. (The results have been included in the chapter on the production of ethanol from biomass.)

LEGISLATION

In the Spring of 1977, the Nebraska Legislature voted to raise the three-cent tax credit on Gasohol to a five-cent tax credit on the first 20 million gallons of Gasohol sold per year.

In May of 1977, Nebraska members of Congress introduced significant amendments to the 1977 Farm Bill. This legislation provides for four loan guarantees up to $15 million dollars each for the construction of industrial hydrocarbon or alcohol plants in the United States. The amendment also provides for $24 million worth of grants to be allocated to land grant colleges and universities for research on the processes and by-products of hydrocarbon or agricultural alcohol production. The seed for this legislation was planted by the Nebraska Gasohol Committee. The USDA is implementing this program.

The federal farm bill legislation contributed more than anything else in bringing together many midwestern states on the subject of Gasohol. Many states began to contact the Nebraska Gasohol office due to the legislation. Seeing a strong demand for this information, the Committee felt it was time to demonstrate its work on Gasohol for the benefit of the agricultural, automotive, and petroleum interests. The Committee decided to hold the Nebraska-Midwest Regional Gasohol Conference on November 1 and 2, 1977. The conference was expanded to incorporate reports on Gasohol-related projects in other states.

FORMATION OF NATIONAL GASOHOL COMMISSION

The Nebraska-Midwest Regional Gasohol Conference was attended by representatives from 26 states and 3 provinces of Canada. A significant resolution passed by the participants of the Gasohol conference recommended that:

> "The Nebraska-Midwest Gasohol Conference go on record in
> support of the creation of a National Gasohol Commission to
> coordinate research, develop resources, encourage public interest,
> and to provide public information on the production and use of
> grain alcohol, and invite grain producing states to join by par-
> ticipating through legislative, administrative, farm organizations
> and civic groups, and that the National Gasohol Commission
> hold an organizational meeting by March 1, 1978 for the pur-
> poses of creating by-laws and establishing criteria for member-
> ship."

The first organizational meeting of the National Gasohol Commission took place in Lincoln, Nebraska on January 24, 1978 at the invitation of the Nebraska Gasohol Committee. It was at this meeting that the Commission was officially formed. Representatives from ten states participated in this formation. Representatives were asked to go back to their respective states and obtain funds, state appropriations or otherwise, to finance the Commission. A second meeting was held again in Lincoln on March 7. Representatives from five new states participated at this meeting. A total of 16 states aspired to become official members of the Commission as of July 1978. The basic purposes of the National Gasohol Commission are to:

(1) Coordinate and disseminate existing information on Gasohol and agriculturally derived alcohol fuels among the member states;

(2) Develop and promote state and federal legislative incentives for Gasohol; and

(3) Develop and coordinate any further research on Gasohol and agriculturally derived alcohol fuels needed so as to avoid research duplication among the member states.

It is highly probable that the headquarters for the Commission will be located in Lincoln, Nebraska.

Probably one of the most significant developments was the opening of a service station in Lincoln, Nebraska selling Gasohol, which has been pumping about 8,000 gallons of Gasohol per week since the end of February, 1978.

THE FUTURE

The Committee is looking ahead in the next year or so to developing small feasible agricultural alcohol producing units that can be located in small rural communities or on the farm. The Committee will be concentrating also on the utilization of farm stover and other forms of biomass for alcohol fuels.

These Committee projects and progressive thinking are the reasons why Nebraska is called the national leader of the Gasohol movement.

ETHANOL PRODUCTION
AND USE IN BRAZIL

The information in this chapter is based on "Energetics, Economics and Prospects of Fuel Alcohols in Brasil", by A.V. de Carvalho, Jr., W.N. Milfont, Jr., V. Yang and S.C. Trindade of Centro de Technologia Promon (CTP), Brazil and "The Use of Ethanol from Biomass as an Alternative Fuel in Brazil", by H. Heitland, H.W. Czaschke and N. Pinto of Volkswagen do Brasil, both in *Proceedings: International Symposium on Alcohol Fuel Technology— Methanol and Ethanol,* November 21-23, 1977, Wolfsburg, Federal Republic of Germany (NTIS CONF-771175), English translation published by the U.S. Department of Energy, July 1978.

INTRODUCTION

Brazil currently imports about 40% of its total primary energy consumption of 100 million tons of oil equivalent.

Components of total primary energy are: petroleum (43%), wood and charcoal (25%), hydro (24%), sugarcane bagasse (4.5%) and coal (3.5%). About 82% of oil and 67% of coal are imported (1).

Annual growth rates of primary energy consumption have been high over the last 10 years: 7.3% for total energy, 10.6% for oil. Domestic oil production has lagged behind, resulting in a severe strain on the balance of payments since 1974.

Several measures have been taken to correct this situation, including intensified exploration for oil, coal and uranium and development of nonconventional domestic energy sources such as fuel alcohols.

Since the 1930s, sugarcane molasses alcohol (ethanol) has been blended with gasoline in amounts up to 30% by volume. This policy is aimed at stabilizing the sugarcane industry, which could harvest a steady amount of sugarcane and yet produce more or less sugar according to the market situation. Molasses, a residue from sugar manufacture, would then be produced in a varying amount and converted into fermentation ethanol which would find an assured market in the gasoline pool. From the available statistics one finds that the amounts of ethanol in blends varied with the region of the country and the time of the year, and seldom went above 10% by volume.

More recently the Brazilian Government has sponsored research in the use of fermentation alcohol in Otto and diesel engines and gas turbines, and development of processes to produce ethanol from other sources such as starch-containing materials (cassava, babassu, etc.) and cellulose-containing materials (wood and crop residues).

Brazil is probably one of the few countries in the world where there is enough land to produce food, feed, fuels and chemicals. Its land area (8,500,000 km^2) is larger than the whole of Europe, and spreads over favorable latitudes.

A National Alcohol Program (PNA or Proalcool) was established in November 1975 to promote a rapid increase in fermentation ethanol production from sugarcane, cassava and other renewable raw materials. The program provides subsidized financing for new ethanol plants and expansion and revamping of existing distilleries.

ALCOHOLS AS FUELS IN BRAZIL

In Brazil, several experiments with the use of straight ethanol as a fuel for Otto engines have been conducted since the 1920s (2)(3).

The use of ethanol-gasoline blends in Brazil was started by 1931, when, by law, imported gasoline was to be blended with 5% by volume of ethanol obtained in the domestic market when available (2).

Anhydrous ethanol production was linked to the sugar industry and followed the irregular pattern shown in Figure 9.1. Ethanol added to gasoline over 1967 through 1977, as shown in Table 9.1, followed a similar behavior. Blending of ethanol with gasoline in recent years was done during the truck loading operation (mainly in the State of Sao Paulo). The proportion of ethanol was set in the range of 10 to 15% by volume as specified by the National Petroleum Council (CNP). Therefore, data in Table 9.1 represent yearly averages.

Occasional erratic performance (vapor lock, etc.) of Brazilian engines fueled with high ethanol blends (up to 30% volume) led to experiments by the automotive industry in Brazil. Experimental studies on straight ethanol were also conducted at universities and research institutes.

In recent years, Air Force Technical Center (CTA) supported by the CNP and the Ministry of Industry and Commerce (MIC) investigated the performance of Brazilian engines fueled with ethanol-gasoline blends. The overall conclusion was that all Brazilian vehicles were fit to burn gasoline blends containing up to 20% ethanol without any engine modification.

Figure 9.1: Ethanol Production in Brazil—1950 Through 1975 (4)

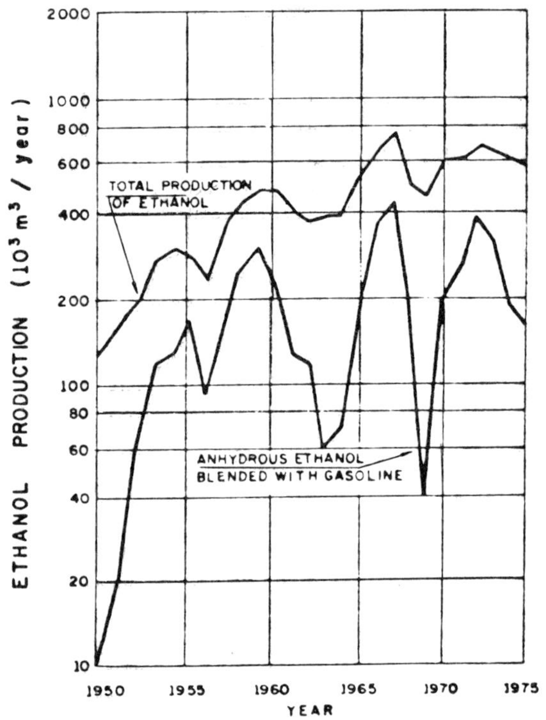

Source: NTIS CONF-771175

Table 9.1: Addition of Anhydrous Ethanol to Regular Gasoline (5)

Year	Anhydrous Ethanol (10^3 m^3)	Regular Gasoline + Ethanol (10^3 m^3)	. . Annual Average (% vol). . . National	State of S. Paulo
1967	437.2	7,145	6.2	13.5
1968	191.3	8,052	2.3	5.1
1969	31.8	8,492	0.3	0.4
1970	183.6	9,340	1.9	4.6
1971	253.8	10,075	2.5	5.8
1972	391.1	11,217	3.5	8.6
1973	308.8	12,224	2.5	7.0
1974	190.2	13,586	1.4	3.1
1975	162.2	14,355	1.1	2.4
1976	171.2	14,546	1.2	2.6
1977*	850.0	15,003	5.6	10.0

*Estimate.

Source: NTIS CONF-771175

Also, the optimum blend on the basis of fuel volumetric consumption contained 10 to 15% ethanol. CTA has also investigated the use of straight ethanol as fuel for gas turbines and diesel engines (fumigation with carburetors). Straight ethanol engine development led to several technical and public demonstrations by CTA, including the so-called "National Integration Tour". In this nationwide tour, three test vehicles covered over 8,000 km on straight ethanol (96%) between the latitude of 20°S and the equator. To cope with the new situation, the Brazilian automotive industries (Chrysler, General Motors, Volkswagen, Mercedes-Benz, Fiat and others) developed their own straight ethanol engine prototypes.

PROSPECTS OF ALCOHOLS AS FUELS

In Brazil, over the last four years only about 30% of the total ethanol production was blended with gasoline (6). The remainder was used for industrial purposes, exports and other uses. This pattern should change drastically due to the Proalcool.

Figure 9.2: Consumption of Oil-Derived Fuels in Brazil—1966 Through 1976 (1)(5)

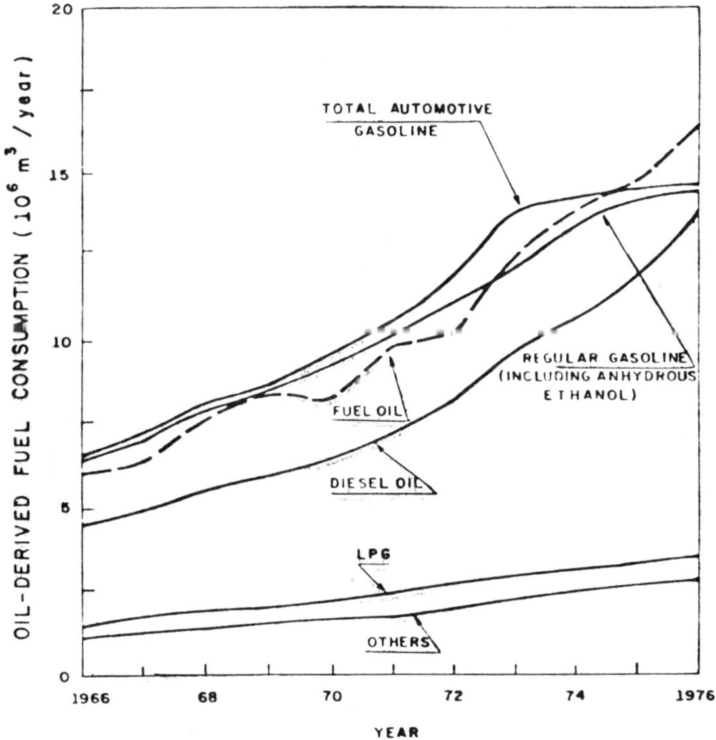

Source: NTIS CONF-771175

The Brazilian consumption of oil-derived fuels in the last decade (Figure 9.2) shows that substantial savings in oil imports can be obtained by reducing consumption of fuel oil, gasoline and diesel.

Gasoline and diesel oil, as expected, are consumed mainly in transportation. Despite the large distances involved, highway transportation growth was considerable over the last 10 years (15% annually). This occurred in detriment of railway and ship transportation and urban mass transport systems.

The 141 new ethanol distilleries authorized by Proalcool up to August 1977 will lead to an increased supply of 3.2 million cubic meters ethanol per year by 1983, if they are implemented as initially planned. The total investment for these 141 projects reaches $9,000,000 (U.S.).

Figure 9.3 shows a projected production of ethanol and methanol, and, for reference, the projected gasoline consumption for 1977 through 1985.

Figure 9.3: Projected Alcohols Production and Automotive Gasoline Demand in Brazil—1977 Through 1985

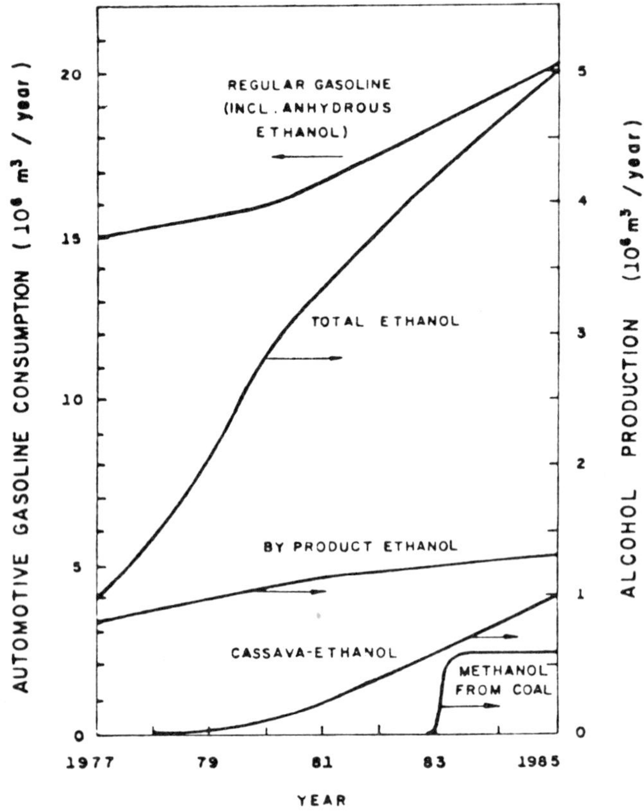

Source: NTIS CONF-771175

The following premises were assumed (4):

> In 1985, total production of ethanol would reach 5 million
> cubic meters;

> By 1980, about 80% of the Proalcool production approved by
> June 1977 would be in operation;

> In 1985, production of cassava ethanol would attain 1 million
> cubic meters;

> Cassava ethanol production would be slow in taking off due to
> expectation over the first project's performance and to the
> longer implementation period;

> The first coal gasification plant in Brazil could produce
> 1,500 tons methanol per day (although it is planned to produce
> ammonia); and

> Regular gasoline consumption growth rate (including ethanol)
> is: 2% pa, 1977 through 1980; and 5% pa, 1980 through 1985.

Otto Engines with Ethanol Blends

Two types of automotive gasolines, A and B, are marketed in Brazil. Type A
is the regular gasoline with 73 MON minimum rate. Consequently, most Brazilian Otto engines have a compression ratio of about 7:1. The addition of 5%
volume ethanol to this type of gasoline results in an increase of 2 MON units.
Such relatively high increases result from the fact that alcohols blending octane
values (BOV) are higher for lower octane number gasolines.

Type B is the Brazilian premium automotive gasoline with an 82 MON minimum
rating. This type of gasoline is not commercially blended with ethanol. Its
consumption has decreased due to the current pricing policy for oil-derived fuels.
In 1976, type B represented only 1.2% of total gasoline consumption in Brazil.

Most Brazilian engines still operate with a rich mixture (excess fuel relative
to the stoichiometric air-fuel ratio). Therefore, driveability and fuel economy
are not expected to be greatly affected with alcohol additions up to 20% volume.
The so-called "alcohol leaning effect" is the primary cause for reduction of HC
and CO emissions, but NO_x and aldehydes emissions are expected to increase (7).

Until 1978 Brazil did not have automobile emission standards; however, specific
HC and CO exhaust emission legislation for used vehicles (Otto engines) was
expected for 1978, based on yearly compulsory tuning following manufacturer's
recommendations. Emission legislation for new vehicles will probably be established by 1979-1980 based on the U.S. 1975 Federal Test Procedure.

Since the alcohol content in gasoline affects emission levels, the alcohol content
should be kept constant through the year, at least in the main cities where the
automotive pollution is more critical and its control more needed. The city of
Sao Paulo and others have already started consuming blends with steady ethanol
contents. Today, there are about 20 blending stations all over the country.

From June 1977 through May 1978, it is expected that all regular gasoline
marketed in the city of Sao Paulo (1,300,000 vehicles equipped with Otto

engines) will contain 18 to 20% ethanol. The need of assuring precise variation limits led to testing several blending techniques.

The possibility of producing low-lead, low-octane gasoline at some refineries is under consideration, provided that the ethanol content could be guaranteed to reach the specified octane rating at the filling station. The environmental impact and vehicle performance resulting from this widespread use of uniform ethanol-gasoline blend is now being determined in Sao Paulo by the EPA of the State of Sao Paulo.

Otto Engines with Straight Ethanol

In August 1977, 35 Otto vehicles converted to straight ethanol (96%) by CTA were incorporated into the maintenance fleets of two Brazilian utility companies. Table 9.2 shows a somewhat optimistic estimate of the number of fleet vehicles to be equipped with straight ethanol engines through 1980.

Table 9.2: Projected Straight Ethanol Fleet Vehicles in Brazil—1977 Through 1980

| Year | . . Number of Vehicles . . | | 96% Ethanol Consumption $(10^3 \text{ m}^3/\text{year})$ |
	Cars	Busses	
1977	1,130	20	0.9
1978	5,300	140	11.2
1979	10,300	1,140	41.8
1980	15,300	2,140	85.6

Source: NTIS CONF-771175

Practical Experience with Pure Ethanol Engines: Operating Behavior — Air- and water-cooled pure ethanol engines were developed under a cooperative ethanol research program by Volkswagen do Brasil. The development of the 1.6 liter, 4-cylinder, opposed-piston VWB engine took place in two steps:

(1)   The gasoline-operated engine was modified for pure ethanol operation. The compression ratio was kept at 7.2:1; and

(2)   The compression ratio was raised to 11:1 in order to increase the efficiency and performance and to improve emissions.

Intense heating of the intake system by hot exhaust gas was necessary in order to improve mixture distribution and increase the mixture temperature. Intake system modifications were also necessary in order to provide for optimum distribution of the alcohol/air mixture to the individual cylinders. The carburetor was to be modified to permit 65% higher alcohol flow-through versus gasoline operation because of the low calorific value of alcohol compared with gasoline (26,780 kJ/kg for ethanol versus 43,950 kJ/kg for gasoline). The airflow had to be reduced by 38%.

Mixture distribution posed most of the problems during the engine test-stand phase of the experimental program. Additional problems were posed by the ignition system. Intense mixture heating, ignition retard, higher ignition voltage (26 kV) and longer sparks solved most of the problems.

Fuel and Energy Consumption During Normal Driving — Alcohol-powered engines had 4% less fuel consumption when the compression ratio was raised from 7.2:1 to 11:1. It is obvious that the low calorific value of ethanol leads to a higher volumetric fuel consumption in terms of liters per 100 km in road traffic at a normal compression ratio of 7.2:1. General energy savings of roughly 20% are obtained for alcohol-powered engines versus gasoline-powered engines.

Engine Behavior — A comparison between gasoline and ethanol operation revealed that the ethanol-operated engine produced only slightly higher engine output at high speeds in full load driving when the compression ratio remained unchanged. However, an increase in the compression ratio from 7.2:1 to 11.0:1 resulted in an increase of the maximum output from 35.5 kW (48 hp) to 41.8 kW (57 hp), i.e., an 18.7% improvement. The use of ethanol may also result in an increase of 20.5% of the maximum torque. The primary effects are a better internal cooling of the engine, higher volumetric efficiency of the fuel/air mixture, and improved combustion at higher compression ratios.

Energy savings of up to 30% were obtained with high compression ratio ethanol engines. Improved energy utilization was obtained at higher compression ratios during full load operation.

Driveability — Driveability tests were performed on vehicles equipped with 1.5 liter, air-cooled VW ethanol engines. The results showed that minor hesitation may occur during acceleration from low speeds. Driveability is good at high engine speeds and vehicle velocities and at ambient temperatures around 15°C.

Higher ignition voltages and longer sparks were found to result in substantially improved driveability. Test vehcile driveability was significantly improved when 5 or 10 volume percent of gasoline was added to the ethanol.

High performance ignition systems were used to improve the cold start behavior of ethanol engines. This was demonstrated by the increase in ignition voltage from 19.5 to 26 kV during the tests. An excellent way to improve cold-start behavior is the blending of 5 or 10 volume percent gasoline into the ethanol. Low ambient temperatures (below +5°C) require straight gasoline for cold-start. Therefore, Volkswagen do Brasil developed a simple cold-start system consisting of a 2 liter gasoline tank, an electric pump, a solenoid valve, an injection nozzle, and the necessary injection lines.

Other Results — During the early stages of this work, ethanol-powered engines always were destroyed when the intake air was heated excessively and engine knock occurred. The problem was solved when preheating was reduced, especially under full load operation, and when a sufficiently safe margin from the knock limit was reached. The highest temperature of engine parts was 290°C at the cylinder head during full load operation.

Volkswagen do Brasil and Volkswagen Research are carrying out endurance tests on three test vehicles with air-cooled engines (in Brazil) and on one test vehicle with a water-cooled engine in Germany.

While the results are very promising, work continues on some problems such as better mixture distribution, avoidance of high-speed knocking, improved cold-start behavior and warm-up phase driveability.

Diesel Engines

Very few researchers in Brazil worked on alcohols as fuel for diesel engines. The pricing policy in this country for oil-derived fuels kept the diesel oil price relatively low in comparison to gasoline price, thus discouraging interest in alternative fuel research. However, the high rate of diesel oil consumption (9% per year for 1973 through 1976) may change this situation.

Several experiments and/or studies on ethanol fumigation, engine conversion to Otto cycle, fuel additives to increase cetane number, and individual and simultaneous direct injection of ethanol-diesel oil are now underway by the Brazilian automotive industry and CTA. The utilization of diesel-vegetable oil (for example, soybean, cotton, babassu) blends are also being investigated.

The use of alcohols as substitutes for diesel oil in Brazil present several advantages, mainly with respect to the improvement of air quality parameters in congested urban environment. The sulfur content of the diesel oil can reach 1.3% by weight; therefore, its replacement by sulfur-free alcohol should abate SO_2 emissions.

ENERGETICS AND ECONOMICS OF CASSAVA AND SUGARCANE FUEL ETHANOL

Among the fuel alcohols under consideration in Brazil, ethanol produced from cassava starch presents the most interesting opportunity. This is due to the absence as yet of large scale commercial experience and to the productivity possibilities of agronomic and industrial processing aspects of the cassava-ethanol technology.

Also social factors, such as income distribution, play a major role. On the other hand, energetics and economics have been a controversial matter. Sugarcane will be taken as reference since it has been the sole source of ethanol in Brazil.

Energetics

The net energy ratio (NER) analysis (8) is carried out by considering a system with well-defined boundaries and applying energy equivalence criteria for the energy inputs and outputs. NER results should be carefully analyzed since they may vary significantly, depending on the boundaries and energy equivalence factors.

Figure 9.4 shows the system composed of cassava farms and ethanol distillery where the source of wood used as fuel in the process is included in the system boundaries.

The equivalent energy values of all economically significant inputs and outputs of the cassava system considered (150 cubic meters per day ethanol distillery, external supply of electric power, wood as fuel for process steam generation) are shown in Table 9.3.

Figure 9.4: Boundaries and Energy Flows of Cassava Farms/Distillery/Commercial Forest System

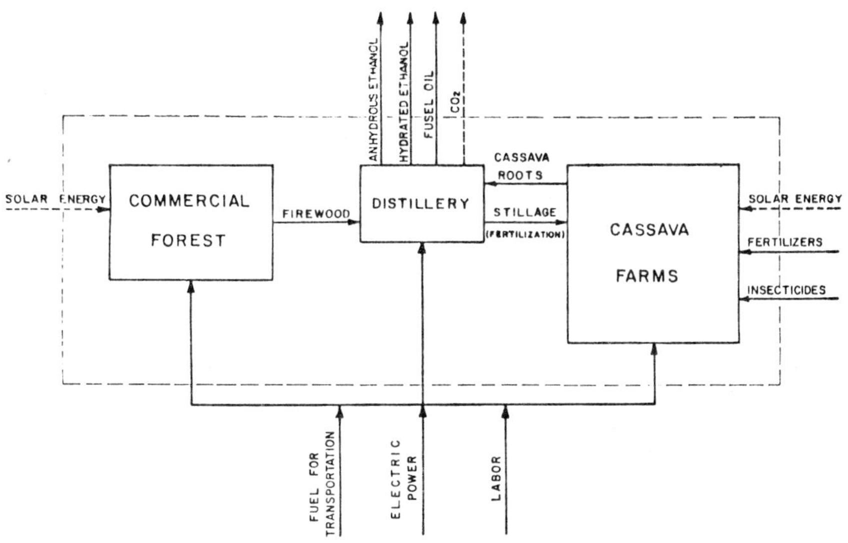

Source: NTIS CONF-771175

Table 9.3: Energy Inputs and Outputs of Cassava Farms/Distillery System*

Flows Crossing System Boundaries	Consumption or Production	Equivalent Energy (10^3 kcal)
Inputs		
Distillery		
Electric power, kWh	448.00	385.33
Sodium hydroxide, kg	2.93	21.10
Benzene, kg	0.90	8.73
Molasses, kg	5.66	6.51
Ammonium sulfate, kg	22.55	4.51
Superphosphate, kg	22.70	4.09
Labor + others	—	nil
Cassava farms		
Diesel, kg	27.42	274.20
Ammonium sulfate, kg	63.00	12.60
Potassium chloride, kg	15.00	11.33
Insecticides, kg	9.96	1.79
Superphosphate, kg	7.50	1.35
Electric power, kWh	0.59	0.51
Labor, man-day	47.50	0.14
Farms- and forest-distillery transportation		
Diesel (stillage), kg	12.26	122.60
Diesel (cassava roots), kg	5.48	54.80
Diesel (wood), kg	3.54	35.40
Total inputs	—	944.99

(continued)

Table 9.3:　(continued)

Flows Crossing System Boundaries	Consumption or Production	Equivalent Energy (10^3 kcal)
Outputs		
Anhydrous ethanol, kg	790.00	5,271.67
Hydrated ethanol, kg	41.60	265.53
Fusel oil, kg	4.80	30.64
Total outputs	—	5,567.84

*Forest included in the system boundaries, external supply of electric power.
Basis: 1 m^3 anhydrous ethanol and 150 m^3/day ethanol distillery.

Source:　NTIS CONF-771175

Table 9.4 shows similar data for sugarcane (150 cubic meters per day ethanol independent distillery, total on-site generation of electric power, sugarcane bagasse as fuel for steam generation).

Table 9.4:　Energy Inputs and Outputs of Sugarcane Farms/Distillery System*

Flows Crossing System Boundaries	Consumption or Production	Equivalent Energy (10^3 kcal)
Inputs		
Distillery		
Benzene, kg	0.90	8.73
Ammonium sulfate, kg.	22.55	4.51
Superphosphate, kg	22.70	4.09
Labor + others	—	nil
Sugarcane farms		
Diesel, kg	37.34	373.40
Ammonium sulfate, kg	121.00	24.20
Potassium chloride, kg	24.50	18.50
Superphosphate, kg	10.50	1.89
Herbicides, kg	0.69	0.12
Insecticides, kg	0.11	0.02
Labor, man-day	7.55	0.02
Farms-distillery transportation		
Diesel, stillage, kg	12.77	127.70
Diesel, sugarcane, kg	9.73	97.30
Total inputs	—	660.48
Outputs		
Anhydrous ethanol, kg	790.00	5,271.67
Hydrated ethanol, kg	48.00	306.38
Fusel oil, kg	2.00	12.77
Total outputs	—	5,590.82

*Total on-site generation of electric power. Basis: 1 m^3 anhydrous ethanol
and 150 m^3/day ethanol distillery.

Source:　NTIS CONF-771175

Diesel oil was considered as fuel for transportation instead of ethanol, because most sugarcane ethanol distilleries (considered as reference) do not employ ethanol engines. The use of ethanol as fuel for transportation will result in substantial fossil fuel savings and NER gains.

NER results for the two cases mentioned above are shown in Table 9.5, as well as the final results for the case of total on-site generation of electric power for the cassava distillery.

Table 9.5: Energetics of Ethanol Production from Sugarcane and Cassava (Basis: 1 m³ Anhydrous Ethanol)

| Raw Material | Case* | Output | Energy (10^6 kcal) | | | | NER = Output/ Input |
| | | | Input | | | | |
			Agriculture	Distillery	Transportation	Total	
Sugarcane	A	5.59	0.42	0.017	0.26	0.70	8.0
Cassava	B	5.57	0.30	0.43	0.21	0.94	5.9
Cassava	C	5.57	0.30	0.045	0.27	0.62	9.0

*A is total on-site generation of electric power (sugarcane bagasse as fuel). B is external supply of electric power (wood as fuel). C is total on-site generation of electric power (wood as fuel).

Source: NTIS CONF-771175

Economics

Tables 9.6 and 9.7 show the composition, under similar basis, of the cost of anhydrous ethanol produced in independent distilleries (not connected with sugar mills) from sugarcane (via sugarcane juice) and from cassava.

These costs correspond to the calculated selling prices of fuel ethanol ex-distillery, with taxes levied considered in the same way as for the market price established by the Brazilian government, i.e., the so-called "acquisition value from the producer" (9).

Taxes levied on this value are much lower than on ethanol price for other industrial purposes. All cost figures refer to the third quarter, 1977.

A 150 cubic meter per day distillery was chosen since this capacity represents an average size distillery considering the Proalcool approved projects to date.

The cost of by-product ethanol (from molasses-based distilleries) was not considered for comparison, since it is dependent on the cane sugar economy and cannot be clearly established. Additionally, emphasis of the Proalcool in the near future will be on ethanol from independent distilleries.

Note that capital investment for cassava and sugarcane distilleries of the same capacity are of the same order of magnitude. Additional equipment for conversion of starch to sugar and related to electric power generation, accounts for the higher initial fixed investment in cassava distilleries.

Table 9.6: Economics of Production of Anhydrous Ethanol from Sugarcane Juice*

	Value	
Investment		
Fixed investment (10^6 U.S.\$)	13.1	
Working capital (10^6 U.S.\$)**	2.0	
	(U.S.\$/m^3)	(%)
Composition of selling price		
Feedstock, sugarcane at \$10.9 (U.S.)/t	164	49.3
Chemicals and utilities	5	1.5
By-products***	16	4.8
Labor	11	3.3
Maintenance materials, operating supplies, insurance and administrative expenses	24	7.2
Value added taxes†	28	8.4
Income tax	21	6.3
Depreciation††	47	14.1
Net operating profit††	49	14.7
Calculated selling price as fuel, ex-distillery	333	100.0

*Basis: 150 m^3/day (27,000 m^3/yr) distillery of anhydrous ethanol operating 180 days/yr. [Exchange rates: \$15.00(Cr)/\$1.00(U.S.).]

**Includes ethanol inventory corresponding to 50 days operating at manufacturing cost.

***Difference between cost of direct application of stillage as fertilizer and credit of sales of hydrated ethanol and fusel oil.

†Major component of this item (93%) corresponds to the value of taxes levied on the feedstock.

††Return on investment of 12%/yr, DCF, based on the annual sum of depreciation plus net operating profit, and 15 yr operational life for the distillery.

Source: NTIS CONF-771175

Table 9.7: Economics of Production of Anhydrous Ethanol from Cassava*

	External Supply of Electric Power		Total On-Site Power Generation
Investment			
Fixed investment (10^6 U.S.\$)	15.76		19.46
Working capital (10^6 U.S.\$)	1.06		1.08
	(U.S.\$/m^3)	(%)	(U.S.\$/m^3)
Composition of selling price			
Feedstock, cassava roots at \$29.2 (U.S.)/t	200	59.1	200
Enzymes and chemicals	29	8.6	29
Utilities			
Water	—	—	—
Electric power at \$28.5 (U.S.)/MWh	13 ⎫	7.4	—
Wood at \$7.0 (U.S./t)	12 ⎭		31
By-products' credit**	18	5.3	20

(continued)

Table 9.7: (continued)

	(U.S.$/m^3)	(%)	(U.S.$/m^3)
Labor	9	2.7	9
Maintenance materials, operating supplies, insurance and administrative expenses	17	5.0	20
Value added taxes***	3	1.0	4
Income tax	12	3.5	16
Depreciation†	32	9.4	38
Net operating profit†	29	8.6	36
Calculated selling price as fuel, ex-distillery	338	100.0	363

*Basis: 150 m^3/day (49,500 m^3/yr) distillery of anhydrous ethanol operating 330 days/yr. [Exchange rate: $15.00(Cr)/$1.00(U.S.).]

**Difference between cost of direct application of stillage as fertilizer, and credit of sales of hydrated ethanol and fusel oil.

***Value of social tax only. Feedstock is considered exempt of taxes and does not contribute to this item.

†Return on investment of 12%/yr, DCF, based on the annual sum of depreciation plus net operating profit and 15 yr operational life for the distillery.

Source: NTIS CONF-771175

Sugarcane distilleries, on the other hand, incur a large commitment of working capital due to the limited operating period as a result of agronomic constraint and legal requirement for ethanol storage and supply. Under present legislation, distilleries are required to have an ethanol storage capacity adequate for an even continuous ethanol supply all year round, including the sugarcane off-season.

It is noteworthy that the agricultural feedstock in both types of distilleries is the major cost component, accounting for 50 to 60% of the total estimated ethanol selling price. The economics of fermentation ethanol is therefore highly sensitive to agricultural productivity and industrial yield.

Cost of ethanol per unit energy content is considerably higher than that of gasoline due to the relatively low heating value of ethanol. Available experimental data with blends indicate, however, that from the standpoint of vehicle performance, measured in kilometers per liter, ethanol is roughly equivalent to gasoline on a volumetric basis.

The estimated fuel ethanol selling prices in U.S. dollars (ex-distillery) at $333/m^3 ($16.7/10^6 Btu) from sugarcane, and $338/m^3 ($16.9/10^6 Btu) and $363/m^3 ($18.1/10^6 Btu) from cassava with and without electric power generation, respectively, are higher than the fixed market price, presently around $290/m^3 ($14.5/10^6 Btu).

Also, all the estimated prices are between 81 and 83% of the gasoline retail price on a volume basis, at the gasoline filling station, presently $413/m^3 ($13.8/10^6 Btu), allowing a small margin for covering the blending and distribution costs.

It is readily apparent that political decisions can have a significant impact on the feasibility of fermentation of ethanol by independent distilleries. This can be done by setting the selling price closer to the real cost and the government absorbing part of the blending and distribution costs to preserve a profit margin to ethanol manufacturers and fuel distributors and retailers.

CONCLUSIONS

Brazil is one of the few countries in the world which has a combination of resources and needs such that large-scale production of fermentation alcohol is viable. The National Alcohol Program (Proalcool) served to provide the initial momentum for the development of this alternate energy source.

Ethanol should account for the largest portion of the Brazilian alcohol pool in the near term, since it is based on a well-established technology and existing industrial infrastructure.

Alcohol application should be heavily geared toward Otto and diesel engines for automotive transport.

Energetics analysis of cassava and sugarcane ethanol production has shown that, in both cases, a net positive return on energy investment is achieved, considering current Brazilian agricultural and industrial technologies.

Presently, the estimated costs (in U.S. dollars) of cassava-ethanol ($338/m^3 and $363/m^3) and sugarcane ethanol ($333/m^3) exceed the price established by the government ($290/m^3). The estimated error of the calculated selling prices is ±15%. Analysis of the composition of these costs indicate that feedstock accounts for a substantial share of the total.

In addition, these estimated costs rank close to that of gasoline at the gasoline station ($413/m^3), allowing a small margin to cover the costs of alcohol blending and distribution. Thus, in the short term, political decisions can alter the economics of fermentation ethanol. In the medium term, upcoming agricultural productivity and industrial yield improvements together with probable increases in the price of petroleum are expected to make fermentation ethanol increasingly more competitive in Brazil.

REFERENCES

(1) Ministerio das Minas e Energia, Brasil, "Balanco Energetico Nacional" (National Energy Balance), Brasilia (1977).

(2) Oliveira, E.S., "Alcool Motor e Motores a Explosao", (Power Alcohol and Internal Combustion Engines), Rio de Janeiro, Instituto de Tecnologia, 356 p. (1937).

(3) Semana de Tecnologia Industrial, Rio de Janeiro, *Etanol: Combustivel e Materia-Prima,* Rio de Janeiro, MIC/STI, 459 p. (1976).

(4) Centro de Tecnoligia Promon, "Alcohol from Cassava—Technical Economic Feasibility, Plant Implementation Schedule and Technological Sensitivity of a Typical Distillery" (A multiclient study) (1977).

(5) Faissal, L., "Mercados, Comercializacao e Precos de Alcool" (Ethanol Market, Commerce and Prices), presented at Simposio Sobre Producao de Alcool no Nordeste, 1, Fortaleza (Brasil) (1977), MINTER/SEPLAN/SUDENE/BNB, 38 p. (1977).

(6) Instituto do Acucar e do Alcool, "IAA-Relatorio 1976", (IAA-1976 Annual Report), Rio de Janeiro (1977).

(7) Furey, R.L. and Jackson, M.W., "Exhaust and Evaporative Emissions from a Brazilian Chevrolet Fueled with Ethanol-Gasoline Blends", *Intersociety Energy Conversion Engineering Conference*, 12, Washington, D.C. pp. 54-61.

(8) Varisco, D.C., "Net Energy and Synthetic Fuels", presented at the 81st National AIChE Meeting, Kansas City, Missouri (April 11-14 1976).

(9) Instituto do Acucar e do Alcool, "Ato No. 19/77", Rio de Janeiro (June 2, 1977).

MOTOR VEHICLE OPERATION WITH ETHANOL AND ETHANOL BLENDS

VARIATIONS IN ENGINE PERFORMANCE WITH CHANGES IN FUEL/AIR RATIO—A DISCUSSION

> Information in this section is based on "Variations in Automotive Engine Performance and Emissions with Changes in Air/Fuel Ratio When Operating on Alcohol Fuels," by E.E. Ecklund of the U.S. Department of Energy, in *Comparative Automotive Engine Operation When Fueled with Ethanol and Methanol* (DOE HCP/W1737-01), prepared for the DOE, May 1978.

Alcohols generally provide better fuel energy economy and lower total exhaust emissions than gasoline when used in internal combustion engines. This has been demonstrated in many engine and vehicle tests, but basically applies to straight or neat alcohols. With regard to fuel economy and exhaust emissions, a 10 volume percent alcohol blend in gasoline (whether ethanol or methanol) exhibits no significant advantage or disadvantage as compared to gasoline.

These statements about blends are, of course, statistical generalizations, but they are amply supported by the great bulk of experimental evidence at hand. Certainly, experimental results will vary slightly from time to time, from test to test, engine to engine, and fuel to fuel. In general, however, an alcohol blend will not provide any improvement in fuel economy or exhaust emissions that could not be had with gasoline and appropriate engine tuning changes. To see why this is so, and why road test results sometimes differ from this general rule, it helps to review the exhaust emissions from spark ignition engines, the ways in which they are formed, and the influence of engine construction and operating variables on their formation. With this basic understanding, it becomes possible to deduce the results of a change in fuel, from straight gasoline to a low level (say 10% by volume) alcohol blend.

Emission and Fuel Economy Fundamentals

For a baseline case, consider a typical, multicylinder automotive engine operated on straight commercial gasoline. The combustion emissions from this engine-fuel system will fall into one of five major categories: (1) carbon monoxide (CO), (2) various hydrocarbon compounds (HC), (3) oxides of nitrogen (NO_x), (4) oxides of sulfur (SO_x) and (5) particulates and smoke.

Of these, only the first three are regulated, and the last two are much less significant than the others. Very small amounts of sulfur are present in gasoline and some will be burned (oxidized) to sulfates which are classified together as SO_x. These vehicle sulfate emissions are not a significant contribution to the general air pollution problem. Particulates are mainly salts of lead or other metals used as antiknock additives. These are expected to decline with time as their use in gasoline is phased out.

Spark ignition engines emit smoke only when in poor mechanical condition (oil burning) or when grossly maladjusted (much excess fuel).

In examining the three remaining regulated exhaust emissions, CO, HC, and NO_x, we find that a crucial factor in their formation characteristics is the relative mixture of fuel and air (fuel-air ratio or F/A) that is metered to the engine. Based on its chemical composition, each fuel has a precise F/A requirement, expressed in pounds of fuel per pound of air, at which the theoretically correct amount of air is available to provide sufficient oxygen (O_2) to convert all of the carbon in the fuel to carbon dioxide (CO_2) and all of the hydrogen to water vapor (H_2O) with neither fuel nor oxygen left over.

The chemically correct or stoichiometric F/A is about 0.069 for a typical gasoline. If the F/A is numerically lower than 0.069, the mixture is said to be lean (fuel lean) and unreacted oxygen will be left over. If the F/A is numerically larger than 0.069, the mixture is said to be rich and excess fuel will be left over, partially unburned.

Now let us turn our attention to the major exhaust emissions, one by one, and see how variations in F/A influence their concentration in the engine exhaust. Figure 10.1 is a graph which shows relative amounts of exhaust emissions as a function of F/A.

The amount of carbon monoxide is determined almost entirely by F/A. Mixtures richer than stoichiometric inevitably lead to higher CO emissions as indicated in Figure 10.1. With excess fuel present relative to the amount of air (oxygen) available, some carbon atoms will simply be unable to burn completely to CO_2 and instead will partially oxidize to CO. This partial oxidation process increases as the mixture gets richer. As the mixture gets leaner, some CO will always be present in the exhaust because fuel and air are never perfectly mixed and apportioned in all cylinders. But CO emissions can be minimized by operating the engine at leaner-than-stoichiometric conditions.

Hydrocarbon (HC) emissions from gasoline combustion are mostly unburned fuel (UBF). Gasoline is a complex mixture of hydrocarbon molecules of varying structures and sizes. If an engine is operated at a rich F/A, not all of the fuel can be burned completely in the combustion environment, so rich mixtures raise HC emissions. Perhaps the most important HC-producing mechanism is flame "quenching." As a flame front at very high temperature (2000 to 5000°F) approaches a relatively cool (200° to 300°F) combustion chamber wall, heat is extracted from it. At some point, usually a few thousandths of an inch from the wall, enough heat is taken from the flame and from the mixture near the wall to put out or "quench" the flame. The total "quench layer" over the entire surface area of the chamber may contain an appreciable volume of hydrocarbons (gasoline) that never become hot enough to burn before being expelled in the exhaust. Thus, no engine which burns its fuel in a premixed condition with air can ever have zero HC emissions.

Figure 10.1: Efficiency, Power, and Emissions as a Function of Fuel-Air Ratio

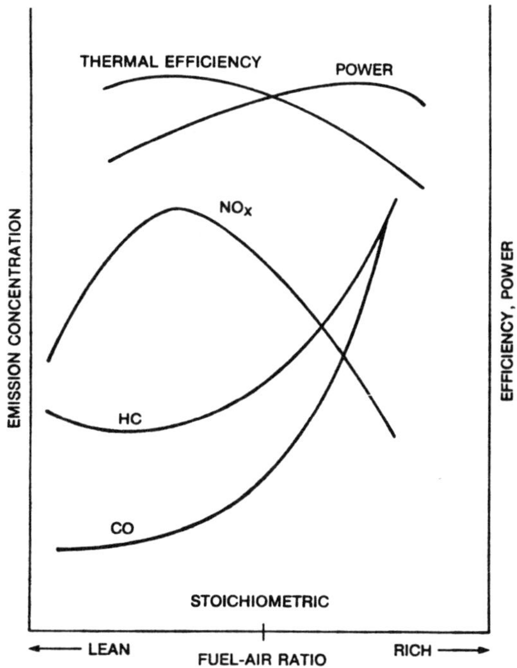

Source: DOE HCP/W1737-01

HC emissions decrease as the mixture becomes leaner, as shown in Figure 10.1, until the lean misfire limit is approached. Then they start to increase. If mixtures become so lean that a strong, self-sustaining flame cannot be established or cannot propagate through the entire combustion volume in the time available, partially or completely unburned fuel will escape through the exhaust valve.

In the figure HC and CO both tend toward optimum low values at F/A somewhat leaner than stoichiometric. Depending on the engine and the gasoline, these F/As range from about 0.065 to perhaps 0.056. Hydrocarbons reach their minimum where CO is very low. Fuel economy is maximum somewhere in this range also, and for the same basic reason: the engine is making the most efficient use of its gasoline, burning it as completely as it can and extracting a maximum of the chemical energy which is converted to heat. Thus thermal efficiency is at a maximum.

The maverick emission species is NO_x. It results when normally inert nitrogen (N_2), present in air at about 3.76 times the concentration of oxygen, chemically unites with that oxygen. High temperatures are needed to make this happen, generally above 2000°F. Such temperatures are easily reached and exceeded in gasoline combustion, and while high temperatures help reduce HC and CO to minimum concentrations, they produce more nitrogen oxides, or NO_x. The hottest flames theoretically are stoichiometric mixture flames, but it is evident that NO_x peaks at a somewhat leaner F/A. The increase in free oxygen exerts a stronger influence on NO_x formation than the slowly falling combustion temperatures as mixtures are leaned slightly from around 0.069 F/A. When the F/A is leaned progressively away from this value, NO_x begins falling as excess air "soaks up" the heat of combustion and lowers peak flame temperatures.

Figure 10.1 also shows the effects of F/A on engine power output. Power output is maximum at a slightly rich F/A mixture, because flame speeds are highest and chemical equilibria most favorable under these conditions. Power drops off as the mixture gets richer than this, due to incomplete combustion of the fuel.

The combination of these factors has a significant influence on automotive design. As Figure 10.1 shows, by operating at very lean gasoline-air mixtures all emissions can be reduced at least to near-minimum levels. This trend was followed beginning in the late 1960s and early 1970s by virtually all automobile manufacturers. At first HC and CO were the target, but as the trend continued and NO_x emissions standards were enacted, it was found that engines were reaching the limits of their lean-mixture tolerance as shown in Figure 10.1 by the steeper increase in HC emissions. In practice, the result was somewhat worse NO_x emissions and rough-running engines, the victims of lean misfire, surges and hesitations.

In the mid-70s, as the manufacturers learned to extend the lean mixture tolerance limits of gasoline engines and to lower combustion temperatures with inert recycled exhaust gas for NO_x control, automobile power and smoothness improved. When the catalytic converter was developed, it reinforced this trend toward smoother operation by allowing engines to return to richer air-fuel ratios CO and HC engine emissions increased, but because these could be afterburned in the catalytic converter, engine performance and efficiency were improved and all exhaust emissions minimized in the catalytic converter before they reached the atmosphere.

Alcohol/Gasoline Blends

With these perspectives on gasoline engine behavior and the historical trends in emissions control in mind, the effects of replacing straight gasoline with alcohol fuels, either 10% alcohol-gasoline blends or straight (neat) alcohols should be considered.

Due to the fact that alcohols contain oxygen, they have very different stoichiometric F/As than gasoline, i.e., about 0.155 methanol and 0.111 for ethanol.

When alcohol is added to gasoline to make a 10% blend, the physical qualities of the blend are not changed significantly from those of gasoline. A carburetor will meter the blend the same as it would meter straight gasoline. It will deliver to the engine a blend-air mixture at about the same F/A as it was calibrated for on gasoline. The specific F/A will depend on the model year of the car. But if 10% of the blend is ethanol, the blend stoichiometric F/A will be 0.0716.

A new parameter, called equivalence ratio (Φ), is often used in place of F/A to facilitate comparisons between different fuels. The equivalence ratio is defined as the ratio of the actual F/A to the stoichiometric F/A for a specific fuel. Therefore, for a stoichiometric mixture the equivalence ratio is 1.0, for lean F/As the equivalence ratio is less than 1.0, and for rich F/As the equivalence ratio is greater than 1.0.

Going back to the example, if the carburetor is calibrated for gasoline at an equivalence ratio of 1.0 and a blend is substituted which contains 10 volume percent ethanol in gasoline, the equivalence ratio becomes:

$$\frac{0.069 \text{ (actual F/A)}}{0.0716 \text{ (stoichiometric F/A for the blend)}} = 0.96$$

Therefore the F/A for the blend is leaner than that for gasoline at a fixed carburetor calibration.

This phenomenon is called the blend leaning effect and explains why adding alcohol to gasoline does little that cannot be done with straight gasoline by mechanical carburetor adjustment, i.e., leaning out the mixture to a lower equivalence ratio. How the engine emission and fuel economy respond to this blend leaning will depend on how rich or lean the engine was running on straight gasoline.

Increases or decreases are possible for NO_x and HC, but if the engine does not become so lean that it stalls, CO can only decrease.

If compared to gasoline at the same equivalence ratio, however, as would be the case if the carburetor were readjusted in the rich direction to maintain the same equivalence ratio after going to a blend, the blend provides essentially identical HC and CO emissions and fuel economy. Small reductions in NO_x usually occur due to other special characteristics of alcohols (such as a cooler burning flame). For a 10% blend, these reductions are roughly 10% of those possible with neat alcohols, (as will be discussed next) and thus are quite small.

Straight Alcohols

Combustion of straight alcohols produces substantially less NO_x than does gasoline at a corresponding equivalence ratio, because of the markedly lower combustion temperatures. Because alcohols can be burned at lower equivalence ratios than gasoline without lapsing into misfire and other lean combustion problems, further NO_x and CO reductions are available and corresponding engine efficiency increases can be obtained. Unburned hydrocarbon fuel (UBF) emissions are still present, but they include some different chemical species than those found in gasoline combustion emissions because alcohols themselves are from a different hydrocarbon chemical family (sometimes know as partially oxidized hydrocarbons). Operated at the same equivalence ratio, an engine will produce very similar HC and CO emissions levels on straight gasoline and straight alcohols, with lower NO_x and higher efficiency from the straight alcohols.

Due to chemical characteristics of the fuel and the related combustion phenomena, the amount of NO_x is reduced by about a half to two-thirds with straight alcohol. Thermal efficiency is increased in the order of 10%. These relationships occur at all comparable equivalence ratios. Ethanol has about two-thirds the energy content of gasoline on an equal volume basis. The volume of ethanol required is half again that of gasoline, all other things being equal.

Fortunately, engines can be operated reliably (without lean misfire, etc.) on alcohol at equivalence ratios that are leaner than gasoline, i.e., the reliable lean operating limit equivalence ratio for alcohols is leaner than the reliable lean operating limit equivalence ratio for gasoline. This has the positive effect of improving volumetric fuel economy when operating on alcohol to a level that is greater than two-thirds (for ethanol) the gasoline fuel economy level noted.

Other Factors

Further improvements in fuel economy can be gained by raising the compression ratio of an engine designed to use alcohols to a point where it can capture the octane benefit provided by straight alcohols. The octane rating for ethanol is much higher than those of today's commercially available gasolines. Octane rating is a measure of knock (or ping) resistance, and has a relationship to the compression ratio of an engine. Octane rating in itself provides no fuel economy benefit. A higher compression engine, which requires higher octane fuel, is more efficient than the same engine with lower compression ratio, and thus provides greater fuel economy, other variables being held equal. However, putting a high octane fuel in an engine that does not require it gives no fuel economy benefit by itself.

The leaning effect encountered when substituting an alcohol/gasoline blend in place of gasoline is of potential concern with regard to new vehicles. The automotive manufacturers must yearly certify their vehicles (with regard to regulated emissions and fuel economy) using gasoline. If these same emission and fuel economy regulations were to apply both to gasoline and alcohol/gasoline blends, it is questionable whether they could be achieved without new vehicle/engine design considerations.

Another related factor is the question of whether or not alcohol/gasoline blends can be marketed in compliance with the Clean Air Act unless extensive and expensive qualification tests are conducted to verify that engine/vehicle emission control equipment will not deteriorate in the lifetime of vehicle use due to the effects of the blends.

Although ethanol might be considered a better fuel than methanol because the problems in use are not as severe, the performance characteristics of ethanol almost universally fall between those of gasoline and methanol, whether those differences are positive or negative. In general, ethanol and methanol performances are sufficiently similar that no differentiation is required. If provisions are made to accommodate methanol fuel, ethanol can also be accommodated.

Summary

In summary, the general relationship between the amount of air (oxygen) and the amount of fuel is somewhat complex. However, it is well-known and understood by engine and fuel experts and important to the design of engine fuel management systems. Switching from strictly petroleum-based fuels to others such as alcohol or alcohol/gasoline blends will alter the fuel/air relationship in a fashion similar to that of adjusting the carburetor. On an equal equivalence ratio basis, straight alcohols will appreciably and favorably alter NO_x emissions and engine thermal efficiency, compared to gasoline. With alcohol/gasoline blends, these same characteristic changes are roughly proportional to the percent alcohol content. For a 10 volume percent alcohol/90 volume percent gasoline blend, levels of regulated emissions and fuel economy are not significantly different from those of gasoline.

ENGINE PERFORMANCE TEST DATA—A DISCUSSION

> The information in this section is based on *Status of Alcohol Fuels Utilization Technology for Highway Transportation*, prepared by Mueller Associates, Inc. for the U.S. Department of Energy (DOE Report HCP/M2923-01), June 1978.

Exhaust Emissions

Experimental work performed at the University of Miami (1) compared the exhaust emissions of a dynamometer-mounted, 151 CID General Motors Pontiac four-cylinder engine (CR = 8.1) when operated on Indolene clear reference fuel and a 20 volume percent ethanol/Indolene blend. An identical test series was performed using a 20 volume percent methanol/Indolene blend so a direct comparison of the two alcohols' performances (as blends) could be obtained.

All tests were made at 2,000 engine rpm and 16 in of mercury manifold vacuum. Ignition timing was set advanced 2° (crank) from the value predicted by an experimentally determined MBT (mininum spark advance for best torque) vs fuel-air equivalence ratio curve for all three fuels.

The last step was taken to ensure operation near peak efficiency since previous University of Miami work (2) had noted sharp engine efficiency drop-offs as ignition timing was retarded from MBT. The following results were obtained: (a) at any constant equivalence ratio in the range examined, carbon monoxide emissions for all three fuels were identical; (b) NO_x emissions for the ethanol blend were between those for the base fuel and those of the methanol blend, with both blends offering NO_x reductions at near-stoichiometric equivalence ratios; and (c) hydrocarbon emissions for the ethanol blend were reduced from baseline at any equivalence ratio in the test range, but not as much as by the methanol blend.

The curves of Figures 10.2, 10.3 and 10.4 have been drawn in preparation of this report using the data points reported by the University of Miami.

The University of Santa Clara conducted a limited series of similar tests with neat ethanol (and neat methanol) using their 2.3-liter Ford Pinto dynamometer engine (CR = 8.44) and an Indolene performance baseline developed on it (3). These tests were also made at 2,000 rpm, but at WOT (wide-open throttle). Ignition timing was MBT. Ethanol was inducted using the Dresserator critical flow metering device and the Indolene with the OEM (original equipment manufacturer) carburetor. Both units were used to induct methanol.

Figures 10.5 and 10.6 display the results of these tests conducted at the University of Santa Clara. (It should be noted that the authors of DOE-EC-77-X-01-2933 have taken the liberty of deleting scattered data points taken using the WHB carburetion system, a proprietary shock wave fuel metering and preparation device, as these did not relate directly to data taken with other induction devices and were too few in number to permit a meaningful curve to be drawn.)

Trends suggested by these tests were as follows: (a) NO_x emissions were reduced from the Indolene baseline by both ethanol and methanol. However, at equivalence ratios in the very lean and very rich ranges, ethanol increased NO_x emissions above the Indolene values, whereas methanol decreased NO_x at all equivalence ratios examined. Ethanol never reduced NO_x as much as methanol; and (b) Unburned fuel (UBF) emissions with ethanol were very close to those with Indolene, becoming slightly lower at richer equivalence ratios.

Figure 10.2:   Steady State Part Load CO Emissions vs Fuel-Air Equivalence Ratio for an Ethanol/Gasoline Blend and Indolene

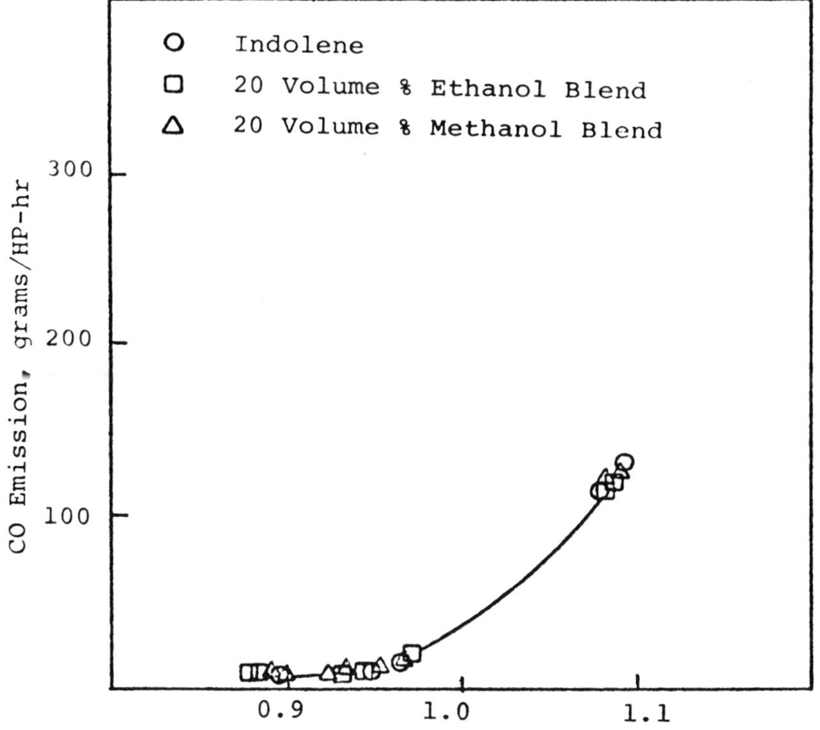

Source: DOE HCP/M2923-01

Figure 10.3: Steady State Part Load NO$_x$ Emissions vs Fuel-Air Equivalence
Ratio for an Ethanol/Gasoline Blend and Indolene

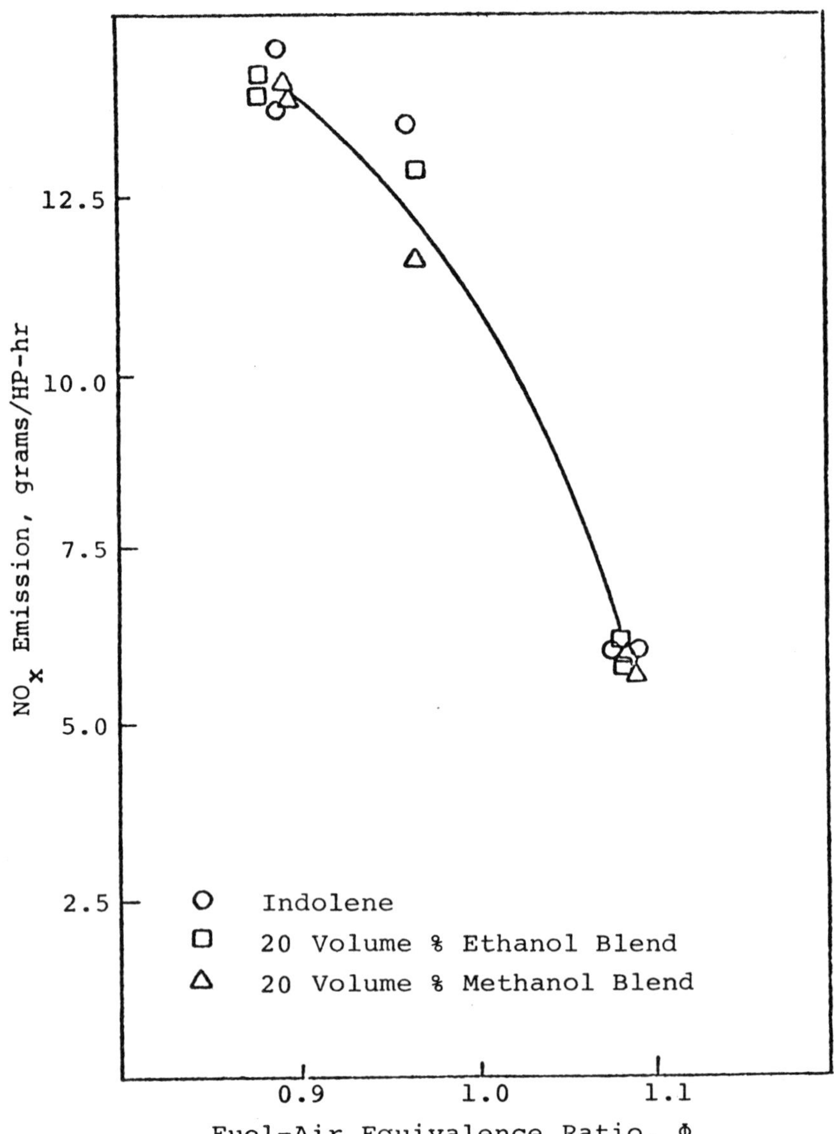

Source: DOE HCP/M2923-01

Figure 10.4: Steady State Part Load FID Hydrocarbon Emissions vs Fuel-Air Equivalence Ratio for an Ethanol/Gasoline Blend and Indolene

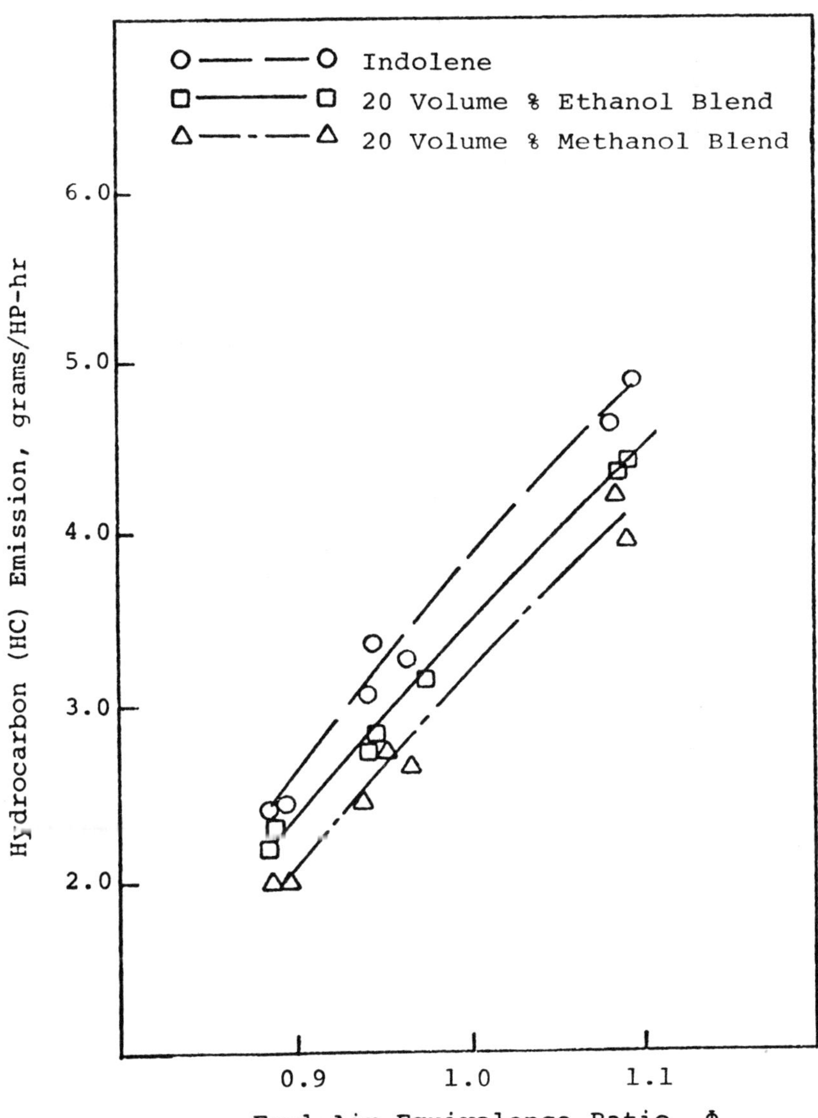

Source: DOE HCP/M2923-01

Since different induction equipment was used for the three test fuels, caution must be exercised in interpreting these results. The most directly comparable are those generated with the Dresserator for ethanol and methanol. Neither set of alcohol data can be compared directly to the Indolene test results since the stock carburetor was the only metering device used with Indolene.

General Motors (4) has reported results of EPA CVS-3 testing using a 1974 (Brazilian) Chevrolet 151 CID sedan operated on 5, 10 and 20 volume percent blends of ethanol with gasolines of two different volatilities. Emissions measured during the GM test showed reduced HC and CO, but increased NO_x and aldehydes.

Figure 10.5: Steady State WOT NO_x Emissions vs Fuel-Air Equivalence Ratio for Neat Alcohols and Indolene (3)

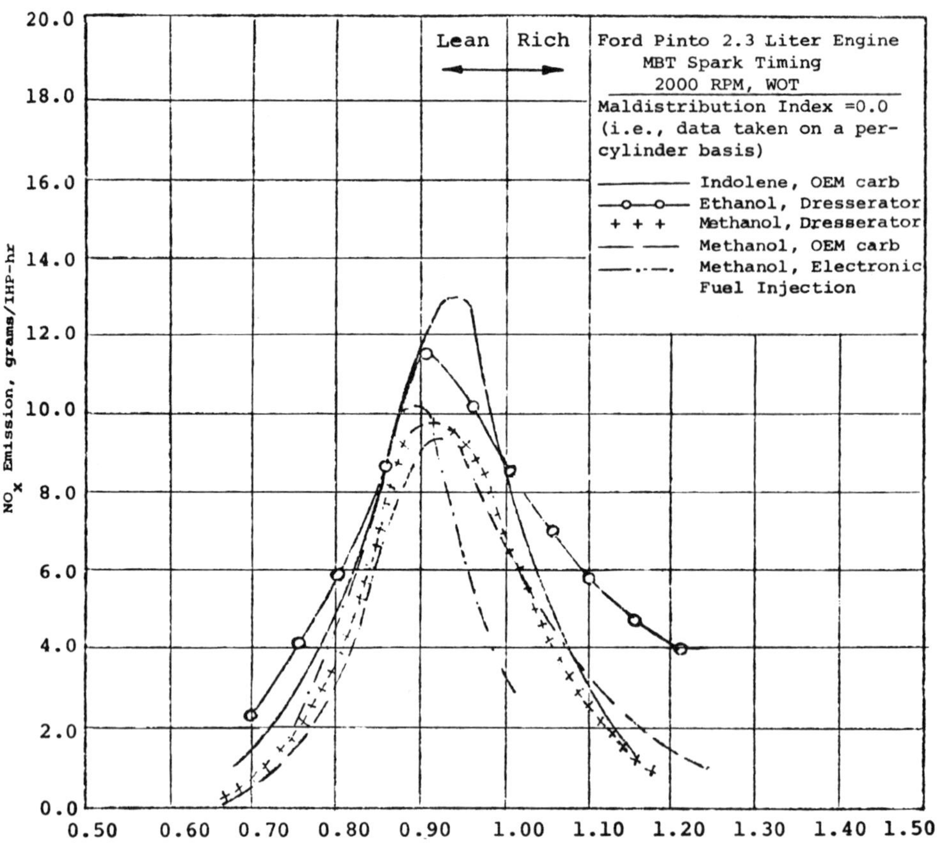

Source: DOE HCP/M2923-01

Figure 10.6: Steady State, WOT UBF Emissions vs Fuel-Air Equivalance Ratio for Neat Alcohols and Indolene (3)

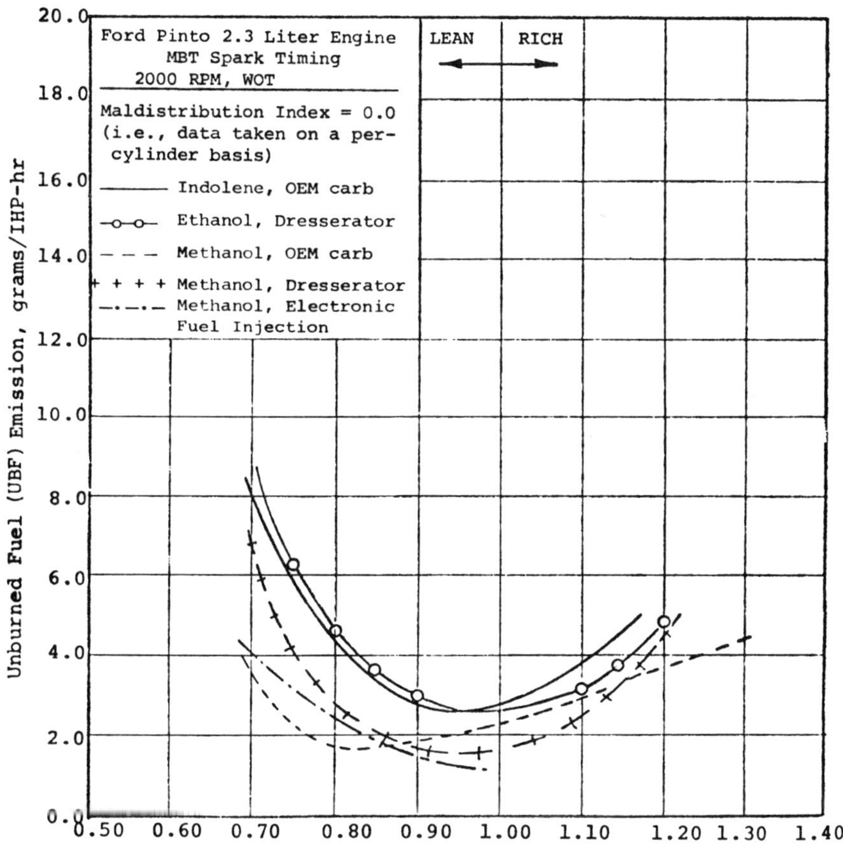

Source: DOE HCP/M2923-01

Evaporative emissions were slightly higher with ethanol blends than with straight gasoline. The test vehicle had no emission controls and the carburetor was adjusted to manufacturer's specification (which is fuel rich to provide acceptable driveability on up to 25% ethanol blends) and left at that setting for the duration of the testing.

Emission results shown in Figures 10.7 and 10.8 indicate a roughly linear variation with blend level for each pollutant, as ethanol's blend leaning effect adjusts the equivalence ratio toward leaner values.

Evaporative emissions from the Brazilian vehicle (4) were measured using the SHED technique (described in SAE J-171a). Running evaporative losses were

 Ethyl Alcohol Production and Use as a Motor Fuel

Figure 10.7: HC and CO Emissions with Ethanol/Gasoline Blends (4)

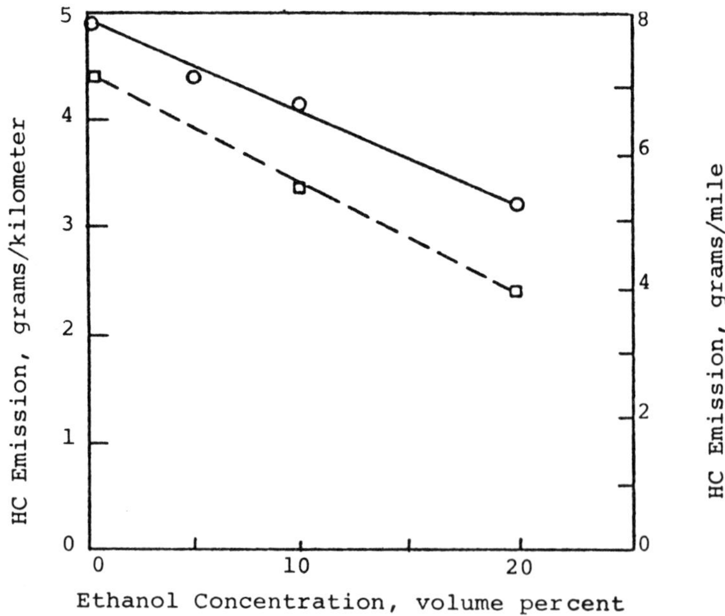

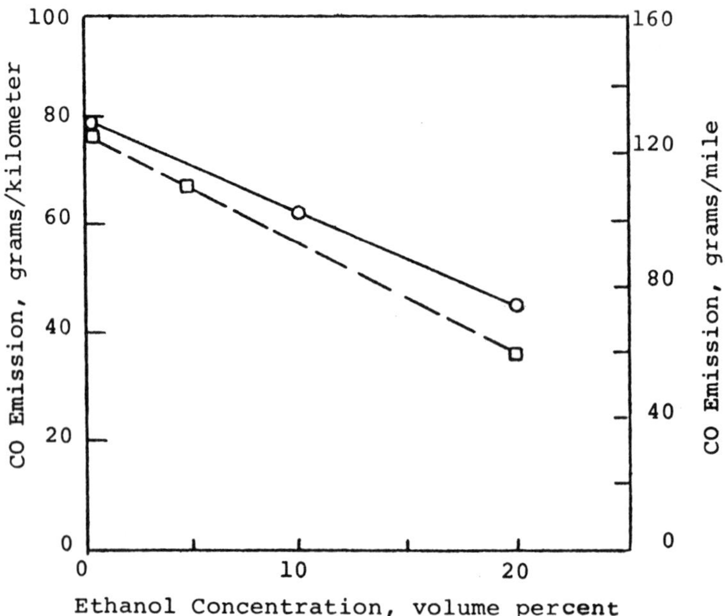

Source: DOE HCP/M2923-01

Figure 10.8: NO_x and Aldehyde Emissions with Ethanol/Gasoline Blends (4)

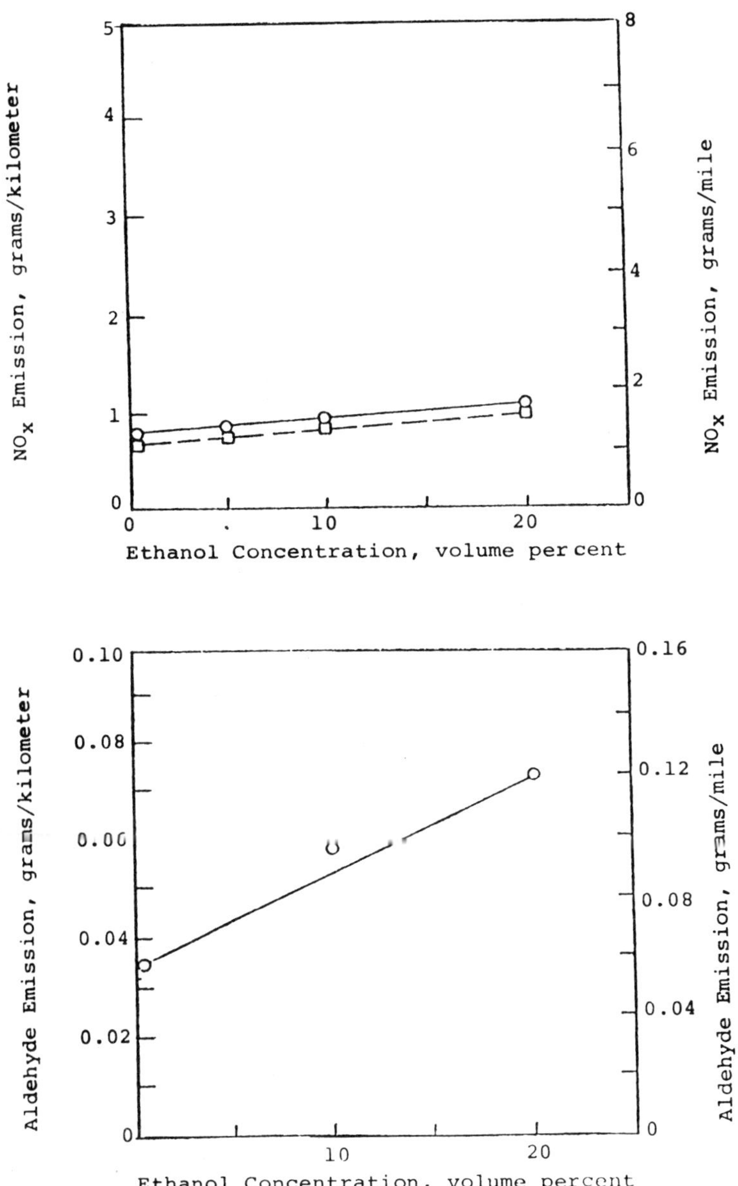

Source: DOE HCP/M2923-01

measured only for emissions occurring during a one-hour hot soak (after dynamometer operation) and during the simulated diurnal fuel temperature rise period. Results appear in Table 10.1 below.

Table 10.1: Evaporative Emissions—SHED Test (4)

Fuel		Diurnal*	Hot Soak	Total
(1)	Gasoline	14.65	3.44	18.09
		14.49	3.13	17.62
	Average	14.57	3.28	17.85
(2)	10% ethanol	12.59	8.29	20.88
		12.81	2.91**	15.72**
	Average	12.70	5.60	18.30
(3)	10% ethanol-adjusted	12.84	7.22	20.05
	volatility base gasoline	11.24	6.15	17.39
	Average	12.04	6.68	18.72

The header "Emissions (g/test)" spans the Diurnal, Hot Soak and Total columns.

*15.6° to 29°C temperature rise.

**Probable error, reflected also in the second test total of fuel (2) and in the average values for hot soak and total emissions of fuel (2). Test repeatability was good with gasoline, only fair with fuel (3). Repeatability was good on the diurnal portion with fuel 2, and very poor on the hot soak portion.

Source: DOE HCP/M2923-01

Because of its higher vapor pressure, fuel 2 was expected to give higher evaporative emissions than gasoline. However, its (fuel 2) diurnal emissions were lower than those of gasoline. The blend made with adjusted volatility gasoline also gave lower evaporative emissions than gasoline.

Within the United States, a program on ethanol blends that has generated great interest is the Nebraska "Gasohol" (trademark of the Nebraska Agricultural Products Industrial Utilization Committee) effort based on a 10 volume percent anhydrous ethanol/gasoline blend with the intent that the alcohol ultimately be derived from grain.

A road test designed to accumulate data on fuel consumption, wear, performance and emissions was conducted beginning in December 1974 (5). Forty-six vehicles were fueled with combinations of 10% ethanol/unleaded gasoline, 10% ethanol/regular gasoline, regular gasoline and unleaded gasoline.

Two of the Nebraska vehicles were tested over the 1976 EPA CVS emission test cycle at the Bartlesville Energy Research Center in June, 1977 (6). Emissions tests on gasoline and Gasohol show a CO reduction of 30% during Gasohol-fueled operation, a result of the blend leaning effect.

Emissions of NO_x and ˙HC were not significantly affected, but aldehydes were 15% higher. Neither vehicle's carburetor was adjusted during the testing, but one had been operating on base gasoline and the other on Gasohol in the Nebraska program before being submitted for testing at BERC (see Table 10.2).

Table 10.2: Exhaust Emissions (EPA CVS-3 Test Cycle) Gasohol vs Base Fuel (6)

| | . . . . .Exhaust Emissions, g/mile . . . . | | | | |
	CO	HC	NO$_x$	Aldehydes	Unburned Ethanol
	. . . . Gasohol Vehicle.				
Gasohol	25.9±7.5	2.8±0.1	2.1±0.3	0.24±0.0	0.14±0
Base fuel	38.1±0.1	2.7±0.0	1.8±0.0	0.23±0.0	—
	. . . . Control Vehicle.				
Gasohol	15.6±0.2	1.8±0.1	2.4±0.1	0.17±0.0	0.09±0
Base fuel	23.1±2.3	2.0±0.4	2.7±0.1	0.15±0.0	—

Source: DOE HCP/M2923-01

Performance and Fuel Economy

Bench tests comparing neat ethanol, neat methanol and a 15% methanol blend, all to gasoline, have been reported by Volkswagenwerk AG. Figures 10.9 and 10.10 show, respectively, the comparisons for power output and energy consumption. The data are for engines whose ignition and carburetion calibrations were adjusted to account for the fuel used (7).

Figure 10.9: Comparative Power Output (7)

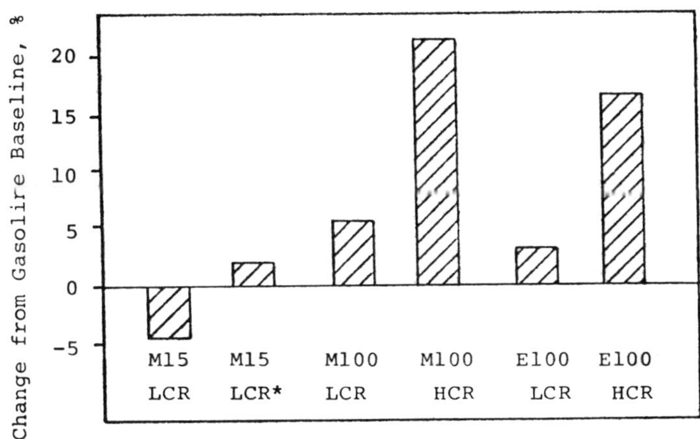

E100:   neat ethanol (5% water)

M15:   15 volume % methanol blend

M100:   neat methanol

LCR: low compression ratio engine

HCR: high compression ratio engine

*:   carburetor adjusted to restore equivalence
         ratio

Source: DOE HCP/M2923-01

Figure 10.10: Comparative Specific Energy Consumption (7)

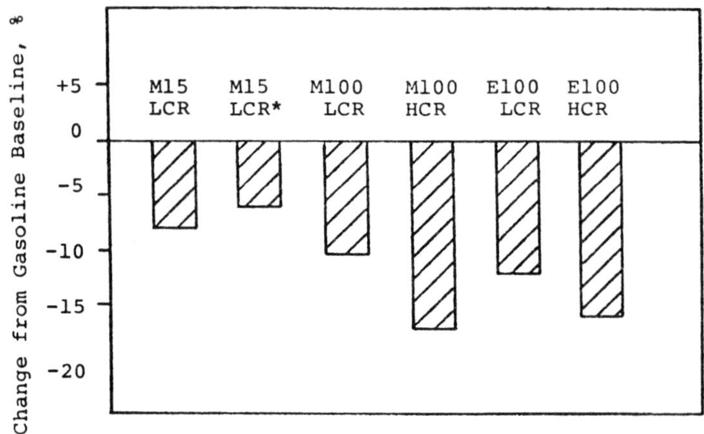

E100:   neat ethanol (5% water)

M15:   15 volume % methanol blend

M100:   neat methanol

LCR: low compression ratio engine

HCR: high compression ratio engine

*:   carburetor adjusted to restore equivalence
         ratio

Source: DOE HCP/M2923-01

The University of Santa Clara has also performed a small amount of work to
quantify the efficiency of ethanol-fueled engines (3). They report improved
thermal efficiency for the neat ethanol-fueled engine over its performance on
Indolene. The improvement was not as large as that gained by using methanol,
except near the stoichiometric equivalence ratio. Here again, the data demand
careful interpretation.

The baseline Indolene data are not directly comparable to ethanol and methanol
data due to the differing experimental induction systems used. The curves
(Figure 10.11) for the two alcohols are more comparable, however, and are
generally consistent with the tendency of ethanol to provide performance and
efficiency gains over gasoline, but smaller gains than are available from methanol.
(Again, the authors have deleted information on the WHB induction system
which was not explicitly used to induct ethanol.)

Unofficial results of Brazilian tests during a neat ethyl alcohol demonstration
program indicated losses in volumetric fuel economy (liters/kilometer) to be
only 5 to 10% above that for gasoline (8). Brazilian regular gasoline has a very
low octane rating (73 MON) and the addition of alcohol significantly increases

octane rating. In Brazil, ethanol content in blends varies from none to 30% during the year at any given gas station. National yearly averages (1972-76) have been 1 to 4% ethanol, however (8).

In conjunction with the vehicle emission tests performed by General Motors on the Brazilian vehicle (4), fuel economy determinations for ethanol/gasoline operation were also made.

Figure 10.11:   Indicated Thermal Efficiency vs Fuel-Air Equivalence Ratio for Neat Alcohols and Indolene (3)

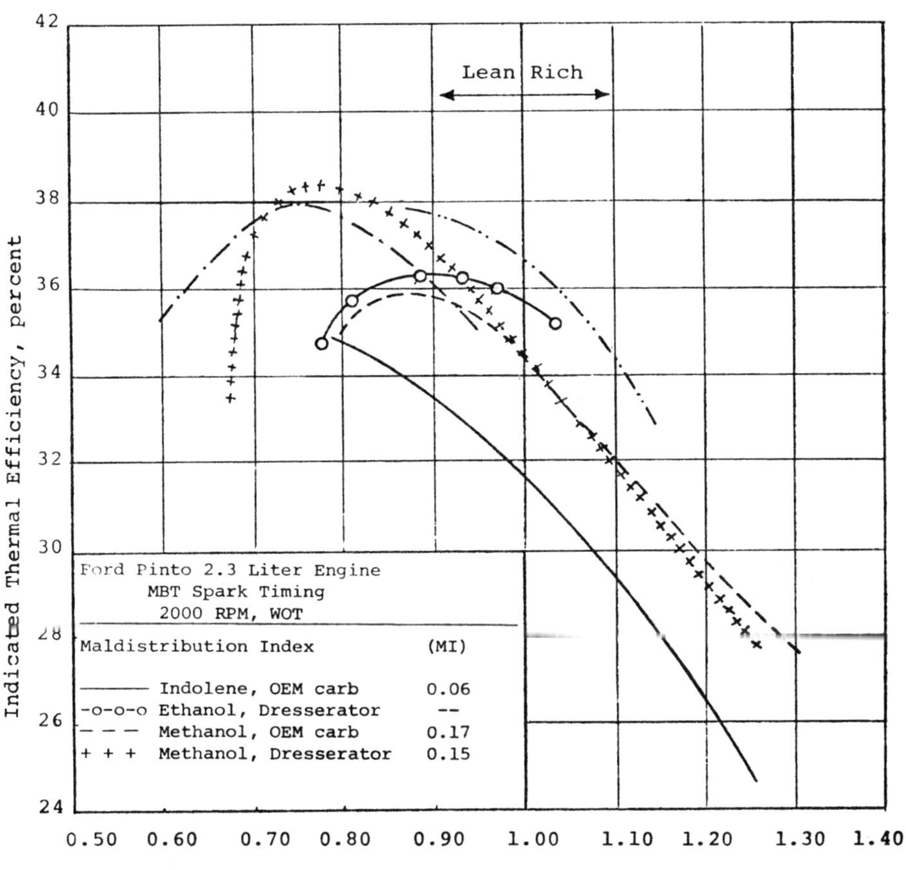

Source: DOE HCP/M2923-01

Fuel economy figures, calculated from exhaust gas products, are shown in Figure 10.12. As blend level increases to 20%, the leaning effect of the ethyl alcohol is slightly more important than its reduced energy content. Using a blend with a lower volatility gasoline (dashed line in Figure 10.12) provides a larger gain in fuel economy as blend level is increased. This vapor pressure effect on fuel economy is not explained.

Figure 10.12: Fuel and Energy Economy Comparison of Ethanol/Gasoline Blends and Gasoline (4)

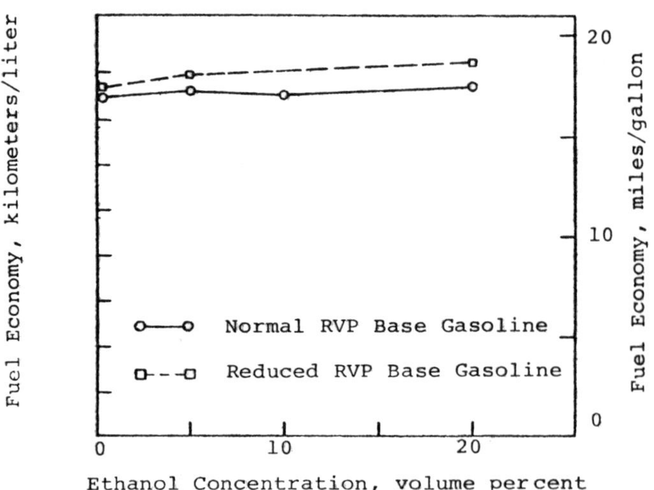

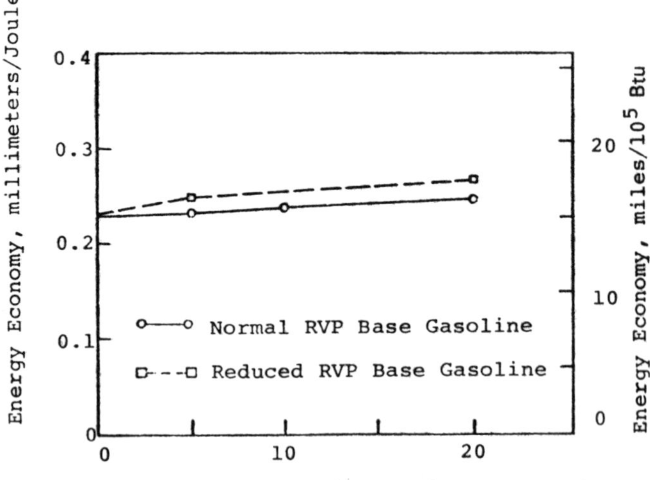

Source: DOE HCP/M2923-01

Likewise, for the Nebraska Gasohol vehicles tested at BERC, fuel economy determinations were made from results of the EPA CVS driving cycle. Volumetric fuel economy was not significantly different between base gasoline and the 10%

ethanol blend. A slight (2%) decrease in volumetric fuel economy was noted for highway driving (6). In the Nebraska program itself, volumetric blend fuel economy has been reported to be 5.3% improved over that obtained with unleaded gasoline (5). Small improvements (2 to 3%) were noted in energy economy for the Gasohol at BERC (6) with the Gasohol program (5) reporting an 8.7% improvement in miles per Btu performance.

The results of the University of Miami's performance evaluations (1) are displayed in Figures 10.13 and 10.14. The curves, as in Figures 10.2, 10.3 and 10.4, have been drawn during preparation of this report. Data points are as reported by the University.

Figure 10.13: Steady State Part Throttle Torque vs Fuel-Air Equivalence Ratio for an Ethanol/Gasoline Blend and Indolene

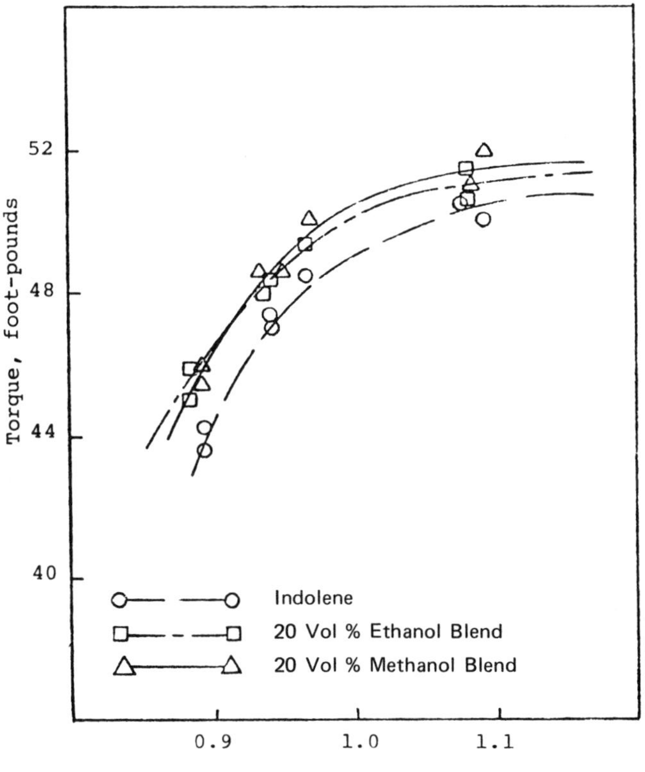

Source: DOE HCP/M2923-01

Note that the alcohol blends provide increased torque and thermal efficiency over gasoline at all equivalence ratios investigated. The ethanol blend is less helpful in these regards than the methanol blend.

Figure 10.14: Steady State Part Throttle Brake Thermal Efficiency vs Fuel-Air
Equivalence Ratio for an Ethanol/Gasoline Blend and Indolene

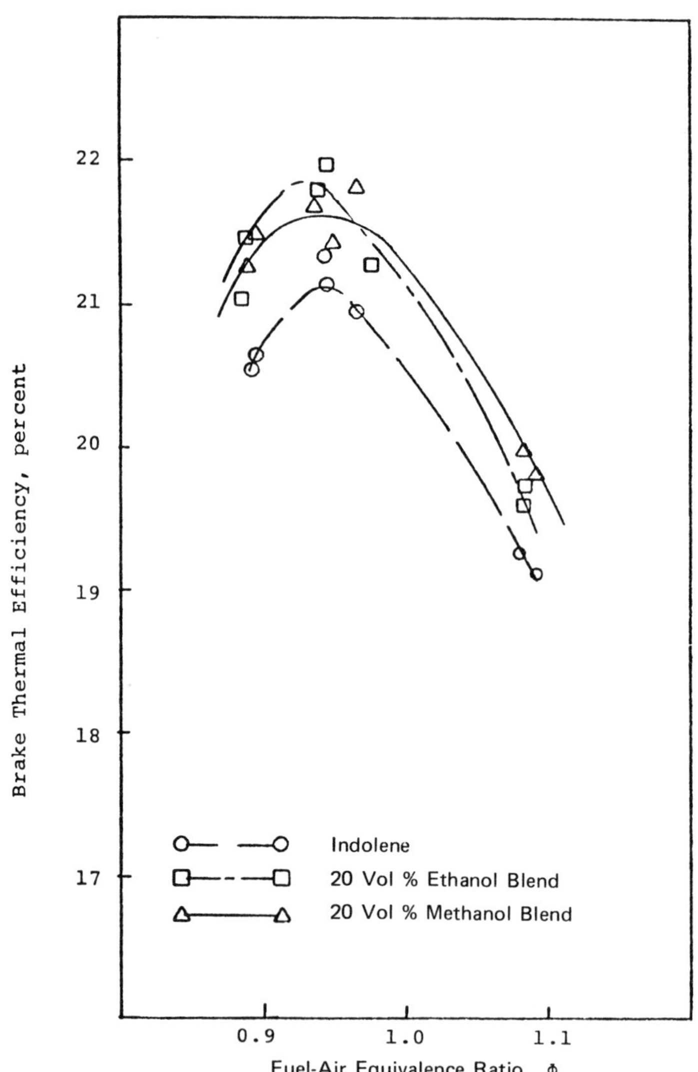

Source: DOE HCP/M2923-01

Driveability

Volkswagen has briefly reported on tests using neat ethanol in three test vehicles
having 1.5 liter, air-cooled VW engines (7). Problems experienced, and yet to
be solved, included cold-start difficulties and warm-up drieveability. Addition

of 5 to 10% gasoline to the ethanol significantly improved driveability as did increased ignition voltage and spark duration.

The Nebraska Gasohol road test program includes the distribution of questionnaires to fleet drivers and evaluation of their responses. No problems have been reported from vapor lock at ambient temperatures to 100°F and altitudes to 5,000 ft. Cold-start problems likewise are reported to have been nil. Phase separation has not been encountered at winter ambient temperatures as low as –30°F (9).

Materials Compatibility

No materials compatibility problems have been reported in either the Volkswagen do Brasil (neat ethanol) or the Nebraska Gasohol (10% ethanol/gasoline blend) programs. Periodic checks are performed in the Nebraska program of spark plug condition, cylinder compression pressure, bore condition, valve and valve seat condition (10). No unusual wear or engine deterioration has been reported (10).

However, it has been reported that all vehicles used in Brazil employ ethanol- and gasoline-compatible materials, but that the introductory production from a Brazilian Fiat plant (based on the Italian design of the time) experienced unspecified incompatibilities and were all recalled for corrective action (11). Since ethanol has been added to Brazilian motor fuels on an aperiodic basis over the last 40 years, other automotive manufacturers have routinely designed for ethanol compatibility.

Engine/Fuel System/Vehicle Design

No explicit design accommodations to the special characteristics of neat ethanol or ethanol/gasoline blend fuels have been reported necessary in the Volkswagen do Brasil, Nebraska or GM (Brazil) tests. For equivalent range, a neat ethanol-fueled vehicle would need to incorporate a fuel tank enlarged in proportion to the difference in heating values of ethanol and gasoline. The phase stability of ethanol/gasoline/water suggests that the specific problems engendered by phase separation may be relatively absent.

General Motors (4) noted increased evaporative emissions in ethanol-blend-fueled operation. This suggests the need for an assessment of evaporative emission control systems.

Fuels Characterization

Volatility: Ethanol, as a single compound, has a unique boiling point, unlike gasoline which boils over a broad range of temperatures. Ethanol has an effect on the Reid Vapor Pressure (RVP) and distillation curve. Although this problem is not as severe as with methanol, it has a disproportionate effect on the front end volatility of these blends.

Figure 10.15 illustrates the ASTM D-86 distillation curve for Nebraska Gasohol (12). Note the small divergence of this curve from that of the unleaded gasoline in the 0 to 5% distilled range.

Figure 10.15: ASTM D-86 Distillation Curves for Nebraska Gasohol and Unleaded Gasoline (12)

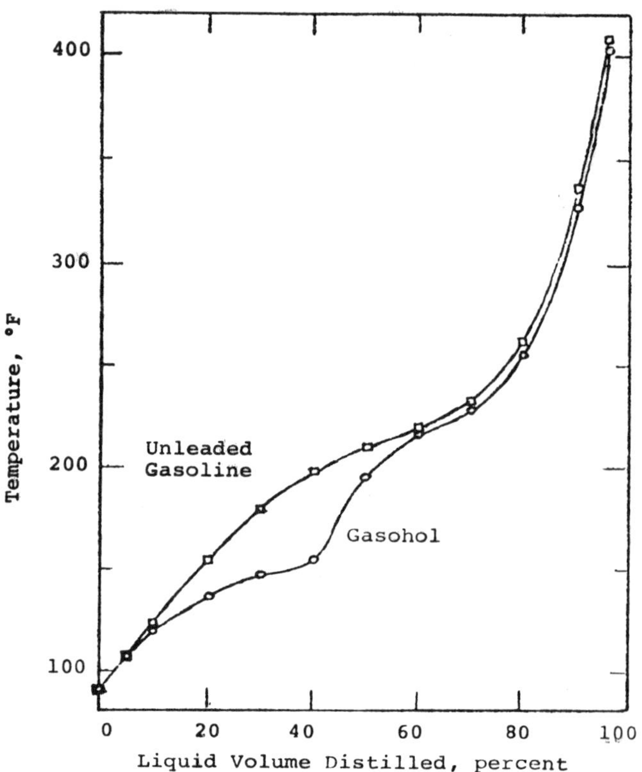

Source: DOE HCP/M2923-01

Professor William Scheller (University of Nebraska) has expressed the opinion
that this factor may explain the absence of vapor lock problems in the Gasohol
road test program (12). Likewise, the more volatile nature of the Gasohol
in the 10 to 60% range is believed a possible explanation of the good winter
driveability and cold startability. Table 10.3 compares certain properties of unleaded gasoline and Gasohol.

Table 10.3: Comparative Properties of Unleaded Gasoline and Gasohol (12)

Inspection	ASTM Test	Unleaded Gasoline	Gasohol
Liquid volume, % ethanol	—	0	11.8
API gravity	D-287	69.1	66.5
Reid vapor pressure (RVP), psi	D-323	12.0	11.8
Copper strip	D-130	1-A	1-A
Saturates, liquid volume %	D-1319	85.0	—
Olefins, liquid volume %	D-1319	1.5	—

(continued)

Table 10.3: (continued)

Inspection	ASTM Test	Unleaded Gasoline	Gasohol
Aromatics, liquid volume %	D-1319	13.5	—
Research octane number (RON)	D-2699	92.9	97.7
Motor octane number (MON)	D-2700	87.0	88.2
(RON + MON)/2	—	90.0	93.0

Source: DOE HCP/M2923-01

For the purposes of comparison, note Figure 10.16 from the GM Brazilian work. Here the volatility of a 10 volume percent ethanol blend is slightly, but distinctly higher than that of the base gasoline (4). Figures 10.17 (4) and 10.18 (13) are curves showing RVP increase versus ethanol concentration for blends. Note that small concentrations of ethanol provide significant RVP increases, but substantially smaller ones than methanol at equal concentrations.

Figure 10.16: ASTM Distillation Curves for Gasoline and a 10 Volume Percent Ethanol/Gasoline Blend (4)

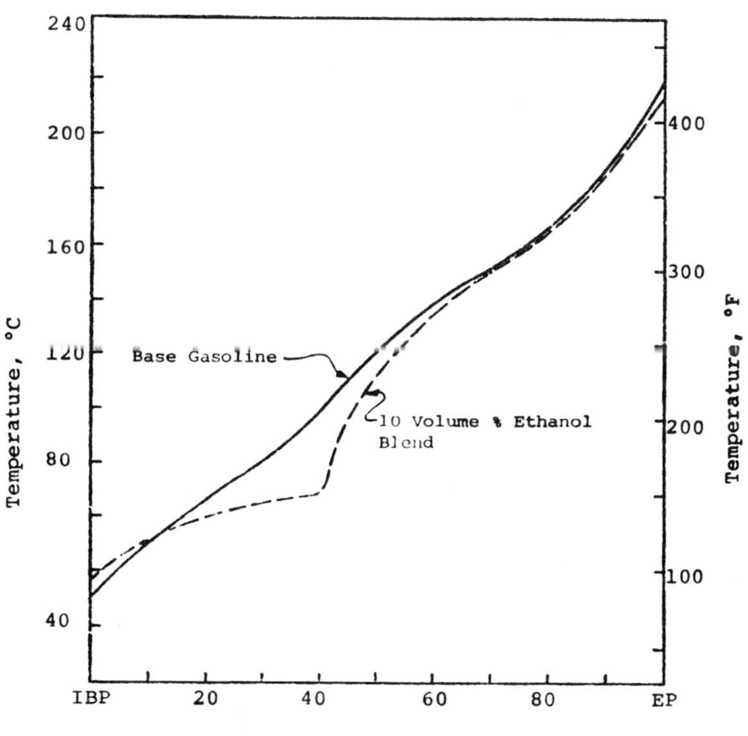

Source: DOE HCP/M2923-01

Figure 10.17: Reid Vapor Pressure vs Ethanol Concentration for Brazilian Blend Fuels (4)

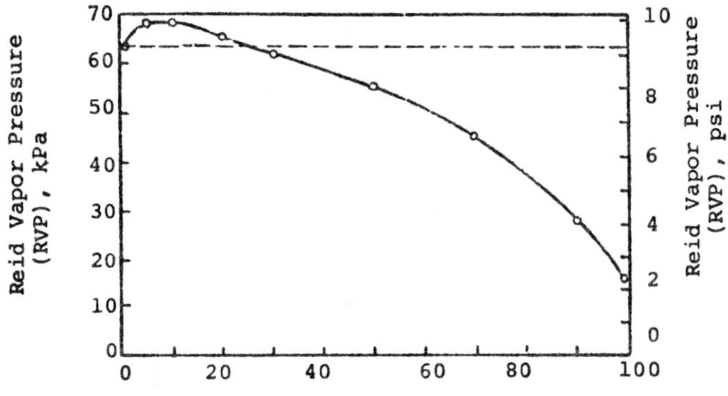

Source: DOE HCP/M2923-01

Figure 10.18: Reid Vapor Pressure Increase of Gasolines vs Concentration of Several Alcohols (13)

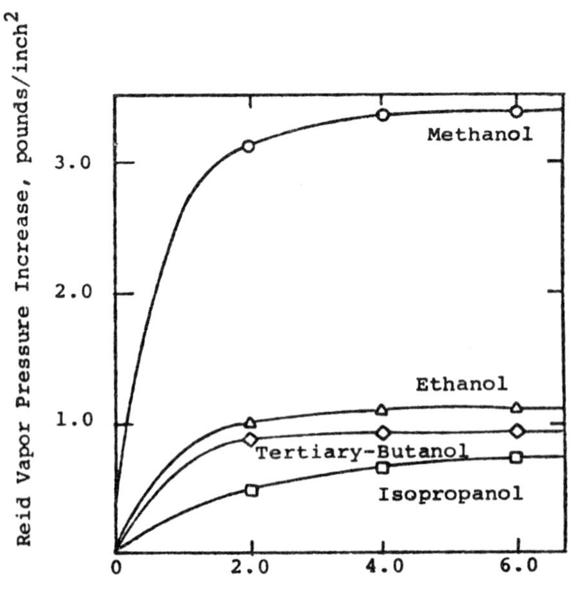

Source: DOE HCP/M2923-01

Table 10.4 contains information on the general properties (especially the volatility properties) of the gasolines and ethanol/gasoline blends used in the GM (Brazil) work (4).

Table 10.4: Properties of Brazilian Test Fuels (4)

	Gasoline	Ethanol 5%	Ethanol 10%	Ethanol 20%	Ethanol Adjusted Volume 5%	Ethanol Adjusted Volume 10%	Ethanol Adjusted Volume 20%
Reid vapor pressure							
kPa	62.7	68.9	68.3	65.5	63.4	62.7	64.1
psi	9.1	10.0	9.9	9.5	9.2	9.1	9.3
ASTM distillation, °C at percent evaporation							
IBP	31.5	36.5	38	39	38.5	40	39.5
10%	51.5	51.5	53.5	56	52.5	55	55.5
20%	65	57.5	59.5	62.5	58.5	61	62.5
30%	80	67.5	65	68	69.5	66	67.5
40%	99.5	95	70	72	97	73	71.5
50%	120	118	109	74.5	119	112.5	74
70%	152.5	154	152	144.5	153	150	142.5
90%	185.5	186.5	185.5	182.5	187	186	183.5
EP	218.5	221.5	216	216.5	214.5	217	219
Formula	$CH_{1.95}$	$CH_{1.97}O_{0.02}$	$CH_{2.02}O_{0.03}$	$CH_{2.12}O_{0.07}$	–	–	–
Stoichiometric A/F	14.66	14.34	14.05	13.47	–	–	–
Specific gravity at 15.6°C	0.738	0.741	0.743	0.748	0.743	0.745	0.747
RON	81.2	83.6	86.9	92.0	–	–	–
MON	77.3	77.2	78.1	80.0	–	–	–
Lead content							
g/l	0.52	–	–	–	–	–	–
g/gal	1.95	–	–	–	–	–	–
Sulfur, wt %	0.01	–	–	–	–	–	–
Hydrocarbon type							
% paraffins	72	–	–	–	–	–	–
% olefins	13	–	–	–	–	–	–
% aromatics	15	–	–	–	–	–	–

Source: DOE HCP/M2923-01

Solubility and Water Sensitivity: Ethanol/gasoline blends are known to be somewhat more tolerant of water than methanol/gasoline blends, but in practical terms, the difference is small. A 25% ethanol/gasoline blend will tolerate about 1% water at 70°F, while a similar methanol/gasoline blend can tolerate about 0.4% (14). Likewise, the solubility of (anhydrous) ethanol in hydrocarbons is higher than that of methanol (14). This behavior had led some researchers to employ ethanol as a methanol/hydrocarbon cosolvent.

Octane Rating: Table 10.5 gives the RON and MON ratings for various ethanol/gasoline blends. The octane rating of ethanol has been reported by various researchers. The values fall in the range of 91 to 105 RON.

Table 10.5: Octane Comparisons of Various Ethanol/Gasoline Blends (15)

Blend Composition	RON	MON
Base gasoline	92.7	83.4
+5 vol % ethanol	94.3	84.9
+10 vol % ethanol	96.3	86.2
+20 vol % ethanol	100.3	88.6
+30 vol % ethanol	103.6	90.7

Source: DOE HCP/M2923-01

Dual Fuel Diesel Application

A small amount of work has been reported by the Indian Institute of Technology on the use of ethanol in diesel pilot-fueled engines. Tests were conducted on a Kirloskar AVI single cylinder engine rated for 5 brake horsepower (BHP) or a governed speed of 1,500 rpm (speed at which tests were conducted).

Test conditions covered various diesel fuel flow rates corresponding to between 20 and 100% of the full load, diesel fuel-only rate. In a single cylinder engine with a CR of 16.5, hydrocarbon emissions rose rapidly with increasing ethanol addition and NO_x emissions fell. At 20:1 CR, HC emissions were lowered and NO_x was substantially unchanged (16). Ignition delay increases were noted as distillate flow decreased and/or as ethanol flow increased. End gas knock was encountered as ethanol addition increased at constant distillate fuel rate (16). Exhaust gas temperatures fell with ethanol addition. Thermal efficiency for all dual fuel modes was lower at part load than that of the straight distillate-fueled engine.

Advancing the injection timing (4° to 6° crank) beyond its rated BHP (brake horsepower) point of 27° before top dead center (BTDC) allowed the pilot distillate to be decreased from 50 to 30%, the ethanol accounting for 80% of the total heat release at the lower pilot rates (16).

Variable injection timing was felt to be an ultimate necessity, but was not expected to provide a total solution, due to lengthy ignition delays induced by alcohol addition. At 20:1 CR, ignition delay was comparable to straight distillate fuel and thus fixed injection timing was thought to be adequate. Additives (nitromethane and aniline) were effective at 0.5 to 1% concentrations in reducing ignition delay and HC emissions and raising thermal efficiency. NO_x emissions, however, rose.

Additional studies on alcohol/diesel blends as emulsions and solutions are in preliminary stages as reported in "Engine Experiments with Alcohol Fuel Blends" by C.A. Moses of Southwest Research Institute at the *Alcohol Fuels Technology Third International Symposium—Volume III,* Asilomar, California, May 1979.

Results on higher concentrations of ethanol-diesel fuel solutions (as high as 30% ethanol) show good combustion at all load/speed conditions; 50% concentrations could be used at low speeds.

Environmental, Health and Safety Considerations

Environmental: There is little information on the environmental effects which may result from the utilization of ethanol as a fuel other than combustion emissions. It is reasonable to assume that damage from spills or other acute exposure would be significantly less than for similar methanol and gasoline spills since denatured ethanol is not as toxic. However, information on acute or low-level, chronic exposure is required for ethanol to predict its environmental impact.

Health: Although ethanol is not highly toxic, discomfitures such as coughing, eye irritation and headaches will occur if it is inhaled for continuous periods under poor ventilation conditions. The Threshold Limit Value for ethanol is 1,000 ppm (1,900 mg/m^3). The intoxication effects of ethanol have been observed and studied at length. Ingestion of alcohol causes intoxication at levels below 100 g. Larger intake will cause stupor and poisoning (17).

The production and use of ethanol is very rigidly controlled in the United States by the Treasury Department's Bureau of Alcohol, Tobacco and Firearms. Specific formulations have been established for the major classes of denatured ethanol. If neat ethanol were used as a fuel, it would be necessary to ensure that it not be easily separated from the constituents added for denaturing purposes (17). For example, it is possible to separate ethanol from gasoline in a mixture by the addition of water. The separated ethanol could be further purified by passing it through a bed of activated charcoal.

Safety: Ethanol is a flammable liquid. Its flash point is significantly above that for gasoline, but its flammability limits are considerably wider, 3.3 to 19% in air by volume. The safety aspects of ethanol/gasoline blends would be about the same as for gasoline alone (17). As with gasoline, there are safety hazards in the form of explosions or fires associated with the handling and storage of ethanol. If neat ethanol is used extensively, it will also be used for peripheral purposes and like gasoline, may often be used or stored improperly.

Table 10.6 presents the hazard identification system established by the National Fire Protection Association (NFPA) and the recommended fire extinguishing agents. The recommended hazard identification system "provides simple, readily recognizable and easily understood markings which will give at a glance a general idea of the inherent hazards of any material and the order of severity of these hazards as they relate to fire prevention, exposure and control" (18).

Table 10.6: Combustion and Safety Hazard Comparison of Alcohols and Gasoline (18)(19)

	.NFPA Hazard Identification Signals*.			Extinguishing
	Health	Fire	Reactivity	Agents**
Gasoline	1	3	0	b, d
Methanol	1	3	0	c, d
Ethanol	0	3	0	a, c, d

*4 indicates a severe hazard; 0 indicates no special hazard.

**a is water, b is foam, c is alcohol foam, and d is carbon dioxide or dry chemical.

Source: DOE EC-77-X-01-2923

ADDITIONAL ENGINE TEST PROGRAMS

EPA Gasohol Test Program

> The material in this section is based on *Gasohol Test Program* (NTIS PB-290-569), prepared by R. Lawrence of the Technology Assessment and Evaluation Branch of the U.S. Environmental Protection Agency, Motor Vehicle Emission Laboratory, December 1978.

A request to permit the use of 10% ethanol in gasoline ("gasohol") was considered by EPA-Mobile Source Enforcement Division (MSED). The Emission Control Technology Division (ECTD) in Ann Arbor, MI was requested to assist MSED by testing ten vehicles on two gasoline fuels and three gasohol fuels.

Test Procedure: The test procedure agreed upon for the ECTD program was to test each vehicle twice on each fuel using the standard FTP (Federal Test Procedure) with SHED procedure as used for certification tests. Some modifications were necessary to allow for cannister weights to be taken before and after the Diurnal Breathing Loss (DBL) test and after the Hot Soak (HS) test.

Void test criteria normally applied to certification tests were waived for some tests where engineering judgment could be used to verify that the test results were valid for the purpose of this program. Typical examples of this include tests where a heat build for diurnal emissions might be $1°F$ out of tolerance or tests where an exhaust emission analyzer might respan 3 to 4% low when the tolerance is ±2%.

Some portions of the FTP were made more restrictive to provide more repeatable SHED results.

The overnight soak tolerance of 12 to 36 hours was adjusted to 12 to 24 hours.

Two preconditioning driving cycles with a one hour hot soak between them and refueling prior to each cycle were required each time the fuel type was changed.

The complete test procedure is shown below. Six vehicles could be run each day using two SHED and two chassis dynamometers. The vehicles were separated into two groups. The first group followed the fuel sequence of 1, 2, 3, 4, 5, (6), 3 with duplicate tests each time. The second group followed the fuel sequence of 3, 4, 5, (6), 3 and then fuels 1 and 2 if time permitted. Nearly all vehicles did receive duplicate tests on all fuels.

Gasohol Test Sequence

   (1)   Drain and refuel to 20% tank capacity
   (2)   Run 1 LA-4 cycle
        a. Check idle CO and rpm first time on each fuel
   (3)   Hot soak one hour (key off to key on)
   (4)   Drain and refuel to 40% tank capacity
   (5)   Run 1 LA-4 cycle
   (6)   Soak 12 to 24 hours at $68°$ to $86°F$ (key off to key on)
   (7)   Run 1 FTP with SHED:
        a. Drain and refuel to 40% tank capacity (leave fuel cap off)

       b.. Move vehicle to SHED
       c. Weigh cannister
       d. Check cannister lines
       e. Perform one-hour diurnal heat build (fuel cap on at 60°F)
       f. Immediately after heat build:
           remove heat blanket
           weigh cannister
           reinstall cannister and check cannister lines
       g. Run 3-bag FTP emissions test within 15 to 60 minutes of end of
           diurnal test
       h. Run one-hour hot soak immediately following emissions test
       i. Weigh cannister immediately following hot soak test

  (8)   Precondition for next test:
       a. If within 24 hours of FTP key off go to step (4)
       b. If longer than 24 hours since FTP key off go to step (1)
       c. If changing fuel type go to step (1)

  (9)   Two tests for each fuel type with following sequence:
          1, 2, 3, 4, 6, (5), 3 for group 1 vehicles
          3, 4, 6, (5), 3, 1, 2 for group 2 vehicles
          (6 fuel runs x 2 tests each x 11 vehicles = 132 tests)

Fuels: Five fuels were chosen for comparison as follows.

  Fuel 1: Indolene [RVP (Reid Vapor Pressure) = 9.0]
  Fuel 2: 90% Indolene (same fuel batch as fuel 1) plus 10% ethanol (RVP = 9.2)
  Fuel 3: Summer grade gasoline (SG) (RVP = 10.0)
  Fuel 4: 90% fuel 3 plus 10% ethanol (RVP 10.7)
  Fuel 5: Blended gasohol containing 10% ethanol and approximating the RVP and
        distillation characteristics of fuel 3

The reason for running fuels 1 and 2 was to show the changes in emissions which result when the certification fuel is combined with ethanol thus increasing fuel volatility.

Fuel 3 was selected as a base fuel which might be representative of national average summer grade fuel.

Fuel 4 shows the effect on emssions when ethanol is added to fuel 3 as might be done by a fuel retailer or distributor. Fuel 4 volatility is higher than fuel 3.

Fuel 5 was to be a gasohol blend with RVP and distillation curve similar to fuel 3. It is not known if this fuel is representative of what a commercial gasohol would be if it were blended by the refiner to meet market requirements. Because ethanol significantly alters the distillation curve it was difficult to blend a gasohol fuel to meet the distillation curve of a gasoline fuel.

All fuels were ordered by MSED from Howell Hydrocarbons. However, because of time constraints EPA-ECTD started testing on in-house Indolene (fuel 1) and blended fuel 2 using fuel 1 and locally purchased ethanol. Fuels 3, 4, and 5 were supplied by Howell Hydrocarbons. Fuel 5 was found to be out of tolerance and was not used. It was replaced and the replacement was designated fuel 6 (RVP = 10.0).

A fuel sample was drawn from a fuel cart each time the cart was refueled. Since the fuel cart capacity is 50 gallons there is at least one fuel sample for every 55 gallon drum of fuel supplied. Figure 10.19 shows typical distillation data of the five fuels used.

Figure 10.19: Distillation Curves of Test Fuels

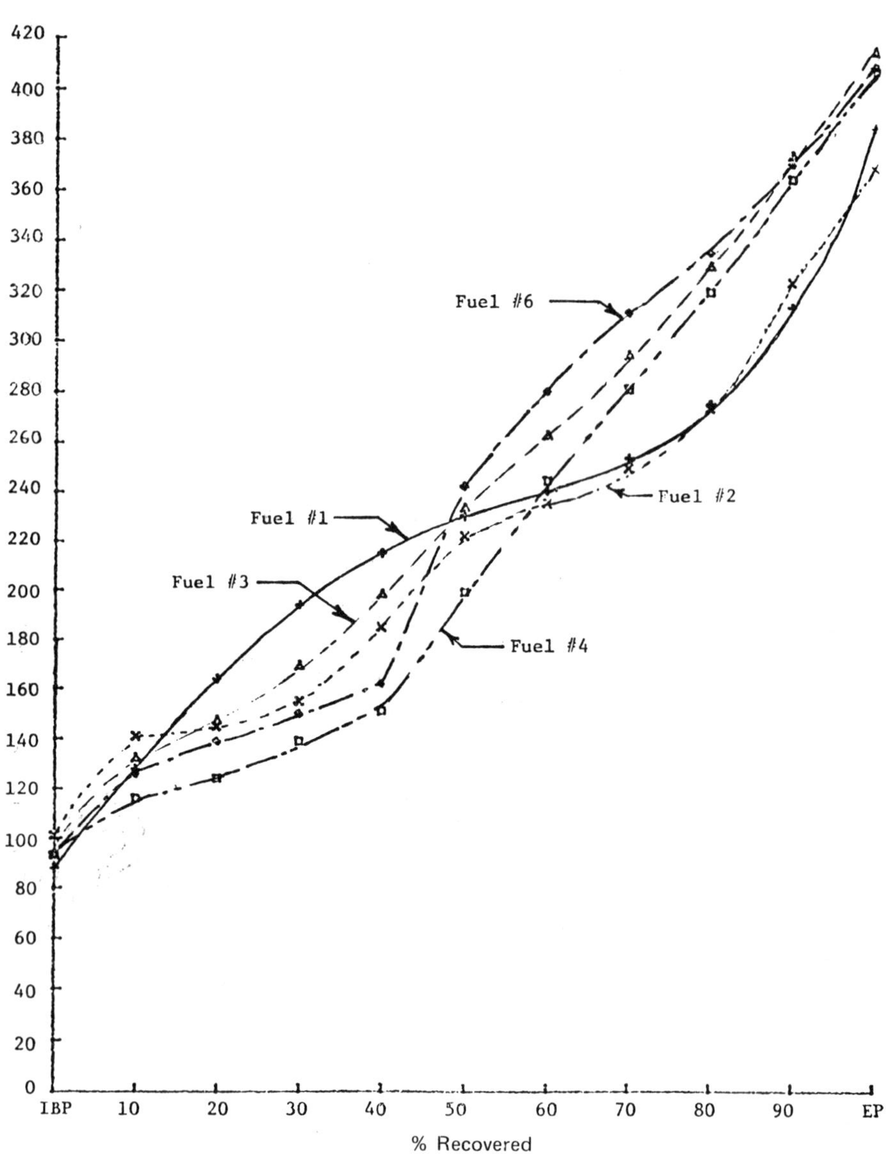

Source: NTIS PB-290-569

By comparing fuel 2 with fuel 1 or fuel 4 with fuel 3 the increase in volatility caused by the addition of 10% ethanol can be seen. Fuel 6 compared with fuel 3 illustrates the difficulty encountered in trying to blend a "gasohol" to the same same distillation curve as a typical gasoline.

Vehicles: All vehicles were supplied by the vehicle manufacturers. Ten 1978 and 1979 vehicles were to be run. To ensure completion of ten vehicles in the required time eleven were requested from manufacturers. All eleven were received and all completed the test sequence.

All eleven vehicles were catalyst equipped—four with three-way catalysts and seven with oxidation catalysts. The vehicles included four from Ford (two three-way catalysts and two oxidation catalysts); four from GM (two three-way catalysts and two oxidation catalysts); two from Chrysler; and one from Toyota.

Data and Discussion: The data have been summarized and emissions on each gasoline and gasohol fuel are shown in Figures 10.20, 10.21 and 10.22. (The total hydrocarbon exhaust plus evaporative emissions for 3.3 trips per day are designated "TOTHC". DBL, HSL, and TLOSS are the diurnal, hot soak, and total evaporative emissions, respectively.)

Emissions and fuel economy for Indolene with 10% ethanol are compared with indolene and the two "commercial" gasohols are compared with the SG base fuel (fuel 3) for all vehicles (Figure 10.23). The gasohol fuels increased both diurnal and hot soak evaporative emissions by 29 to 71%. Total evaporative emissions increased by 49 to 62% on gasohol fuels. Total HC emissions (evaporation plus exhaust) increased by 11 to 32% on gasohol fuels. CO emissions decreased 20 to 34% and fuel economy decreased 1 to 5% on gasohol. NO_x increased 6 to 11% on gasohol fuels.

Evaporative emissions with fuel 6 (blended gasohol) were slightly lower than with fuel 4 but HC and CO exhaust emissions were higher on fuel 6 than on fuel 4. The total HC (exhaust plus evaporation) for 3.3 trips per day were higher. Fuel 4/3 showed an 18% increase and fuel 6/3 showed a 32% increase in total HC emissions.

The fuel inspection data show that gasohol fuels blended by adding ethanol to a base gasoline are more volatile than the base fuel. The increased RVP and front end volatility of gasohol would be expected to increase diurnal and hot soak losses, respectively. The oxygen present in alcohol causes leaner operation and would be expected to decrease exhaust HC and CO, unless other fuel characteristic changes such as density, viscosity, or volatility were dominant.

Vehicle emission data support the above relationships:

    (1)   Higher RVP gave higher diurnal losses.
    (2)   Increased front end volatility (up to 50% point) increased
           hot soak losses.
    (3)   Gasohol generally gave lower HC and CO exhaust emissions
           than gasoline.

One exception was fuel 6—a blended gasohol. Driveability was poor with stumbling, hesitation and backfiring during acceleration on some vehicles.

Figure 10.20: Average Emissions of Eleven 1978-79 Catalyst Equipped Vehicles

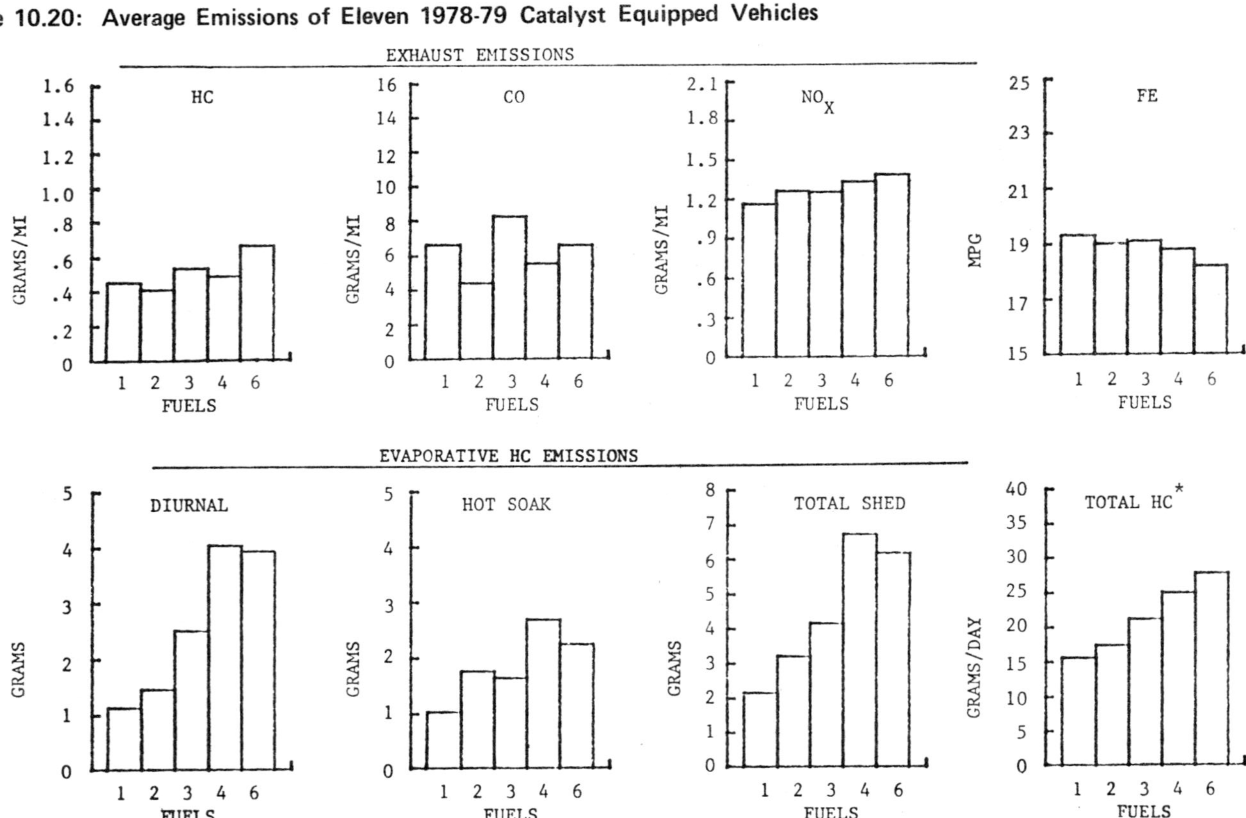

*Exhaust plus evaporative emissions for 3.3 trips per day

Source: NTIS PB-290-569

Figure 10.21: Average Emissions of Four 1978-79 Three-Way Catalyst Equipped Vehicles

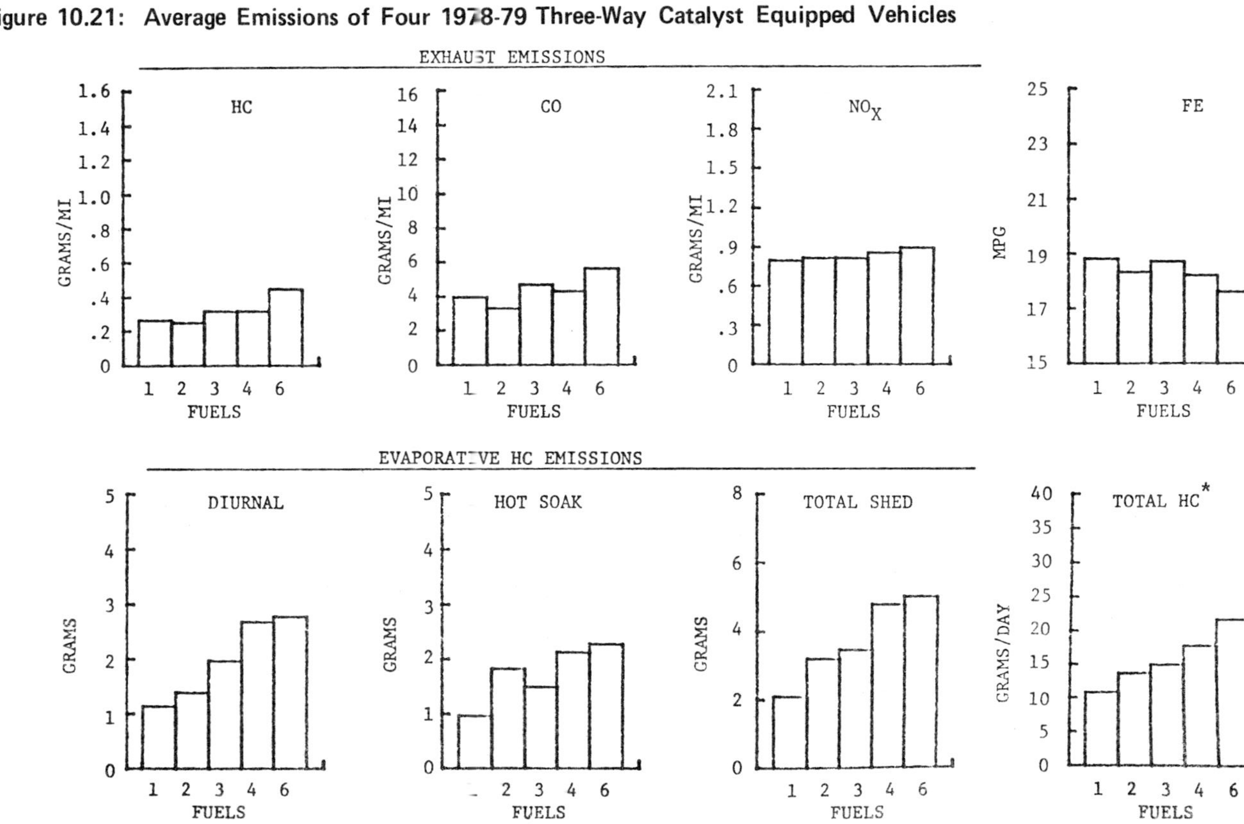

Source: NTIS PB-290-569

Figure 10.22:　Average Emissions of Seven 1978-79 Oxidation Catalyst Equipped Vehicles

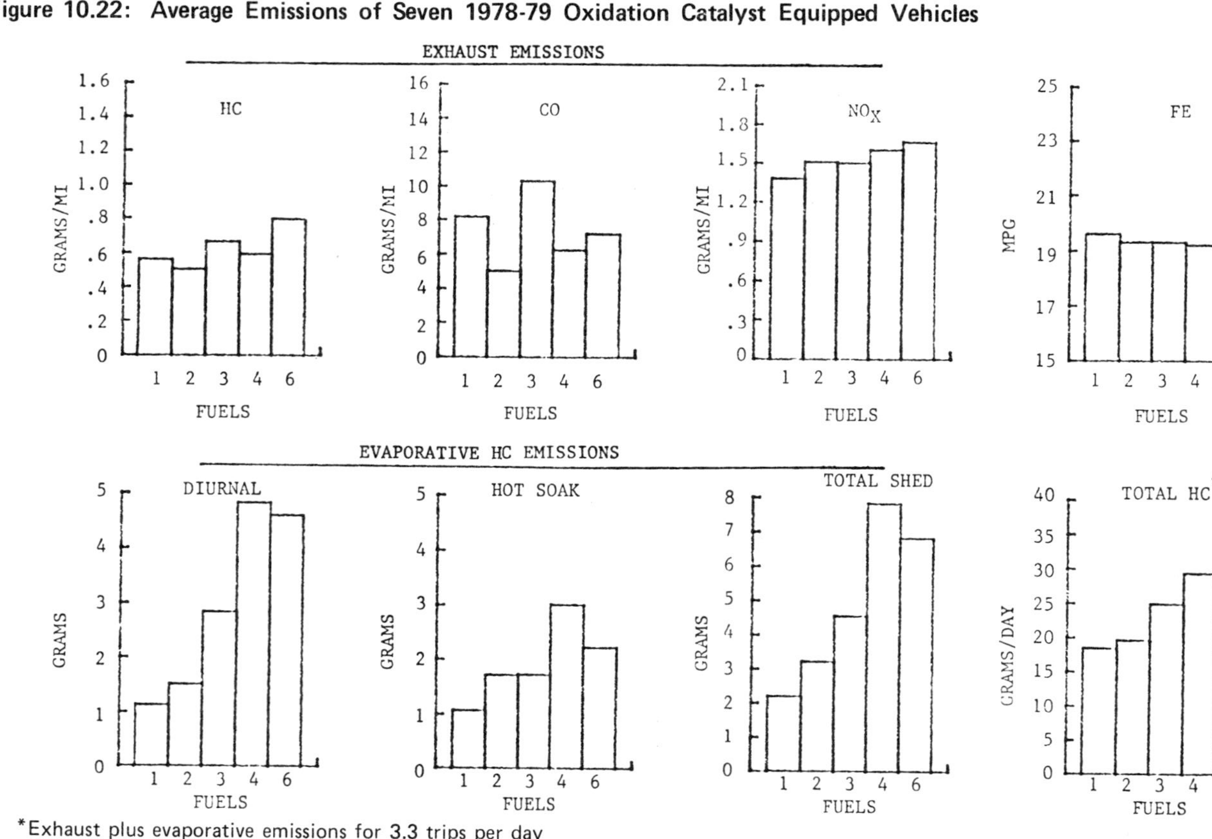

Source:　NTIS PB-290-569

Figure 10.23: Ratios of Average Emissions of Eleven 1978-79 Catalyst Equipped Vehicles

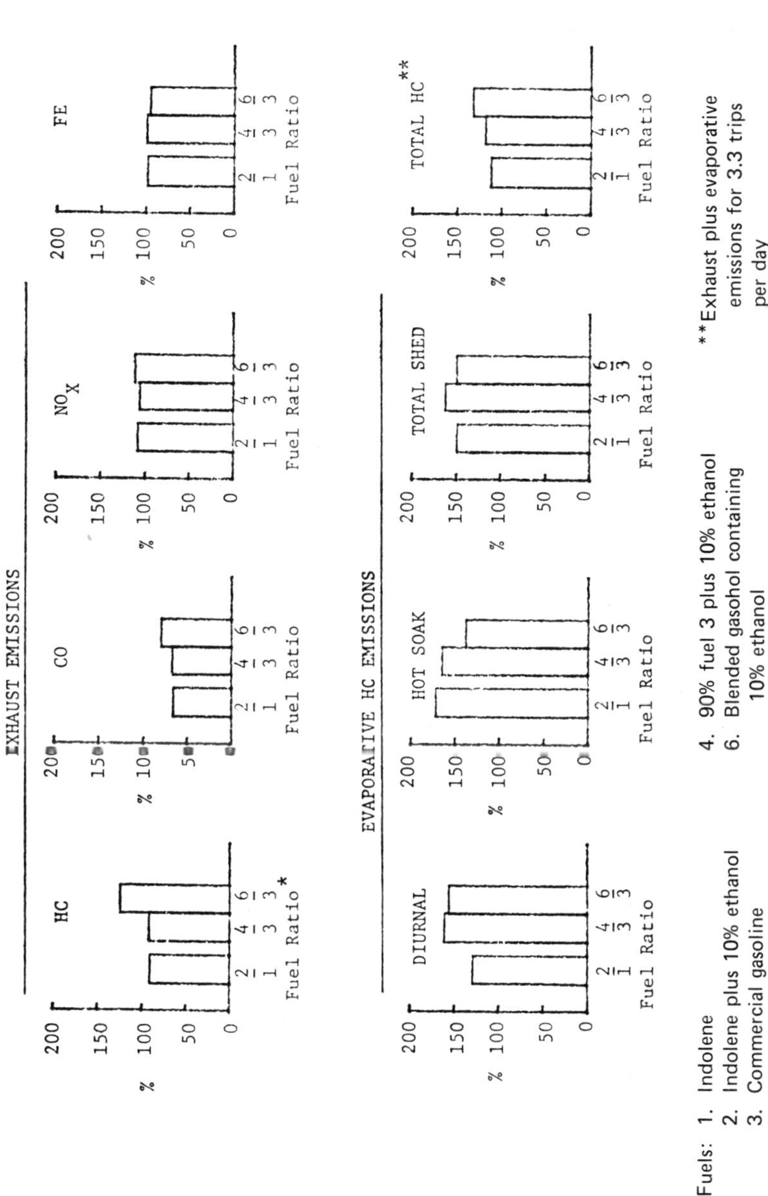

Source: NTIS PB-290-569

The HC emissions were 24% higher on this gasohol than on the base fuel. CO emissions were 20% lower but this is not as great as the 34% decrease in CO emissions seen with the other two gasohols.

The mean emissions, fuel economy (FE), and cannister weights were determined for each vehicle on each fuel. Summaries showing the average of vehicle means for all vehicles; for the TWC (three-way catalyst) equipped vehicles; and for the oxidation catalyst equipped vehicles are shown in Tables 10.7, 10.8 and 10.9, respectively. Here the means for the five fuels, the difference between selected fuels and ratios of the means of selected fuels are given.

The emissions and fuel economy data in these tables were those presented in Figures 10.20, 10.21 and 10.22. The last five columns of each table contain cannister weight data (grams):

DBL	=	diurnal test	DDBL	=	Δ diurnal (ADBL – BDBL)
HSL	=	hot soak test	DTEST	=	Δ test (AHSL – BDBL)
BDBL	=	before diurnal test	TLOSS	=	total evaporative emissions
ADBL	=	after diurnal test	TOTHC	=	exhaust plus evaporative emis-
AHSL	=	after hot soak test			sions for 3.3 trips per day

Cannister weights could not be measured before the hot soak test without interfering with the test.

The cannister weight gains during the diurnal breathing loss test (DDBL) are related to the Reid Vapor Pressures of the fuels. The cannister weights before the diurnal test (BDBL) (which is after a 12 to 24 hour soak) are fuel related but it is not clear which fuel parameter(s) exert the strongest influence.

Vehicle operation on the FTP causes a net decrease in cannister weight from after the diurnal test to after the hot soak test (AHSL – ADBL). This indicates that the cannister is purging during the test and that there was plenty of cannister capacity available during the hot soak test. High hot soak losses then imply that the evaporative emission control systems do not effectively trap hot soak emissions on these vehicles. This is an important consideration since hot soak losses are more significant than diurnal losses from an air quality viewpoint. This is because there is only one diurnal per day but an average of 3.3 hot soaks per day per vehicles in "real world" use.

SHED Alcohol Data: Ethanol measurements were made on some evaporative emissions tests. Capability for ethanol measurement did not exist at EPA-MVEL at the start of this program. The Laboratory Branch, in conjunction with EPA-ORD, was able to provide a gas chromatograph and procedure capable of measuring SHED ethanol concentrations in time to obtain data part way through the program.

Ethanol emissions from gasohol fuels 4 and 6 for the diurnal and hot soak test ranged from 0.1 to 0.6 gram for each test. The average diurnal emissions were 0.26 gram ethanol on 12 tests and the average hot soak emissions were 0.33 gram ethanol on 10 tests. This amounts to 0.6 gram ethanol for a complete test or 1.35 grams ethanol for 3.3 trips (3.3 trips per day = DBL x 1.0 + HSL x 3.3).

Table 10.7: Mean Test Results for Eleven Catalyst Equipped Vehicles

	HC	CO	NO_x	CO_2	FE	DBL	HSL	TLOSS	TOTHC	BDBL	ADBL	AHSL	DDBL	DTEST
		(g/mile)			(mpg)						(g)			
Fuel														
1	0.451	6.61	1.16	470.	19.3	1.12	1.02	2.14	15.60	890.	904.	885.	13.3	-4.9
2	0.410	4.39	1.26	467.	19.0	1.45	1.75	3.20	17.35	891.	907.	892.	16.3	1.1
3	0.535	8.20	1.25	472.	19.1	2.51	1.63	4.15	21.11	900.	919.	898.	19.4	-2.1
4	0.490	5.51	1.33	468.	18.8	4.04	2.68	6.72	24.97	905.	927.	908.	21.5	2.7
6	0.665	6.57	1.33	482.	18.2	3.93	2.24	6.17	27.76	905.	924.	905.	19.3	0.6
Differences between means														
Fuel 2 – fuel 1	-0.041	-2.23	0.09	-3.	-0.4	0.33	0.73	1.06	1.76	1.	4.	7.	3.0	6.0
Fuel 3 – fuel 1	0.084	1.59	0.03	2.	-0.2	1.39	0.61	2.00	5.52	10.	16.	12.	6.1	2.9
Fuel 4 – fuel 3	-0.045	-2.69	0.08	-4.	-0.3	1.53	1.04	2.57	3.86	5.	8.	10.	2.1	4.8
Fuel 6 – fuel 3	0.130	-1.63	0.13	10.	-0.9	1.41	0.61	2.02	6.65	5.	5.	7.	-0.1	2.7
Ratios of means (%)														
Fuel 2/fuel 1	91.	66.	108.	99.	98.	129.	171.	149.	111.	100.	100.	101.	122.	-23.
Fuel 3/fuel 1	119.	124.	107.	100.	98.	224.	160.	193.	135.	101.	102.	101.	145.	42.
Fuel 4/fuel 3	92.	67.	106.	99.	99.	161.	164.	162.	118.	101.	101.	101.	111.	-131.
Fuel 6/fuel 3	124.	80.	111.	102.	95.	156.	137.	149.	132.	101.	101.	101.	100.	-29.

Source: NTIS PB-290-569

Table 10.8: Mean Test Results of Four Three-Way Catalyst Equipped Vehicles

	HC	CO	NO_x	CO_2	FE	DBL	HSL	TLOSS	TOTHC	BDBL	ADBL	AHSL	DDBL	DTEST
		(g/mile)			(mpg)						(g)			
Fuel														
1	0.262	3.94	0.79	479.	18.8	1.13	0.95	2.08	10.75	954.	969.	948.	15.0	-6.4
2	0.247	3.29	0.81	475.	18.3	1.38	1.81	3.19	13.49	958.	976.	959.	18.3	1.4
3	0.314	4.64	0.81	479.	18.7	1.96	1.48	3.45	14.65	971.	993.	968.	21.7	-2.9
4	0314	4.26	0.85	477.	18.2	2.67	2.11	4.79	17.41	977.	1,002.	979.	25.4	2.1
6	0.445	5.57	0.89	488.	17.6	2.77	2.26	5.03	21.24	977.	999.	978.	21.9	1.1
Differences between means														
Fuel 2 – fuel 1	-0.015	-0.65	0.03	-4.	-0.5	0.25	0.86	1.11	2.74	3.	6.	11.	3.3	7.8

(continued)

Table 10.8: (continued)

	HC	CO	NOx	CO₂	FE	DBL	HSL	TLOSS	TOTHC	BDBL	ADBL	AHSL	DDBL	DTEST
	.... (g/mile)				(mpg)	.... (g)								
Fuel 3 – fuel 1	0.053	0.70	0.02	1.	-0.1	0.83	0.53	1.36	3.90	17.	23.	20.	6.7	3.5
Fuel 4 – fuel 3	-0.001	-0.38	0.04	-3.	-0.5	0.71	0.63	1.34	2.76	5.	9.	10.	3.7	5.1
Fuel 6 – fuel 3	0.131	0.93	0.07	9.	-1.1	0.80	0.78	1.58	6.60	6.	6.	10.	0.3	4.0
Ratios of means (%)														
Fuel 2/fuel 1	94.	83.	103.	99.	98.	122.	190.	153.	125.	100.	101.	101.	122.	-21.
Fuel 3/fuel 1	120.	118.	103.	100.	100.	174.	156.	165.	136.	102.	102.	102.	145.	46.
Fuel 4/fuel 3	100.	92.	105.	99.	97.	136.	142.	139.	119.	101.	101.	101.	117.	-72.
Fuel 6/fuel 3	142.	120.	109.	102.	94.	141.	152.	146.	145.	101.	101.	101.	101.	-38.

Source: NTIS PB-290-569

Table 10.9: Mean Test Results of Seven Oxidation Catalyst Equipped Vehicles

	HC	CO	NOx	CO₂	FE	DBL	HSL	TLOSS	TOTHC	BDBL	ADBL	AHSL	DDBL	DTEST
	.... (g/mile)				(mpg)	.... (g)								
Fuel														
1	0.559	8.14	1.38	465.	19.6	1.12	1.06	2.18	18.37	854.	866.	850.	12.3	-4.1
2	0.503	5.01	1.51	463.	19.3	1.50	1.71	3.21	19.56	853.	868.	854.	15.1	1.0
3	0.661	10.24	1.50	468.	19.3	2.83	1.72	4.54	24.81	859.	877.	858.	18.0	-1.6
4	0.590	6.22	1.60	464.	19.2	4.82	3.00	7.82	29.29	865.	884	868.	19.3	3.1
6	0.791	7.14	1.66	479.	18.4	4.59	2.22	6.82	31.49	864.	881.	864.	17.8	0.2
Differences between means														
Fuel 2 – fuel 1	-0.056	-3.13	0.13	-2.	-0.3	0.38	0.65	1.03	1.19	-1.	2.	4.	2.8	5.1
Fuel 3 – fuel 1	0.102	2.10	0.12	3.	-0.4	1.71	0.66	2.37	6.44	5.	11.	8.	5.7	2.5
Fuel 4 – fuel 3	-0.071	-4.02	0.11	-4.	-0.1	1.99	1.28	3.27	4.49	6.	7.	10.	1.2	4.6
Fuel 6 – fuel 3	0.130	-3.10	0.17	11.	-0.8	1.76	0.51	2.27	6.68	4.	4.	6.	-0.2	1.8
Ratios of means (%)														
Fuel 2/fuel 1	90.	62.	110.	99.	98.	134.	162.	147.	107.	100.	100.	101.	122.	-24.
Fuel 3/fuel 1	118.	126.	109.	101.	98.	252.	162.	209.	135.	101.	101.	101.	146.	39.
Fuel 4/fuel 3	89.	61.	107.	99.	99.	170.	175.	172.	118.	101.	101.	101.	107.	-195.
Fuel 6/fuel 3	120.	70.	111.	102.	96.	162.	130.	150.	127.	101.	100.	101.	99.	-14.

Source: NTIS PB-290-569

Correcting the SHED FID (flame ionization detector) for response to ethanol would result in a decrease in SHED HC of about 5% for the gasohol fuels. The ethanol present as determined by the GC would then have to be added to the SHED HC to arrive at the total evaporative HC plus ethanol emissions. This cannot be done directly since HC is given in grams of $CH_{1.85}$ (MW = 13.85) and ethanol is given in grams of C_2H_5OH (MW = 46). The reported HC emissions (evaporative and exhaust) are not corrected for ethanol response of the FID nor for measured ethanol in the sample.

Driveability: Driveability experiments were not run. However, drivers were requested to note any driveability comments on the test data sheets. These comments indicate a slight degradation in driveability on some vehicles on fuels 2, 3 and 4. A more severe degradation in driveability on fuel 6 was noted, with occurrences of backfiring and poor acceleration on several vehicles.

Conclusion: The purpose of this test program was to evaluate the effect on emissions (evaporative and exhaust) that the use of gasohol would have. The data show that gasohol increased total hydrocarbon emissions by 6 to 11% while decreasing CO emissions by 20 to 34% on the eleven 1978 and 1979 vehicles tested.

Driveability on the blended gasohol (fuel 6) degraded to the extent that if commercial fuel like fuel 6 were used it is likely that persons using this fuel would either stop using it or would have their vehicles adjusted to compensate for the different fuel. This would most likely be an air-fuel ratio (A/F) adjustment towards richer operation. Once properly adjusted for gasohol fuel the vehicle exhaust emissions might be expected to be similar to emissions from a vehicle correctly adjusted for and running on gasoline, but evaporative emissions would remain high. However, if a vehicle adjusted for gasohol were then operated on gasoline a rich A/F ratio would result and would likely cause a marked increase in HC and CO emissions while not affecting driveability.

Driveability comments on Indolene plus 10% ethanol compared with Indolene and on SG plus 10% ethanol compared with SG indicated that the driveability was the same in some cases and slightly degraded (hard to start and stalling when cold) in other cases. Thus these "mixed" gasohols did not pose the driveability problem that the "blended" gasohol did.

It is not known if a decreased volatility gasohol could be blended which would not cause an increase in evaporative emissions or degradation in driveability on in-use vehicles. The "blended" gasohol (fuel 6) used in this program did result in increased evaporative emissions over fuel 3 even though its RVP and distillation curve were adjusted close to that of fuel 3.

BERC Testing Program

Information in this section is based on "Alternate Fuels and Data Bank Program," a paper by K. Stamford of the U.S. Department of Energy, Bartlesville Energy Research Center, in the summary report of the *Highway Vehicle Systems Contractors' Coordination Meeting* (DOE CONF-7805102), held in Troy, Michigan, May 1978, issued by the U.S. Department of Energy, September 1978 and on "Ethanol and Methanol Fleet Operation," a paper by K. Stamper

of the U.S. Department of Energy, Bartlesville Energy Research
Center in *Proceedings of Highway Vehicle Systems Contractors'
Coordination Meeting* (DOE CONF-781050), held in Dearborn,
Michigan, October 1978, issued by the U.S. Department of
Energy March 1979.

10% Ethanol Blend Fleet Tests: An experimental program has been set up to
determine the feasibility of using ethanol/gasoline blends to fuel current production automobiles. Five vehicles have been used to determine the performance
characteristics of 10% ethanol/gasoline blends at four ambient temperatures
ranging from 20° to 100°F (Table 10.10). Each test condition is repeated in
triplicate to ensure that representative, repeatable data are taken. Tests are
made using 10% ethanol/90% base gasoline and the clear base gasoline. Measurements of fuel economy and emission rates of CO, NO_x, unburned HC, and
aldehydes are made for each test run on the clear base fuel. In addition to
these, measurements of unburned ethanol are made for each test run on the
10% ethanol/gasoline blend.

Table 10.10: Description of Ethanol Fleet

Vehicle Description	Engine Displacement (CID)	Carb	Test IW (lb)
1975 Dodge Colt*	98	2V	2,750
1976 Chevrolet Impala*	350	2V	4,500
1977 Volvo 242DL**	130	FI	3,000
1977 Pontiac Astre*	151	2V	3,000
1978 Ford Pinto**	140	2V	2,750

*With conventional emission control system
**With three-way catalyst ("California" car)

Source: DOE CONF-7805102

Test results shown in Figure 10.24 represent the test fleet average. Volumetric
fuel economy of the 10% ethanol/gasoline blend is slightly lower than that of
the clear base fuel for all the ambient temperatures run on the urban cycle.
However, the effect is less pronounced at ambient temperatures below 75°F
(Figure 10.24a). The volumetric fuel economy penalty for the 10% ethanol
blend is just as significant in the results of the highway fuel economy tests.
When the fuel economy is compared on an energy basis, the 10% blend shows
a slight increase over the base fuel for all temperatures except 100°F on urban
cycle tests. These results are emphasized in the tests made on the highway cycle
in that slight increases in energy economy can be seen for each of the four ambient temperatures. It is difficult to attach great significance to the slight increase in energy economy numbers for the ethanol blend since there is only a
3.5% difference in the heating value of blend and clear base gasoline. (Analyses
of the test fuels are listed in Table 10.11.)

Carbon monoxide emissions are consistently reduced at all temperatures for the
ethanol blend compared to the base fuel (Figure 10.24b). The major portion
of this reduction comes from the cold start segment of the test cycle.

Figure 10.24: Test Results at Four Ambient Temperatures

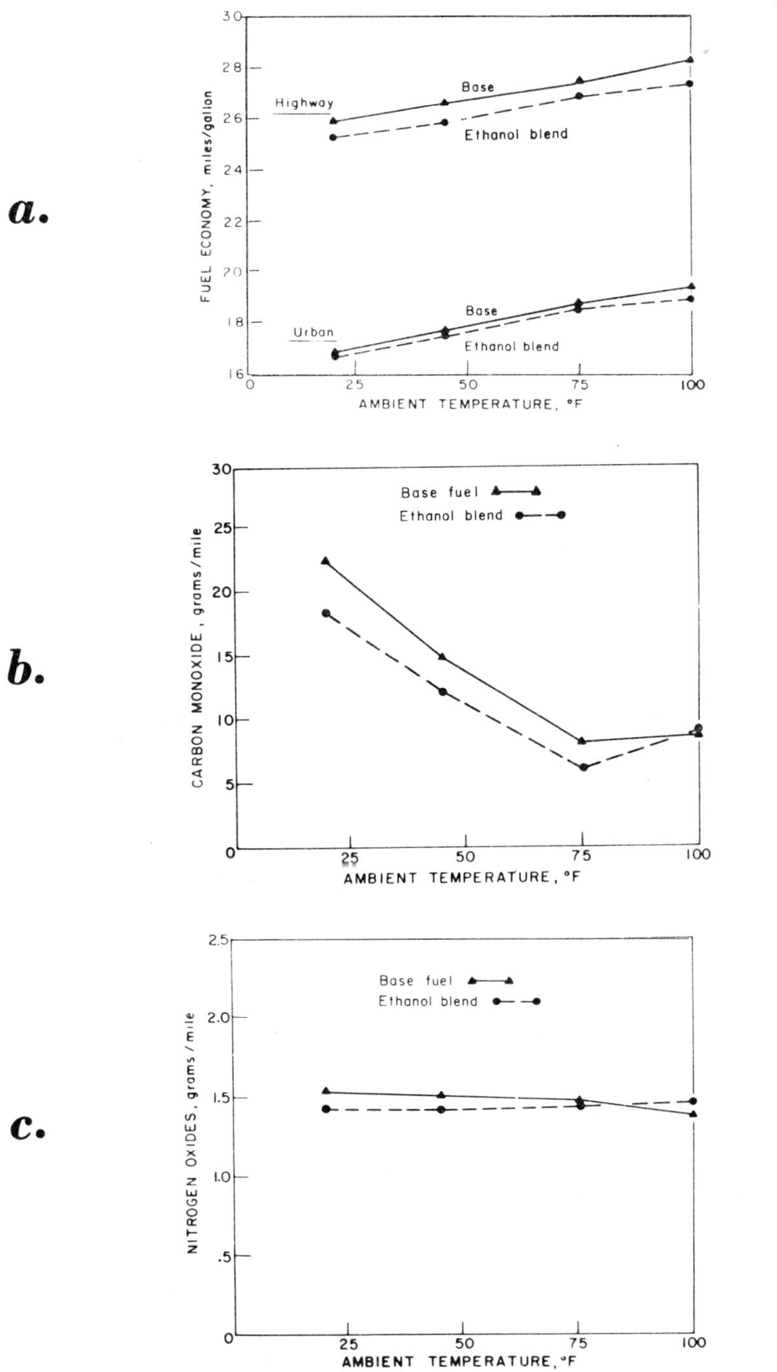

(continued)

Figure 10.24: (continued)

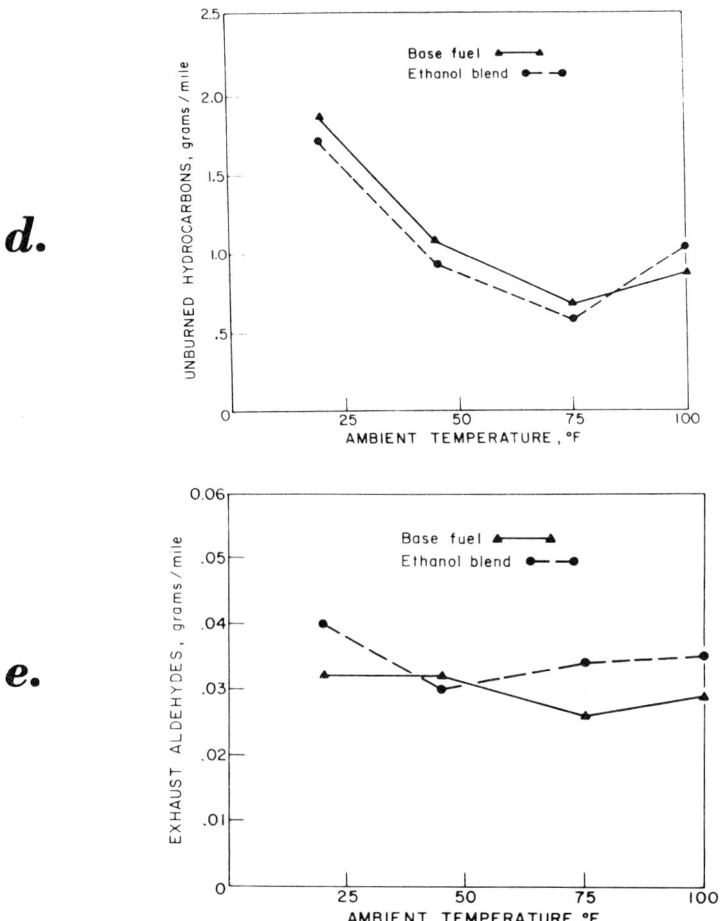

Source: DOE CONF-7805102

Table 10.11: Fuel Analyses

	Base Fuel	Base Fuel plus 10% Ethanol
FIA analysis, %:		
Aromatics	28	–
Olefins	8	–
Saturates	64	–
Distillation, ASTM D86, °F:		
IBP	90	90
5%	111	108
10%	124	118
20%	151	135
30%	184	148

(continued)

Table 10.11: (continued)

	Base Fuel	Base Fuel plus 10% Ethanol
Distillation, ASTM D86, °F:		
40%	210	262
50%	233	217
60%	256	244
70%	282	272
80%	312	301
90%	347	338
95%	380	373
EP	416	410
Specific gravity	0.746	0.751
Reid vapor pressure, psi	9.7	10.9

Source: DOE CONF-7805102

The mass emission rate of NO_x from the fleet CVS (constant volume sampling) emission/fuel economy tests (1975 Federal test procedure) indicates that a slight reduction can be achieved by using the 10% ethanol blend when ambient temperatures are 75°F or below (Figure 10.24c). These data also indicate that there is a slight increase in the NO_x emission rates for the fleet when operating at ambient temperatures of 100°F on the 10% ethanol blend.

The addition of ethanol to the base fuel produced a slight reduction in the unburned HC emissions from the fleet operation over the urban cycle at the three lower ambient temperatures (Figure 10.24d). This reduction comes primarily from the cold start portion of the cycle and suggests that the increased vapor pressure of the blend improves the startability at moderate-to-low temperatures and produces a net decrease in the HC emissions. Tests made at the 100°F ambient temperature indicate that the ethanol/gasoline blend increased the emission rates of unburned HC.

There are no significant effects in the aldehyde emission rates produced by the introduction of ethanol to the base fuel (Figure 10.24e). Typically, the aldehyde emissions constitute less than 6% of the unburned HC mass emissions. The aldehyde emission rates appear to be independent of ambient temperature in the range selected (20° to 100°F).

The emission rates of unburned ethanol show the same ambient temperature dependence as the total unburned HC (Figure 10.25). The mass emissions of unburned ethanol consistently make up approximately 3 to 4% of the total unburned HC emissions for the four ambient temperatures selected.

Road octane tests were run on four of the five test vehicles using three base fuels blended with 0, 5, and 10% ethyl alcohol. The three base fuels had research octane ratings of 81, 86, and 91 RON. The tests were run in accordance with a modified Uniontown technique, except that the tests were conducted using a chassis dynamometer. Triplicate tests were run for each vehicle/ fuel combination.

Figure 10.25: Unburned Ethanol Emissions at Four Ambient Temperatures

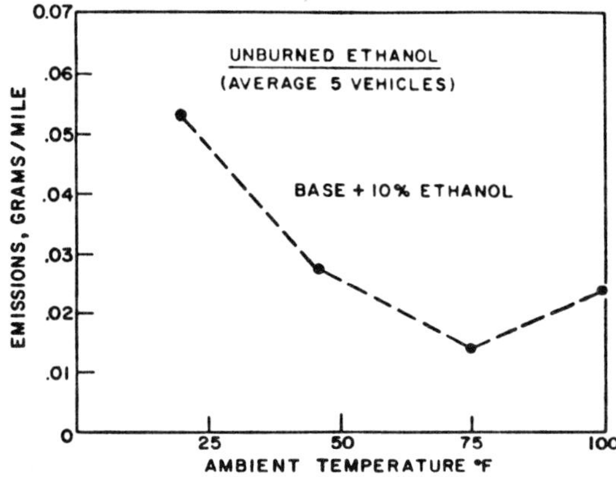

Source: DOE CONF-7805102

The road octane data (Figure 10.26) indicate that, for an 81 RON base fuel, an increase of one-half road octane number per percent ethanol addition can be expected. As the research octane of the base fuel increases, the improvement in the road octane is not as great. The 91 RON base fuel showed an increase of one-third road octane number per percent ethanol addition. The average road octane blending value of ethanol ranged from 122 for the 81 RON base fuel to 117 for the 91 RON base fuel.

Figure 10.26: Road Octane Response to Ethanol in Three Base Fuels

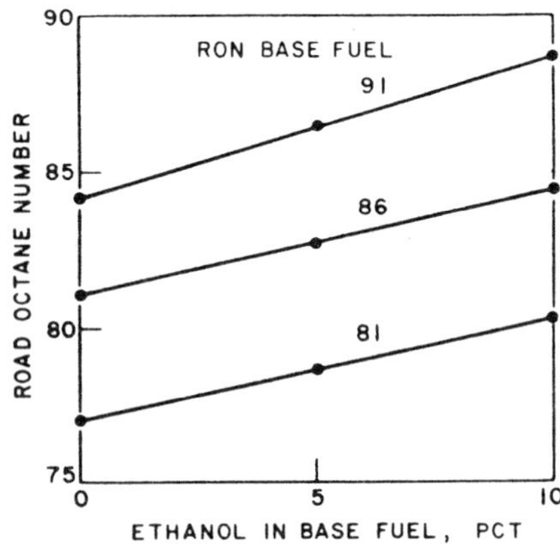

Source: DOE CONF-781050

Physical Properties of Alcohol Blends: A recent study was conducted at BERC to determine the distillation characteristics, vapor pressure, and octane response of several alcohol/gasoline blends. The alcohols which were used in the study

include methanol, ethanol, normal propanol, isobutanol, and two simulated
methyl fuels. The vapor pressure study was conducted with two base gasolines,
one with a 495 mm Hg vapor pressure and the other with a 678 mm Hg vapor
pressure. The alcohols were added in 5, 10, and 15 weight percent concentrations. The methanol blend showed the greatest vapor pressure increase (Figure
10.27a). As the carbon number of the alcohol increased, the vapor pressure of
the blends approached that of an ideal solution according to Raoult's Law. The
methyl fuels were simulated by a mixture of ethanol, n-propanol, and isobutanol
in 2-3-5 proportions. Then 10 and 30 weight percent of this mixture was added
to methanol to produce the fuels referred to as "methyl 10" and "methyl 30".

The vapor pressures of the methyl fuel/gasoline mixtures lie in between those
of the methanol/gasoline blend and the ethanol/gasoline blend. The methyl fuel-
10 shows the greater vapor pressure increase, as could be expected from the
higher concentration of methanol. The high vapor pressure gasoline showed relatively the same results as those with the lower vapor pressure gasoline except
that the absolute increase in vapor pressure was not as large (Figure 10.27b).
Whereas the n-propanol and isobutanol/gasoline blends showed slight increases
in vapor pressure at 5 weight percent concentrations in the lower vapor pressure
gasoline, with the higher vapor pressure base gasoline both show slight decreases
in vapor pressure.

Figure 10.27: Microvapor Pressure Measurements

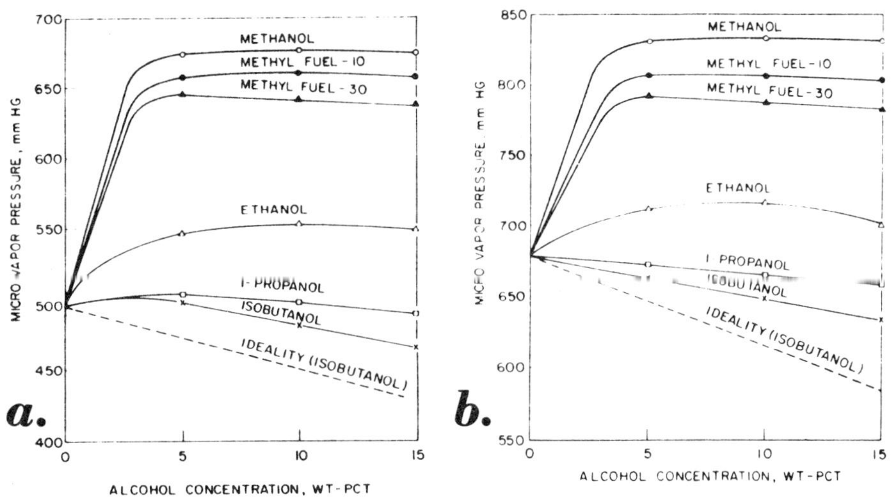

   (a)   Alcohols in solution with a low vapor pressure gasoline
   (b)   Alcohols in solution with a high vapor pressure gasoline

Source: DOE CONF-781050

The distillation characteristics of the alcohol/gasoline blends were determined
for the two gasolines used in the vapor pressure study, each with 5, 10, and
15 weight percent of methanol, ethanol, n-propanol, isobutanol, and the two
methyl fuels. For the sake of brevity, however, only the distillation of the 15%

alcohol solutions in the low vapor pressure (higher boiling point front end) gasoline will be discussed. The results of the other distillations follow the same trend as that of the 15 weight percent alcohol distillation.

The distillation curves are plotted as function of temperature versus volume percent distilled (Figure 10.28). Since the 15 weight percent alcohol addition displaces approximately 15% of the original volume of gasoline, the alcohol/gasoline distillations can be expected to deviate from that of the base gasoline by the fact that the fraction of blended fuel recovered would be expected to be less than that of the base fuel until the boiling point of the blended alcohol is reached.

Figure 10.28: Distillation Characteristics of Alcohols in Solution with a Low Vapor Pressure Gasoline

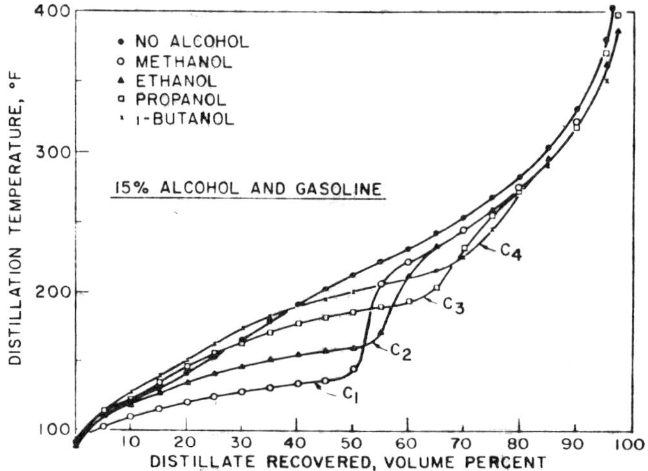

Source: DOE CONF-781050

Above the boiling point of the blended alcohol, the distillation temperature of the blended fuel could be expected to be lower than that of the base gasoline. Therefore, one result of the alcohol addition is to raise the front end of the distillation curve. This tendency is offset by the alcohol forming azeotropes with the hydrocarbon components in the gasoline. Methanol and ethanol form azeotropes with the lower boiling point hydrocarbons. Therefore, the C_1 and C_2 alcohol curves cross that of the base gasoline early in the distillation process.

The normal propanol and isobutanol form azeotropes with higher boiling point hydrocarbon compounds. This results in the distillation curves for the C_3 and C_4 alcohol/gasoline blends being initially higher due to the dilution of the base fuel and cross the distillation curve of the base gasoline prior to boiling point of the alcohol.

The octane measurements of the alcohol/gasoline blends were made by both the motor and research methods using two 1976 CRC octane number requirement fuels. The alcohols and methyl fuels were blended with these fuels in 5, 10, and 15 weight percent concentrations. The research octane number measurements showed that each of the alcohols was effective in raising the octane numbers of the base gasolines (Figure 10.29a). The octane improvement was most pronounced with the lower octane gasoline. Each of the alcohols used in the tests showed a nearly linear relationship between octane number and weight fraction of alcohol in the blend. The largest research octane improvement was found in methanol. Successively less octane improvements were found as the carbon number of the alcohol increased.

Figure 10.29: Octane Ratings

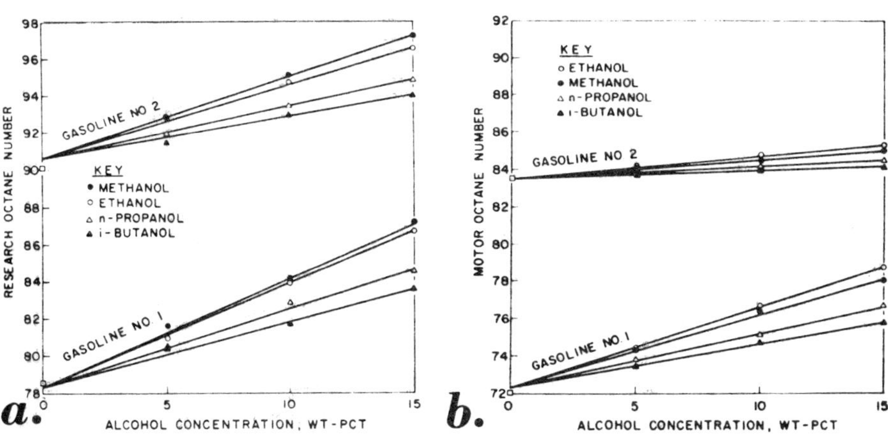

(a)   Research octane rating of alcohols in solution with two 1976 CRC octane requirement gasolines

(b)   Motor octane rating of alcohols with two 1976 CRC octane requirement gasolines

Source: DOE CONF-781050

The motor octane ratings for the alcohols blended with the two base gasolines also show increases for each of the alcohols selected (Figure 10.29b). However, the alcohols are not as effective at improving motor octane as they are research octane. A comparison of the motor octane increase of the base gasolines indicates that the alcohols are much more effective as octane boosting agents in low octane base gasoline. The motor octane measurements show ethanol to be slightly better than methanol for increasing the motor octane in both base gasolines. This situation is reversed from the research octane measurements. The normal propanol and isobutanol are less effective motor octane blending agents per weight percent alcohol in the blend.

The methyl fuels were also evaluated as octane improvers in the two base gasolines. In general, it was found that the octane improvements could be predicted from the weight fraction of alcohols in the methyl fuel and the blending octane value of the individual alcohol components.

Brazilian Engine Calibration Test Program

The information in this section is based on "Brazilian Vehicle
Calibration for Ethanol Fuels" by G.K. Chui, R.D. Anderson
and R.E. Baker of Ford Engineering and Research Staff, USA
and F.B.P. Pinto of Ford of Brazil in the proceedings of the
*Alcohol Fuels Technology Third International Symposium—
Volume II,* Asilomar, California, May 28-31, 1979 (hereinafter
referred to as Asilomar Conference).

Recent energy shortages have demonstrated the need for decreasing dependence
upon petroleum as an automotive fuel source. In Brazil, 70% of the petroleum
used for processing fuels and lubricants is imported, generating a negative effect
on the balance of trade. To alleviate the situation, the Brazilian government has
initiated a national program to supplement gasoline with ethanol. At the present
time a blend of 20% ethyl alcohol and 80% gasoline is being used in the areas
close to the alcohol production centers and the current plan is to have this blend
available nationwide by 1981. In addition, several fleets are already operating
on ethanol alone. The government plans to make available neat ethanol at gas
stations in selected areas by 1981 so that customers can buy vehicles for ethanol
or gasoline. It is the government's objective to have 20% of the 1982 automobile
production volume running on neat ethanol fuel.

Some modifications of engines and vehicles are necessary to use ethanol and
ethanol-gasoline blends. A test program was conducted to determine the extent
to which ethanol and ethanol blends could be accommodated by engine calibration adjustments with minimal hardware modifications. Three Brazilian vehicles
were used in this program: a 1976 1.4 liter sedan, a 1978 1.4 liter sedan and
a 1977 2.3 liter light truck (Table 10.12).

Table 10.12: Descriptions of Brazilian Test Vehicles

	1978 Sedan	1976 Sedan	1977 Truck
Vehicle weight (lb)	2,300	2,300	3,400
Test inertia			
weight class (lb)	2,500	2,000	3,500
Transmission (m)	4	4	4
Axle	4.13:1	4.13:1	5.38:1
Engine (liter)	1.4	1.4	2.3
Compression ratio	8.0:1	8.0:1	7.8:1

Source: Asilomar Conference

Fuel Properties: The base gasoline used in this study was blended from U.S.
stocks to match properties of regular grade gasoline in Brazil and is referred to
as Brazilian gasoline hereafter. Relevant properties are listed in Table 10.13
for this Brazilian gasoline and for the U.S. emissions test fuel Indolene. The
ethanol and blends of 20% ethanol in Brazilian gasoline used in this study are
also described in Table 10.13. The ethanol was denatured and was commercially
available. While this specific composition may not match Brazilian fermentation
ethanol precisely, it was believed to be adequate for this engine test program.

Properties which depend more upon chemical compostion, such as material compatibility or phase stability, could not be evaluated with this ethanol formulation.

Table 10.13: Fuel Properties

| | . . . . Gasoline. . . . . | | 20% Ethanol |
	Brazil	Indolene	Ethanol	80% Brazil Gas
Specific gravity, 60°/60°F	0.747	0.743	0.795	0.754
RVP (100°F), psi	6.1	8.8	2.5	6.5
Distillation, °C				
IBP	38	32	78	42
10%	62	57	--	58
50%	107	107	–	74
90%	179	155	–	176
EP	218	196	–	207
RON	75	97	~110	87
MON	71	88	89	77
FIA, % vol				
Aromatics	25	30	–	20
Olefins	8	7	–	7
Saturates	67	63	–	53
OH	–	–	100*	20
Lead, g/l	0.129	0.004	–	0.103
Sulfur, wt %	0.105	0.02	–	0.084
H/C	1.87	1.86	3.0	2.00
O/C	–	–	0.5	0.05
Heat of combustion, LHV, kJ/kg	~44,100	~44,100	~26,950	–
Heat of vaporization, kJ/kg	~280	~280	~920	–
Stoichiometric A/F	14.57	14.6	8.98	13.76

*Fuel grade ethanol composition: 93% ethanol, 4% methanol, 1% ethyl acetate, 1% gasoline and 1% methyl ethyl ketone.

Source: Asilomar Conference

Properties which make ethanol distinct from gasoline as a liquid fuel include lower heat of combustion, different stoichiometric air-fuel ratio, higher heat of vaporization and a single boiling point rather than a full distillation range. The low heat of combustion implies that a larger amount of fuel must be used to provide the same total energy in the combustion chamber. If driving range must be maintained, an ethanol fueled vehicle will require a larger fuel tank to accommodate this difference in heat of combustion. Substituting ethanol for gasoline without carburetor adjustments does not change the metered air-fuel ratio a great deal since the densities are similar.

However, since the stoichiometric air-fuel ratio is considerably lower for ethanol than for gasoline, the equivalence ratio is very different. Adding ethanol to gasoline causes the equivalence ratio to become progressively leaner. Many of the effects observed with blends or neat ethanol can be attributed to this change in equivalence ratio. Fuel supply and metering systems must be increased in capacity and throughput when alcohols are used in order to maintain equivalent engine operation. Liquid fuels having high heats of vaporization, such as ethanol, may not be completely vaporized in the intake system unless additional heat is supplied from external sources. Liquid fuel in the intake system and in the combustion chamber can create air fuel distribution and combustion problems. This is perceived as poor driveability, particularly in cold weather.

The single boiling point characteristic of ethanol aggravates the problem of mixture preparation. Gasoline contains components with a wide range of volatility so that an acceptable amount of vaporization for combustion can be obtained over all operating conditions without creating unmanageable amounts of vapor in the liquid handling systems. In contrast, ethanol is difficult to evaporate in sufficient quantities at low temperature, but once high temperatures are achieved, it may produce too much vapor in the liquid system, causing fuel pump vapor lock. Such problems often can be resolved by appropriate design revisions to accommodate unique fuel properties.

The vapor pressure of neat ethanol is very low compared with gasoline. However, when blended with gasoline up to about 25%, the more polar ethanol decreases the solubility of light hydrocarbons and thus increases the vapor pressure of the blend.

The intake heat requirement to vaporize ethanol is compounded by its lower heating value. Five times more heat to the intake air-fuel mixture is required to provide the same mixture quality with neat ethanol as with gasoline. This includes a factor of 3.3 for heat of vaporization and a factor of 1.6 for the additional fuel required to provide the same amount of energy. Heat requirements for equivalent mixtures using blends fall between those of gasoline and neat ethanol. Substantial revisions to the intake mixture heat supply are required to achieve equivalent mixture quality using fuels which contain large amounts of alcohol.

Measurement of Unburned Fuel Emissions: Vehicle emission tests were conducted on chassis dynamometers using the federal emission test procedures, which were developed for use with gasoline as the fuel. With ethanol, the exhaust gas of an engine contains unburned ethanol and partial oxidation products of ethanol such as aldehydes.

Additional instrumentation was needed to measure these exhaust gas constituents. The flame ionization detector (FID) commonly used for exhaust hydrocarbons does not respond fully to either alcohols or aldehydes. A correction can be applied only if the relative amounts of the unburned fuel constituents and their FID response are known. Data in the following sections reported as exhaust HC were obtained with an FID and no corrections were applied.

Therefore, these values must be used with caution since the discrepancy increases with ethanol fuel content. FID HC are an adequate representation of total unburned fuel emissions for gasoline and a 20% ethanol blend. At an ethanol content of 80% or higher in the fuel, a significant amount of unburned fuel was not measured by the FID.

Measurements were made of unburned ethanol and aldehydes to determine the magnitude of the underestimation by the FID and to examine the impact of ethanol fuel on air quality. Samples from tests on the 1976 1.4 liter vehicle with Brazilian gasoline, low and high content blends and neat ethanol were collected proportionately from the dilute stream during cruises and CVS-H tests. The ethanol samples were obtained from a cold trap and on a polymer absorbent. The ethanol was detected with a computer controlled gas chromatograph/ mass spectrometer (GC/MS). Aldehydes were collected in dual impingers and analyzed by the 3-methyl-2-benzothiazolone hydrazone (MBTH) method for total aliphatic aldehydes. The aldehydes resulting from ethanol combustion were expected to be primarily aliphatic.

A simplified procedure was available for analysis of the 1978 1.4 liter and 2.3 liter vehicle tests. A Fourier transform infrared spectroscopy method (FTIRS) had been developed to identify and quantify light hydrocarbons and other emission species. Using computer data reduction, the complete IR spectrum can be traversed and quantified readily. An example is given in Table 10.14 for exhaust composition from the 2.3 liter vehicle using neat ethanol.

Table 10.14: Exhaust Composition from Neat Ethanol by FTIRS and Other Methods

| | . . . . FTIRS . . . . | | |
Compound	Amount at 4 Hours (ppm)	Error	Amount at 26 Hours
Water	1.10*	0.2	0.52
Carbon dioxide	1.08*	0.2	1.07
Carbon monoxide	488.	18.4	490.
Heavy hydrocarbons (C_{6+})	0.3**	14.7	0
Nitric oxide	14.4	0.6	4.0
Nitrogen dioxide	46.9	0.6	31.5
Nitrous oxide	0.4	0.2	0.4
Nitrous acid	2.4	0.2	2.2
Hydrogen cyanide	0.1	0.4	0
Ammonia	0	0.2	0.1
Sulfur dioxide	0.3	0.4	0.2
Methane	8.7**	0.2	8.8
Acetylene	4.6**	0.4	4.5
Ethylene	26.0**	0.9	26.0
Ethane	1.8**	0.4	1.7
Propylene	0.5**	1.8	0.6
Isobutane	3.0**	1.8	3.5
Formaldehyde	8.8**	0.2	7.8
Acetaldehyde	57.4**	1.6	55.6
Formic acid	0.6**	0.2	0.2
Methanol	7.6**	0.2	6.4
Ethanol	194.**	0.2	168.
Total NO_x	63.7	–	37.6
Total HC	313.3**	–	280.0
	. . . . Other Methods		
FID HC	240.6**	–	–
NDIR CO	463	–	–
Chemi NO_x	64.6	–	–
NDIR CO_2	1.04*	–	–
MBTH Aldehydes	46.4	–	–

*Percent.
**ppm carbon.

Source: Asilomar Conference

A variety of hydrocarbons are identified along with water, CO, CO_2 and NO_x. Methanol, ethanol, formaldehyde and acetaldehyde are also identified. One source for the methanol is the denaturant. Species with more than five carbon atoms tend to be lost in the C-H stretching absorption when gasoline is present, so the FTIRS is not suitable, in general, to determine total aldehydes. Total aliphatic aldehydes were also measured by MBTH from the same sample. The FTIRS recorded 80% of the total aldehydes in this case.

The FTIRS measurements agree adequately with the CVS bag concentrations of CO and CO_2 by NDIR and of NO_x by chemiluminescence. Assuming no loss of sample in the bag, the comparison of unburned fuel concentration is a measure of the FID response. After adjusting the FTIRS total unburned fuel upward to account for the additional aldehydes observed by MBTH, the FID response was determined to be 0.71 for this test. This is consistent with the published response value for ethanol. The specific amount varies somewhat within the CVS-H tests conducted, and a correction upward of 20 to 40% appears necessary to obtain the total unburned fuel concentration from an FID measurement.

The calculation of unburned fuel mass flow requires an additional correction since the average molecular composition is not $CH_{1.85}$ as in gasoline, but about $CH_3O_{0.5}$, i.e., primarily ethanol and aldehydes. This, together with the adjustment for FID response, requires that unburned fuel mass emissions from a vehicle using neat ethanol, but measured and calculated as for gasoline, must be multiplied by a factor of 2.0 to 2.3 to obtain correct mass emissions. The FID HC data reported in the following sections do not include these adjustments. Sufficient information was not available in all cases and the value measured by the current test procedure was considered useful. However, for quantitative comparison of mass unburned fuel emissions, the adjustments described above are necessary.

The FTIRS is a laboratory instrument and samples must be transported to it from the vehicle test location. CVS sample bags made of Tedlar were used since previous tests had shown these were equivalent to direct sampling from a dilute stream. Significant loss of aldehyde or alcohol sample occurred only when sampling undiluted exhaust which resulted in condensate formation in the bag.

Table 10.14 shows an analysis of a sample extracted from the same bag after overnight storage. Formaldehyde dropped 12%, acetaldehyde 3%, methanol 16% and ethanol 14%. This provided reassurance that substantial quantities of the unburned fuel species were not being lost during transport from the test location to the instrument. Total NO_x and water vapor dropped to half the earlier values.

Minimal Adjustment to Accept Ethanol: To fully take advantage of ethanol as the engine has to be redesigned to utilize its high octane and extended lean limit properties. Such redesign should provide for higher compression ratios and the optimization of ignition timing and air-fuel mixtures. The objective of this study was to determine the extent to which the advantages could be realized by calibration adjustments on current engines with no major hardware revisions. Vehicle tests were conducted with minimal changes necessary to accept ethanol fuels and engine tests were run to develop optimum calibrations.

Direct Substitution of Ethanol Blend — Brazilian vehicles are calibrated to accept either gasoline or a 20% ethanol blend with no adjustments since both are available. Since the stoichiometric air-fuel ratio of ethanol is less than that of gasoline, a direct substitution of an ethanol blend gives a leaner mixture for combustion which, in turn, leads to changes of fuel economy, emissions and vehicle behavior. The effects of replacing Brazilian gasoline with a 20% ethanol blend in the same gasoline were consistent for all three Brazilian vehicles used in this study.

Table 10.15 presents results for the 1978 1.4 liter sedan. Leaning-out of the air-fuel mixture is indicated by a reduction of more than 10% in the fuel-air equivalence ratio, Φ_{FA}. This change was observed at idle and steady speeds as well as in the acceleration modes of the CVS driving cycles.

Table 10.15: 1978 1.4 Liter Sedan Tests

	Brazilian Gasoline	20% Ethanol 80% Brazilian Gasoline	. . . 100% Ethanol . . . Φ_{FA} Idle to Match 20%	Lean Idle
Equivalence ratio, Φ_{FA}				
Idle	0.98	0.86	0.84	0.67
48 kmph	1.07	0.90	1.01	0.92
CVS-H				
1st acceleration	0.97	0.92	0.97	0.92
2nd acceleration	1.00	0.96	0.94	0.91
Fuel economy				
MJ/km basis				
CVS-H	3.12	2.93	2.90	2.69
Road				
City	3.70	3.49	—	—
Suburb	2.60	2.43	—	—
l/100 km basis				
CVS-H	9.5	9.6	13.6	12.6
Road				
City	11.3	11.4	—	—
Suburb	7.0	7.9	—	—
Emissions CVS-H (g/mile)				
HC	3.32	2.44	2.53	2.50
CO	49.37	20.62	27.50	17.00
NO_x	1.86	2.86	1.86	1.66
Formaldehyde	0.05	0.07	0.07	0.4
Acetaldehyde	0.05	0.16	0.34	0.52
Methanol	0.02	0.03	0.14	0.08
Ethanol	—	0.19	2.98	1.19
Performance, sec, 0-96 kmph	18.0	16.9	—	—
Driveability, scale 1-10				
Crowds/cruises	7	5	—	—
WOT/PT acceleration	7	6	—	—

Source: Asilomar Conference

The leaning effects were also reflected in the improvement of energy economy both in the CVS cycle and on the road. Based upon energy consumed per unit distance traveled, the direct substitution of a 20% ethanol blend gave a 6% improvement. On the basis of volume per unit distance, results for the blend were slightly worse than those for gasoline because of the lower energy content in the ethanol.

Emissions from the vehicle also followed the trend of the equivalence ratio. With the 20% ethanol blend, both HC and CO were reduced substantially while NO_x was increased. As shown in Table 10.15, FTIRS aldehydes in the exhaust with the 20% ethanol blend as fuel were slightly higher than with Brazilian gasoline. Unburned ethanol was the more significant of the unregulated emissions when the 20% blend was used.

Performance, as measured by acceleration capability, was improved by direct substitution of gasoline by the 20% ethanol blend. The time needed to accelerate from 0 to 97 km/h was reduced by 1.1 seconds. As no hardware adjustment was made in the vehicle, the improvement could have been due to an overly rich calibration on gasoline and improved volumetric efficiency from the larger heat of vaporization of ethanol.

While the energy economy and power were improved, the driveability of the vehicle deteriorated. Slight surging and hesitation of the vehicle were observed during hot operation. This was probably due to the leaner equivalence ratio. A subsequent slight reduction in carburetor main jet size caused a severe loss in driveability, indicating high sensitivity to equivalence ratio in this range.

Operation on Neat Ethanol — The 1978 1.4 liter test vehicle could be operated on neat ethanol only after the metering jets in the carburetor were enlarged to provide a comparable equivalence ratio. Measurements of intake mixture temperature, A/F distribution and calibration settings helped to define the hardware revisions necessary to best utilize ethanol. The carburetor was adjusted for best economy using neat ethanol through a CVS-H driving cycle.

Cold start CVS tests were not attempted, since substantial intake manifold revisions were necessary before such tests would have been meaningful. While making the carburetor adjustments, comparable driveability was maintained on the chassis dynamometer throughout the emission test driving cycle. A few additional driving modes were added to simulate sensitive operating conditions for driveability. The equivalence ratio at idle was adjusted to match a 20% ethanol blend to compare economy and emissions.

Results in Table 10.15 show that even though the equivalence ratios at all operating conditions were not precisely the same, the CVS-H fuel economy of the two cases were equal on an energy basis but not on a volumetric basis. Data on regulated emissions showed small increases in HC and CO and a significant decrease in NO_x comparing the 20% blend results to neat ethanol. Total aldehydes, as measured by FTIRS, almost doubled while unburned ethanol increased more than ten-fold.

The idle air-fuel mixture misfire limit was leaner with ethanol than with gasoline. A CVS-H test was conducted with an equivalence ratio, Φ_{FA}, at idle equal to 0.67, Table 10.15, which is well below the misfire limit for an engine operated on gasoline. No driveability problems were experienced by the driver through the CVS cycle. The energy economy was improved by 8% while all emissions were reduced except acetaldehyde, which was increased by 50%. If acetaldehyde is of less concern than other emissions, these results illustrate the advantages of the lean burn characteristics of ethanol to the extent they can be realized without intake system modifications to increase mixture heat and improve air-fuel distribution.

Cold start, driveaway evaluations demonstrated high sensitivity to ambient temperature. A cold start was not possible with neat ethanol when the ambient temperature was below 5°C. 10% gasoline added to the fuel bowl extended cold start capability downward to 0°C, and larger quantities were needed at lower ambients. The warm-up period was prolonged compared with gasoline operation, and the choke was required indefinitely to enrich the ethanol/air mixture when the ambient temperature was below 5°C.

Difficulties with operating the ethanol fueled vehicle in cold weather were caused by the high heat of vaporization of ethanol. The only external heat source to the air-fuel mixture on the 1.4 liter engine was from a small contact area between the intake and exhaust manifolds directly under the carburetor. This was insufficient to compensate for the large amount of heat extracted from the intake air to vaporize the ethanol. Intake mixture temperature was depressed resulting in incomplete vaporization.

The temperature depression of the intake mixture is illustrated in Figure 10.30 which shows the temperature change of the air-fuel mixture relative to the carburetor inlet, at several locations and at different vehicle cruise speeds. For neat ethanol, the mixture dropped to the lowest observed temperature across the carburetor.

**Figure 10.30: Temperature Change of Air-Fuel Mixture in Intake Manifold
1.4 Liter 1978 Sedan**

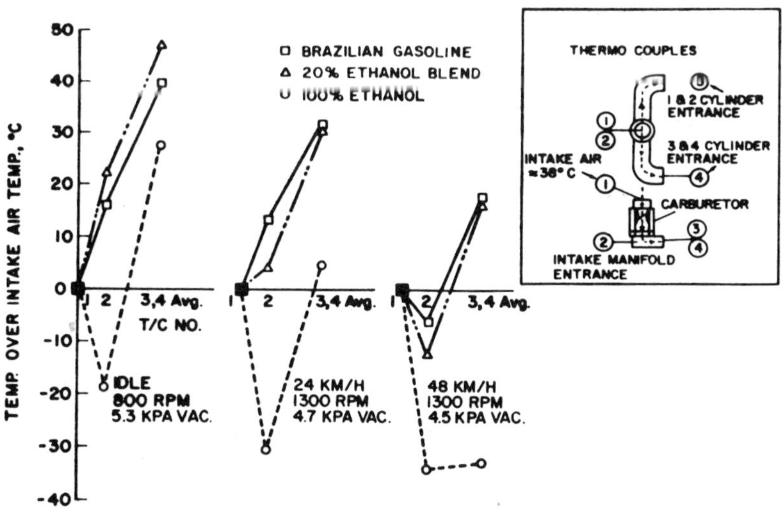

Source: Asilomar Conference

The idle fuel flow was sufficiently low and the residence time sufficiently high so that the mixture recovered more than its loss before entering the cylinders. As the air and fuel flow increased, the supply of external heat became inadequate and the temperature recovery was less. At 48 km/h and above, the initial temperature depression was more than 30°C. With the intake air at about 38°C, icing could occur in the carburetor and temperature recovery of the mixture in the manifold was negligible. In contrast, temperatures of the air fuel mixtures of both the Brazilian gasoline and the 20% ethanol blend were seldom colder than the intake air. Enough external heat was received to fully recover before entering the cylinders.

The problem of vaporization of ethanol also resulted in a more uneven distribution of equivalence ratio among cylinders. Emission samples from cylinders showed the maldistribution increased with vehicle cruise speed from an equivalence ratio spread of 0.06 idle to 0.20 at cruise 80 km/h. This could be attributed to the increase in air and fuel flow rates and the resultant temperature depressions. The distribution is expected to become worse at cold ambient temperatures. The drop in intake mixture temperature and the poor air-fuel distribution limit the extent to which the lean limit properties of ethanol can be utilized in existing engines.

High Ethanol Content Blends — A series of tests were conducted on a 1976 1.4 liter vehicle, to examine the effects of blends with ethanol as the primary component. Blends of 80% and 90% ethanol were compared with a 20% blend and with Brazilian gasoline and neat ethanol. Carburetor adjustments were made to match the equivalence ratio of the 80% blend to the 20% blend. Results of steady-state and CVS-H tests, Table 10.16, were similar to those obtained on the 1978 vehicle, Table 10.15. The 80% and 90% blends were similar to neat ethanol regarding emissions of MBTH aldehydes and unburned ethanol. The presence of gasoline in the fuel improved driveability, especially during warmup.

Equivalence ratio measurements made during the CVS-H tests explain some of the driveability problems experienced on neat ethanol. The equivalence ratio Φ_{FA}, was much leaner than expected at high mass flow rates, causing a significant increase in necessary throttle opening. This might have been due to the formation of vapor in the fuel metering passages. An equivalence ratio of 0.63 was observed with no misfire.

Table 10.16: 1976 1.4 Liter Sedan Emissions

| | .Cruise Tailpipe Emissions (48 km/h). | | | |
	Brazilian Gasoline (BG)	0.20 EtOH 0.80 BG	0.80 EtOH 0.20 BG	0.90 EtOH 0.10 BG	100% EtOH
FID					
HC, ppmc	3,105	1,890	1,552	1,215	675
CO, %	2.74	0.67	1.98	1.50	0.98
NO_x, ppm	—	—	—	—	—
O_2, %	0.8	1.2	0.85	1.1	1.5
CO_2, %	12.0	13.52	12.88	12.88	12.88
MBTH aldehyde, ppm	41.9	47.2	115.8	119.5	128.0
EtOH, ppm	—	32	136	—	168
A/F	13.58	14.06	9.95	9.71	9.27
Φ_{FA}	1.07	0.98	1.04	1.01	0.97
STOIC	14.57	13.76	10.33	9.80	8.98

(continued)

Table 10.16: (continued)

CVS-H Emissions

Fuel	Test	HC (g/mi)	CO (g/mi)	....Economy .... (l/100 km)	(MJ/km)	MBTH Aldehydes as Formaldehyde (g/mi)
Brazilian	1	2.43	60.1	8.74	2.85	0.06
gasoline	2	2.37	57.6	8.43	2.75	0.06
	3	1.79	45.9	9.08	2.97	—
	4	1.84	49.5	8.91	2.90	—
Indolene,	1	1.75	42.3	8.88	2.99	—
clear	2	1.75	52.1	9.01	2.94	—
20% EtOH	1	1.84	32.5	8.84	2.69	0.06
80% BG	2	1.75	32.3	8.78	2.66	0.06
80% EtOH						
20% BG	1	1.69	34.0	12.38	2.91	0.24
100% EtOH	1	1.87	20.1	13.75	2.93	0.26
	2	1.64	17.9	13.68	2.91	0.25

Source: Asilomar Conference

Calibration Development for Neat Ethanol: Results discussed previously deal with the effects of carburetor adjustments with no attempt to change the spark timing. In a separate study, the 2.3 liter engine from a 1977 truck was installed on an engine dynamometer to calibrate spark timing and carburetion simultaneously for best power and best efficiency. Automatic calibrations were then developed and evaluated in the test vehicle.

Engine Calibrations — Engine dynamometer tests were conducted on the 1977 2.3 liter engine for maximum torque to determine the increase possible with ethanol. Spark timing on the production engine must be retarded from MBT because of the low octane of Brazilian gasoline. Figure 10.31 shows knock-limited (KL) torque and thermal efficiency of the engine using Brazilian gasoline and the improvements possible with best fuel and spark timing settings for a 20% blend and for neat ethanol.

The 20% blend was also knock-limited, requiring some spark retard, while MBT was readily obtained with neat ethanol. The torque and efficiency improvements are due primarily to the higher fuel octane. Since the blend included a leaded octane improver at the same level as in the Brazilian test gasoline, removal of lead from the blend will decrease the knock-limited torque.

Maximum part throttle efficiency was determined at several speed/load combinations. Figure 10.32 shows the dependence of efficiency on equivalence ratio with MBT spark timing for an intermediate torque level over a range of engine speeds. Best efficiency was typically between 0.8 to 0.9 Φ_{FA} with enrichment for maximum torque.

Automatic calibrations were developed from the best economy and best torque spark timing and carburetor settings. Figure 10.33 shows centrifugal and vacuum advance curves which closely match most of the points. However, the hardware to implement this calibration was not available for the vehicle test program.

Figure 10.31: Best Torques and Efficiences at WOT–2.3 Liter Engine

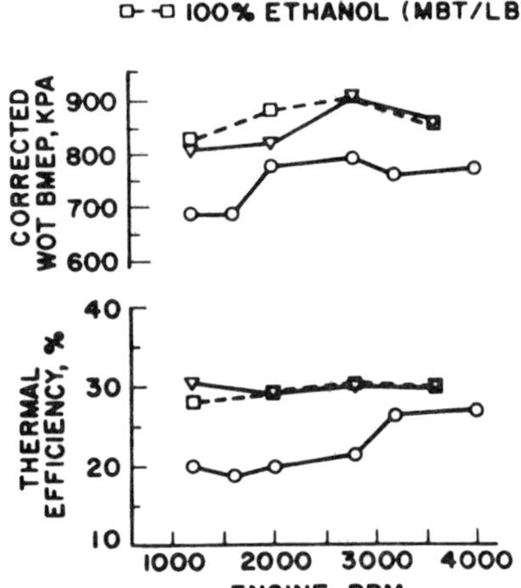

Source: Asilomar Conference

Figure 10.32: Change of Thermal Efficiency with Equivalence Ratio at Part
Throttle, MBT Spark–2.3 Liter Engine

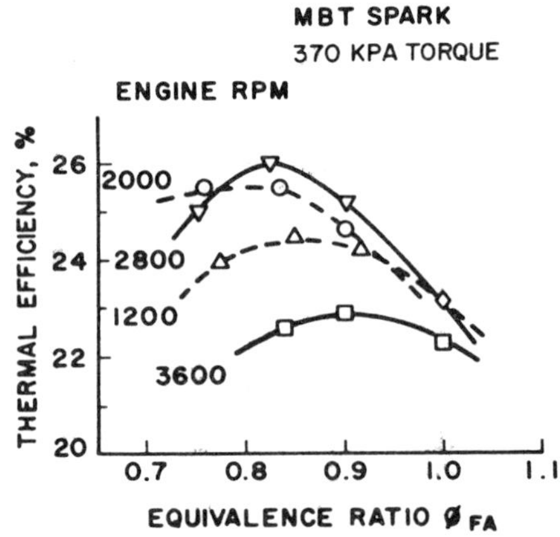

Source: Asilomar Conference

Figure 10.33: Best Economy Spark Timing, 100% Ethanol–2.3 Liter Engine

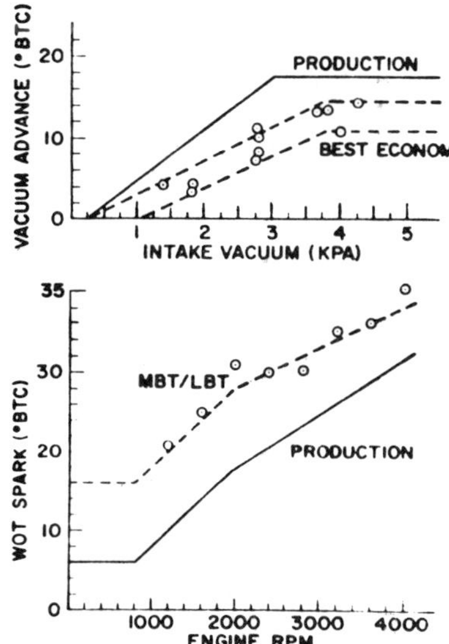

Source: Asilomar Conference

The production curves shown in Figure 10.33 were used in the vehicle tests. Compromises were also made in the carburetor settings. Interaction among the idle, main and power enrichment systems prevented matching the desired equivalence ratios under all conditions; light loads were richer than desired and WOT was somewhat leaner, causing some torque loss, Figure 10.34. Further carburetor modifications could have reduced the compromise, but were not available for the vehicle tests.

Figure 10.34: Comparison of Auto and Best Economy Operations

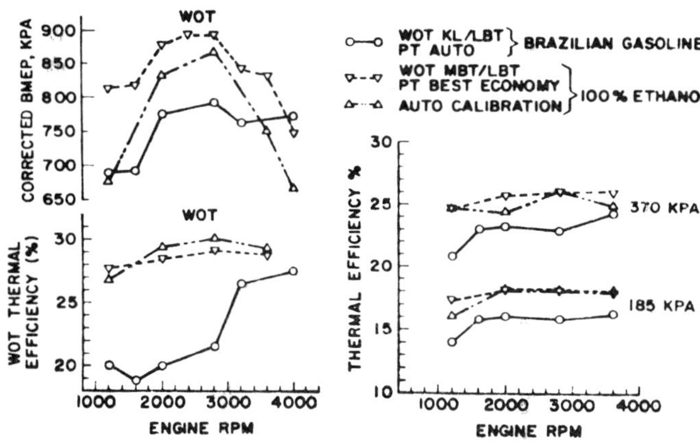

Source: Asilomar Conference

Vehicle Tests — CVS-H tests were conducted on the 2.3 liter engine in a truck test vehicle using Brazilian gasoline with production carburetor and distributor settings, followed by tests with direct substitution of a 20% ethanol blend, Table 10.17. The carburetor developed for neat ethanol was then installed and the CVS-H tests were repeated. Idle and light load equivalence ratio were found to be much richer than expected, as can be seen from the CO results. Carburetor settings were then restored to the best economy values and the idle was set leaner, providing the final data in Table 10.17.

Table 10.17: 2.3 Liter Truck CVS-H Tests

	Brazilian Gasoline	20% Ethanol Blend Substitution	100% Ethanol Automatic . . . Calibration. . . . Power	Economy
Fuel economy				
MJ/km	5.27	4.78	5.30	4.80
l/100 km	16.01	15.68	24.83	22.48
Emissions, g/mi				
HC	5.44	4.65	3.56	3.67
CO	94.47	63.51	89.25	15.14
NO_x	2.72	3.01	1.59	2.85
Formaldehyde	0.87	0.13	0.16	0.31
Acetaldehyde	0	0.10	0.85	1.49
Methanol	0	0.03	0.20	0.29
Ethanol	0	0.11	4.20	5.24

Source: Asilomar Conference

While vehicle performance could be improved by further carburetor and distributor hardware improvements, this fuel economy is believed to be representative of the possible improvements with ethanol. The CVS-H thermal efficiency was improved by 10%, bringing the volumetric fuel economy to within 70% of the base gasoline economy from 65%, as expected from the ratio of the fuel heating values.

Summary and Conclusions: The major effects on engine operation from the use of ethanol fuels can be anticipated from the properties of the fuels. The difference in stoichiometric air-fuel ratio has the effect of shifting the equivalence ratio towards the lean side. Exhaust emissions, fuel economy and driveability reflect this shift. The heating value of the fuel is reduced also, decreasing driving range. The larger heat of vaporization provides some NO_x and octane advantage but it also delays warmup and can drop the intake temperature so low that poor mixture quality and air-fuel distribution cause driveability problems. Volatility characteristics of ethanol are not well suited to current vehicles. However, these problems can be overcome with revisions to the engine and modifications of the fuel.

Most Brazilian vehicles are calibrated to accept either gasoline or a 20% ethanol blend. This calibration must be a compromise between the two fuels and is not optimum for either; operation on gasoline is too rich for best economy and operation on the blend is close to driveability limits.

Operation on neat ethanol was possible after the carburetor metering jets were enlarged to provide a comparable equivalence ratio. Increased power and thermal efficiency were obtained by taking advantage of the higher octane and lean limit properties of ethanol. Lean operation was limited by the depressed intake temperatures, the low vapor/liquid ratio and poor air-fuel distribution. Further improvements obtained by hardware changes to increase compression ratio and improve air-fuel mixture preparation and distribution were beyond the scope of this study.

Optimum calibrations developed on an engine dynamometer were not fully realized in the vehicle due to compromises made by implementing automatic calibrations in the production carburetor and distributor.

Cold start was not possible on neat ethanol at ambient temperatures below 5°C. Even at stabilized engine temperatures, enrichment was required for acceptable driveability at ambient temperatures below 5°C to compensate for the inadequate heat provided by the production manifold. The addition of 10% gasoline to the ethanol extended cold start ability down to 0°C and improved cold driveability.

The use of ethanol fuel decreases NO_x and increases unburned fuel exhaust emissions when compared with gasoline at the same equivalence ratio. CO emissions were similar unless increased by poor air-fuel distribution. The most significant differences in the unburned fuel emissions are the presence of unburned ethanol and substantial amounts of aldehydes, principally acetaldehyde. The impact of these emissions on Brazilian air quality must be determined.

REFERENCES

(1) Adt, R.R., Jr., et al, *Effects of Blending Ethanol with Gasoline on Automotive Engines' Steady State Performance and Regulated Emissions Characteristics,* Topical Report on U.S. DOE contract E(40-1)-5216, Department of Mechanical Engineering, University of Miami, Coral Gables, Florida, (January, 1978).

(2) Adt, R.R., Jr., et al, "The Effect of Up to 30 Volume Percent Methanol Addition per se on the Basic Performance and Exhaust Emissions Characteristics of a Carbureted Spark Ignition Engine," paper presented at International Symposium on Automotive Technology and Automation, Rome, Italy, (September, 1976).

(3) Pefley, R.K., Chairman, Department of Mechanical Engineering, University of Santa Clara, Santa Clara, California, letter to E.E. Ecklund, Chief, Alternative Fuels Utilization Branch, U.S. Department of Energy, (October 13, 1977).

(4) Furey, R.L. and Jackson, M.W., "Exhaust and Evaporative Emissions from a Brazilian Chevrolet Fueled with Ethanol-Gasoline Blends," General Motors Corporation, Warren, Michigan, (1976).

(5) Scheller, W.A., "Tests on Unleaded Gasoline Containing 10% Ethanol—Nebraska Gasohol," paper presented at the International Symposium on Alcohol Fuel Technology, Wolfsburg, FRG, (November 21-23, 1977).

(6) *Gasohol Test Vehicles,* Interim Report prepared by Bartlesville (OK) Energy Research Center, U.S. Department of Energy, (August, 1977).

(7) Bernhardt, W.E., et al, "Recent Progress in Automotive Alcohol Fuel Application," Volkswagenwerk AG, Wolfsburg, FRG, Fourth International Symposium on Automotive Propulsion Systems (NATO/CCMS), Washington, DC, (April, 1977).

(8) Trindade, S.C., Centro de Tecnologia Promon (Brazil), letter to E.E. Ecklund, Chief, Alternative Fuels Utilization Branch, U.S. Department of Energy, (February, 1977).

(9) Scheller, W.A., *Nebraska 2-Million Mile Gasohol Road Test Program,* Sixth Progress Report, University of Nebraska, Lincoln, Nebraska (January, 1977).

(10) Fricke, C., Administrator, (NE) Agricultural Products Industrial Utilization Committee, "Gasohol—Food and Fuel for the Future," *Nebraska Oil Jobber,* (December, 1975).

(11) "Ethanol Materials Compatibility," E.E. Ecklund, Chief, Alternative Fuels Utilization Branch, U.S. Department of Energy, quoting Prof. U.E. Stumpf, Brazilian Center for Aerospace.

(12) Scheller, W.A., Chairman, Department of Chemical Engineering, University of Nebraska (Lincoln), "The Use of Ethanol-Gasoline Mixtures for Automotive Fuels," Symposium on Clean Fuels from Biomass and Wastes (IGT), Orlando, Florida (January 25-28, 1977).

(13) *Alcohols, A Technical Assessment of Their Application as Fuels,* API Publication No. 4261, (July, 1976).

(14) Biller, W.F., "Alcohols as Fuels. A Technical Evaluation," unpublished draft report prepared for American Petroleum Institute, Washington, DC (December, 1974).

(15) Keller, J.L., et al, *Methanol Fuel Modification for Highway Vehicle Use,* Monthly Report 13 for U.S. DOE contract EY-76C-OA-3683, Union Oil Company of California, Brea, California, (September, 1977).

(16) Panchapakesan, N.R., et al, "Factors That Improve the Performance of an Ethanol-Diesel Oil Dual Fuel Engine," Indian Institute of Technology, Madras, India, paper presented at the International Symposium on Alcohol Fuel Technology, Wolfsburg, FRG, (November 21-23, 1977).

(17) Hagey, G. and Parker, A.J., et al, "Methanol and Ethanol Fuels—Environmental, Health and Safety Issues," U.S. Department of Energy and Mueller Associates, Inc., (respectively), paper presented at the International Symposium on Alcohol Fuel Technology, Wolfsburg, FRG, (November 21-23, 1977).

(18) National Fire Protection Association, *Fire Protection Handbook,* 13th Edition, (1969).

(19) Chemical Rubber Company, *Handbook of Laboratory Safety,* (1967).

SOURCES UTILIZED

CAEC-28
The Production of Ethanol from Agricultural Waste—An Economic Evaluation,
prepared by D. Paige and R. Boulton of the University of California, Davis,
for the California Energy Commission, January 1979.

DOE ALO/3729-1
Systems Study of Fuels from Grains and Grasses, prepared by W. Benson, A.
Allen, R. Athey, A. McElroy, M. Davis and M. Bennett of Midwest Research
Institute for the U.S. Department of Energy, February 1978.

DOE CONF-781050
Proceedings of Highway Vehicle Systems Contractors' Coordination Meeting,
Dearborn, Michigan, October 1978, issued by the U.S. Department of Energy,
March 1979.

DOE CONF-7805102
Highway Vehicle Systems Contractors' Coordination Meeting, Troy, Michigan,
May 1978, issued by the U.S. Department of Energy, September 1978.

DOE HCP/ET-2854
Comparative Economic Assessment of Ethanol from Biomass, prepared by The
Mitre Corporation, Metrek Division for the U.S. Department of Energy, Sep-
tember 1978.

DOE HCP/M2923-01
Status of Alcohol Fuels Utilization Technology for Highway Transportation,
prepared by Mueller Associates, Inc. for the U.S. Department of Energy,
June 1978.

DOE HCP/T4101-03
Biomass-Based Alcohol Fuels: The Near-Term Potential for Use with Gasoline,
prepared by W. Park, G. Price and D. Salo of The Mitre Corporation, Metrek
Division, for the U.S. Department of Energy, August 1978.

DOE HCP/W1737-01
Comparative Automotive Engine Operation When Fueled with Ethanol and Methanol, prepared by The University of Santa Clara, The University of Miami and E.E. Ecklund of the U.S. Department of Energy for the U.S. Department of Energy, May 1978.

DOE PE-0012
The Report of the Alcohol Fuels Policy Review, prepared by the U.S. Department of Energy, June 1979.

ERDA E(49-18)-2081
Silvicultural Biomass Farms, Volume I—Summary, prepared by R.E. Inman of The Mitre Corporation, Metrek Division, for the Energy Research and Development Administration, May 1977.

NSF 760009
Survey of Alcohol Fuel Technology. Volume I, prepared by B. Baratz, R. Oulette, W. Park and B. Stokes of Mitre Corporation for the National Science Foundation, November 1975.

NTIS CONF-771175
Proceedings: International Symposium on Alcohol Fuel Technology—Methanol and Ethanol, November 21-23, 1977, Wolfsburg, Federal Republic of Germany, English translation published by the U.S. Department of Energy, July 1978.

NTIS PB-290-569
Gasohol Test Program, by R. Lawrence of the Technology Assessment and Evaluation Branch of the U.S. Environmental Protection Agency, December 1978.

NTIS TID-22781
Fuels from Sugar Crops, edited by R.A. Nathan of Battelle Columbus Laboratories for the U.S. Department of Energy, July 1978.

NTIS TID-29419
Economics of Manufacturing Liquid Fuels from Corn Stover, by D.M. Jenkins, T.S. Reddy and J.R. Harrington of Battelle Columbus Laboratories, October 1978.

"A General History of the Nebraska Grain Alcohol and Gasohol Program," by C.R. Fricke of the Agricultural Products Industrial Utilization Committee, presented at the Symposium on Utilization of Alternate Fuels for Transportation, University of Santa Clara, Santa Clara, California, June 19-23, 1978 and furnished by the National Gasohol Commission.

Alcohol Fuels Technology Third International Symposium—Volumes I, II and III, Asilomar, California, May 1979 (Asilomar Conference).

Chemical Week, August 8, 1979.

"Lowering the Cost of Alcohol," by F.F. Hartline, *Science,* Vol. 206, 41-42, October 5, 1979.

METHANOL TECHNOLOGY & APPLICATION IN MOTOR FUELS 1978

Edited by J. K. Paul

Chemical Technology Review No. 114
Energy Technology Review No. 31

This book contains detailed descriptive information relating to methanol production technology from unusual sources, the utilization of methanol as an automotive fuel, and the conversion of methanol into gasoline. The first chapter is an overview and serves as an introduction to the subject.

The next three chapter are feasibility studies and discuss the production of methanol from coal, solid waste, and natural gas. This book does not discuss in any detail the production of methanol from liquid hydrocarbons, since one of the reasons for the future utilization of methanol in automobiles is to lessen the reliance upon petroleum hydrocarbons. Regarding natural gas, methanol can be produced from natural gas at a distant location, and shipped as a liquid at potentially less cost and less danger than the shipping of liquefied natural gas (LNG). Regarding coal, the cost of methanol appears to be comparable on an energy-equivalent basis to the production of synthetic gasoline and substitute natural gas from coal.

The next two chapters relate to the use of methanol/gasoline blends in automobiles and the use of 100% methanol as a vehicle fuel. Small amounts of methanol can be added to gasoline for use in current engines, however, the use of 100% methanol as a vehicle fuel does require engine changes, including modifications to the carburetor, and the necessity for a greater heat supply to the intake manifold.

The last two chapters relate to the production of gasoline from methanol, and these two chapters are based primarily on the efforts being conducted by the Mobil Corporation in this direction.

This book is based on federally funded studies, U.S. patents and other sources. It does convey the impression of considerable progress in methanol technology.

1. OVERVIEW
Methanol Production
Extraction
Fermentation
Syntheses
From Synthesis Gas
Oxidation of Hydrocarbons
By Irradiation of CO_x
Organic Feedstocks
Inorganic Origins
Raw Materials
  Nonrenewable Sources
  Renewable Sources
Methanol as a Fuel
Combustion Emissions
Suitable Engines

Electrochemical
  Oxidation of Methanol
Human Toxicity:
  Visual Impairment &
  Blindness from Methanol

2. METHANOL FROM COAL
Coal Gasification and
  Methanol Synthesis
DuPont Feasibility Study
Sasol Type Process Study
Badger Conceptual Design

3. METHANOL FROM SOLID WASTE
Conversion of Municipal
  Solid Waste
Pyrolysis Procedures
Suitability Specifications
  for Municipal Wastes
Conversion of Wood Wastes

4. ALASKAN METHANOL
Synthesis & Conversion Plants
Pipeline (Alyeska)
Pumping Power Requirements
Other Transportation
Other Transportation
Methanol Crude Separations
  at Los Angeles
Cost of Supply Gas
Slug Flow Interface
Overall Cost Estimates

5. METHANOL/GASOLINE BLENDS
5%, 10%, 15%, 20% Methanol
Missouri U. Study
Simulation Procedures
Engine Adjustments
Bartlesville Energy Research
  Center Study
Vehicle Optimizations Needed
Performance Mapping
U. of California Study
Engine Configurations
  and Operations
Optimum Amounts in Blends

6. 100% METHANOL MOTOR FUEL
Engine Modifications
Fuel-Air Injection Systems
Cold Start Difficulties
Thermochemical Engines

7. CHEMICAL CONVERSION: METHANOL TO GASOLINE
Mobil Oil Corp. Process
Fixed Bed Pilot Plant
Vehicle Studies with Synthetic
  Gasoline from CH_3OH

8. METHANOL TO GASOLINE
Proprietary Processes
Mobil Oil Corp. Patents
Ethyl Corp. Patents
Others

ISBN 0-8155-0719-4

470 pages

STIRLING ENGINE DESIGN AND FEASIBILITY FOR AUTOMOTIVE USE 1979

Edited by M.J. Collie

Energy Technology Review No. 47

What are the features of the Stirling external combustion engine by which it has advantages over the internal combustion engine? Low noise level, absence of vibration, and clean, almost odorless emissions are assets, but the fact that it can burn almost any fuel is perhaps one of its greatest advantages. Engine wear is also minimized.

Used to pump water and propel fans in the last century, the Stirling engine lay dormant for years. But its revival is at hand, and this apropos book supplies needed data. The first part treats theoretical design, from elementary principles through cycle analysis. It also contains over 800 literature references and a directory of those active in Stirling engine development. The Task I section of Part II describes tests by Ford Motor Company of a vehicle powered by a 170 hp engine; the Task II segment covers a feasibility study of an 80-100 hp engine for automotive use, including improvement potential for emissions and fuel economy.

Chapter headings and **examples of some** subtitles are given below.

PART I: STIRLING ENGINE DESIGN MANUAL

1. MAJOR TYPES
Heat Sources
Solid-Gas Heat Transfer
Gas Transport & Power Take-Off (Seals)
Power Control; Heat Sinking; Working Gas

2. PRESENT & FUTURE APPLICATIONS
Motor Vehicle, Heat Pumping, Biomedical, and Central Station Power

3. AUTOMOTIVE SCALE ENGINES
Philips-Ford—1-98, 4-215, 4-98 Engines
United Stirling—Application, Performance
Design: Seals, Gas Cooler, Gas Heater, Burner & Air Preheater, Power Control
General Motors—Engine Measurements
FFV Engine (Swedish)

4. ENGINE DESIGN METHODS
Stirling Engine Cycle Analysis
First Order Design Methods
Piston-Displacer Engines
Dual Piston Engines

Second Order Design Methods
Basic Power Output
Fluid & Mechanical Friction Losses
Basic Heat Input and Reheat Loss
Conduction—Shuttle, Gas, Solid
Pumping Loss; Temperature Swing Loss
Heat Exchanger Evaluation
Iteration for Effective Gas Temperature
Schmidt Equations
Third Order Design Methods
Fundamental Differential Equations
Comparisons—Urieli, Schock, Vanderburg, Finkelstein, Lewis Research Center

5. THEORY COMPARED TO EXPERIMENT
Allison, MIT Cooling and GPU-3 Engines

6. AUXILIARY DESIGN PROBLEMS

7. SAMPLE DESIGN PROCEDURE

8. REFERENCES

9. DIRECTORY

PART II: FEASIBILITY STUDY

10. TASK I—170 HP ENGINE TESTS
Compared to Internal Combustion Engine
4-215 Engine—Safety, Preheater, Burner, Blower, Heater Head, Drive System
Weight Breakdown of Test Vehicle
Fuel Specifications of Test Vehicle

11. TASK II—80-100 HP ENGINE DESIGN STUDY
Vehicle Size, Weight, Fuels, Accessories, Heating and Cooling, Drivetrain
Automatic Start and Shutdown Control
Power and Temperature Controls
Heater Head Design; Preheater System
Induction & Exhaust Systems—Air Filter, Air Flow Measure Device, Burner Assembly
Sealing and Cooling Systems
Features—Heater Cage, Piston Rods, etc.
Fuel Economy, Emissions, Performance
Alternate Engine Configuration Study

12. APPENDIX
Report on Hydrogen Safety Tests
Engine & Vehicle Package Layout Drawings

ISBN 0-8155-0763-1

470 pages

ELECTRIC AND HYBRID VEHICLES
1979

Edited by M.J. Collie

Energy Technology Review No. 44

This year several hundred electric vehicles, operated by five private companies, will take to the road under an Energy Research and Development Administration program. These will be evaluated for city and suburban driving. Vans similar to those used in England since World War II to carry milk noiselessly "from the dairy to the doorstep" will appear here more often. With such stimulus from government and industry abetted by the petroleum scarcity, this industry should mature in the decade ahead. The number of all-electric vehicles and their hybrid counterparts for passenger and commercial uses should increase a great deal.

This up-to-date assessment of electric vehicles summarizes data on characteristics, costs, maintenance and energy consumption gleaned from actual tests and from the literature and trade. It includes a survey of electric bus operations in 19 foreign and domestic locations and a survey of 11 different types of cars and delivery vans in the United States.

Energy storage devices and power systems were analyzed as alternatives to current automobile propulsion systems, with the following batteries covered: Pb/acid, Ni/Zn, Ni/Fe, Li/Fe sulfide, Na/S, Zn/Cl, Ni/H, Zn/Br, Fe/air, Li/ metal sulfide, and others.

The partial table of contents below lists chapter headings and **examples of some** subtitles. Bibliographies and a glossary complement the text.

PART I. STATE OF THE ART

1. **INTRODUCTION**

2. **INFORMATION AND DATA BASE**

3. **THEORY AND TRACK TESTS**
   User Experience
   Literature Data

4. **ELECTRIC VEHICLE COMPONENTS**
   Tires
   Differentials
   Transmissions
   Traction Motors
   Controllers
   Batteries
   Battery Chargers

5. **HYBRID VEHICLES**
   Types

Operating Modes
Components
Performance Characteristics
APPENDIXES
Batteries

PART II. ELECTROCHEMICAL DEVICES

1. **EVALUATION OF BATTERIES**

2. **DESCRIPTION OF BATTERIES**

PART III. FOREIGN TECHNOLOGY

1. **INTRODUCTION**
   Transportation Energy Requirements
   Energy Sources

2. **FOREIGN R&D**
   Australia
   West Germany
   France
   Italy
   Sweden
   Switzerland
   Japan
   United Kingdom
   USSR
   Eastern European Countries

3. **POWER SOURCES (BATTERIES)**

4. **DRIVE SYSTEMS**
   Traction Systems
   Developments in DC Drives
   Developments in AC Motors

5. **CONTROL SYSTEMS**
   Function of a Controller
   Speed Control Systems
   Systems for Other Controller Functions
   Solid-State Electronic Switching/Controller Devices

6. **MILITARY SYSTEMS**
   Benefits & Drawbacks of Military Use

PART IV. LATEST DEVELOPMENTS AVAILABLE JANUARY 1979

   Near-Term Programs
   Vehicle Product Improvement Program
   Propulsion Systems
   Vehicle Systems Support
   Near-Term Battery Development
   Mechanical Energy Storage
   Foreign and Domestic Status

ISBN 0-8155-0754-2

633 pages